Wildlife Ecology and Conservation in Forest Habitats

Wildlife Ecology and Conservation in Forest Habitats

Guest Editors

Alexey Andreychev
Todd Fredericksen

MDPI

Basel • Beijing • Wuhan • Barcelona • Belgrade • Novi Sad • Cluj • Manchester

Guest Editors
Alexey Andreychev
Faculty of Biotechnology and Biology
National Research Mordovia State University
Saransk
Russia

Todd Fredericksen
School of Natural Sciences and Mathematics
Ferrum College
Ferrum
USA

Editorial Office
MDPI AG
Grosspeteranlage 5
4052 Basel, Switzerland

This is a reprint of the Special Issue, published open access by the journal *Forests* (ISSN 1999-4907), freely accessible at: https://www.mdpi.com/journal/forests/special_issues/QN4T9LV54J.

For citation purposes, cite each article independently as indicated on the article page online and as indicated below:

Lastname, A.A.; Lastname, B.B. Article Title. *Journal Name* **Year**, *Volume Number*, Page Range.

ISBN 978-3-7258-4887-4 (Hbk)
ISBN 978-3-7258-4888-1 (PDF)
https://doi.org/10.3390/books978-3-7258-4888-1

Cover image courtesy of Todd Simon Fredericksen

Contents

About the Editors

Alexey Andreychev

Alexey Andreychev is the head of the Department of General Biology and Ecology at the National Research Mordovian State University. Alexey is interested in animal responses to climate and environmental change, and his current areas of research primarily concern the ecology of mammals and birds of prey in the forest-steppe zone of the Middle Volga region, landscape ecology, vulnerability, and climate change. The main goals of his current research are to assess the factors that facilitate the adaptation of rare rodent and owl species to different natural and urban environments.

Todd Fredericksen

Todd Fredericksen teaches courses in environmental science and biology at Ferrum College. His research interests include the effects of forest management on biodiversity, natural history, the conservation of wildlife species in the Blue Ridge Mountains, and tropical and temperate forest ecology. He is the editor of *Banisteria*, the journal of the Virginia Natural History Society, Editor-in-Chief of the journal *Trees, Forests and People*, and an editor for *Forest Ecology and Management*.

Preface

This collection of articles presents relevant research findings on the impact of various factors on living organisms in forest ecosystems, from butterflies to wild boars, and the ecosystem services provided by forests as nature-based solutions in natural contexts, covering not only provisioning services but also regulating and maintaining services. The impact of environmental change on forest multifunctionality and service provision is also explored. Various wildlife monitoring methods for effective study and management in forest ecosystems are discussed herein.

Alexey Andreychev and Todd Fredericksen
Guest Editors

forests

MDPI

Article

Alien vs. Native—Influence of Fallow Deer (*Dama dama*) Introduction on the Native Roe Deer (*Capreolus capreolus*) Population

Jakub Gryz [1], Dagny Krauze-Gryz [2,*] and Karolina D. Jasińska [2]

1 Department of Forest Ecology, Forest Research Institute, Braci Leśnej 3, Sękocin Stary, 05-090 Raszyn, Poland; j.gryz@ibles.waw.pl

2 Department of Forest Zoology and Wildlife Management, Institute of Forest Sciences, Warsaw University of Life Sciences WULS-SGGW, Nowoursynowska 159, 02-776 Warsaw, Poland; karolina_jasinska@sggw.edu.pl

* Correspondence: dagny_krauze_gryz@sggw.edu.pl

Abstract: Fallow deer is one of the most widespread alien mammals in Europe. We documented the response of the roe deer population shortly after the fallow deer was introduced to a hunting ground in central Poland. Mean roe density dropped from 17.6 ind./100 ha to 10.5 ind./100 ha after the alien species was introduced. In the reference area, where fallow deer was absent, the roe deer density did not change in the analogue study period. At both study sites, mean roe deer productivity before fallow deer introduction was similar (1.6 juv./female). However, in the first study area, the productivity dropped to 1.4, while in the reference study area, it slightly increased to 1.75. The presence of fallow deer influenced roe deer space use negatively, i.e., the number of pellet groups of roe deer decreased with an increase in the number of fallow deer feces. Overall, the introduction of the fallow deer was successful and the population grew quickly. Yet, the economic impact of its introduction was far from satisfactory. At the same time, its negative influence on the roe deer was apparent. This shows that the fallow deer is an alien species threatening local biodiversity.

Keywords: alien invasive species; wildlife management; population density; reproduction rate; interspecific competition

1. Introduction

Alien species are recognized among the biggest threats to global biodiversity [1–4], with alien ungulates being no exception [5]. Nevertheless, predicting the impact of an alien species is not straightforward, as the same species can cause different types and magnitudes of impacts when introduced to different contexts [6]. The common fallow deer (*Dama dama*) is one of the most widespread alien mammals in Europe [7]. Its undoubtedly native geographic range is in Asia Minor [8–10], where fallow deer currently live in just one natural site in south-western Turkey [11,12]. Recent evidence suggests that Anatolia and the Balkans were the sole glacial refugia of the species [13]. In medieval times, the fallow deer became a popular species, being kept in fenced hunting grounds and in parks. In the following centuries, the deer's geographic range expanded steadily because of numerous intentional introductions [9]. Currently, fallow deer live on all continents except Antarctica [14–16]. It is a widespread game, ornamental, and farmed species [8,15–18]. The number of individuals in captivity has already exceeded five million worldwide [19], and the number of free-living fallow deer keeps growing [18].

The exact time of the fallow deer's introduction to Poland is unknown. Most likely, it was in the 17th century, but it may have been as early as the 13th century (review in ref. [20]). In recent decades, the number of fallow deer has grown rapidly for the following reasons: introductions by hunters (in 1987, the national program for the development of fallow deer population in Poland was accepted and supported by the government); escapes from captivity (in 1997, the fallow deer was listed as livestock) [21]; and transborder immigration [20]. In practice, there are no impediments to the introduction of fallow deer. Thus, each year, new populations are established by hunters. Officially, in March 2022, there were 35,540 fallow deer on hunting grounds in all of Poland [22]. Unfortunately, there is no scientifically based system for game monitoring in Poland. Yet, such mostly hunter-reported game statistics may be underestimated by as much as five-fold [23]. Nevertheless, fallow deer populations have clearly increased. In 2001, the fallow deer occurred in 6% of 4965 hunting grounds; in 2020, this figure rose to 19%. Similarly, in 2006, the fallow deer occurred in just one national park [24]; in the next 17 years it was reported in seven more national parks [25–27]. In the 1970s, the fallow deer was present mostly in the western Poland, with some single localities in central Poland. Currently, this species inhabits the whole country [22,27,28]. The number of hunted animals has also increased, i.e., from 1412 ind. in 1990 to 10,117 in the 2021/2022 hunting season [9,22]. The number of captively bred animals (i.e., animals for venison production, further introductions, and scientific, educational, or ornamental purposes) has also grown rapidly. It is estimated that this number increased from less than 200 in 1990 to 16,000 in 2012 [20,29,30]. Importantly, the fallow deer is not included in the List of Invasive Alien Species of European Union concern [31], so its number is expected to grow further.

The fallow deer is a highly adaptable species that thrives in a range of ecosystems and shows a very wide spectrum of food preferences [15]. It also presents great behavioral and social flexibility, including several mating tactics [32]. Fallow deer have proven able to compete for feeding resources with native cervids, with food niche overlap exceeding 50% [33]. Its negative impact on roe deer *Capreolus capreolus* have been clearly documented, e.g., [34,35]. In Italy, this smaller cervid avoided sites heavily penetrated by the fallow deer, and in areas of high density of the fallow deer, roe deer were rare [36]. This alien species was also shown to be able to compete successfully at feeding sites with red deer *Cervus elaphus* [37]. The fallow deer is a host for many species of parasite that might infect native cervids [38], livestock and humans (review in: ref. [16]). It has also been shown to impact native flora and natural habitats negatively [39–45].

Previous studies have documented relations between the fallow and the roe deer in areas where both species have coexisted for a long time, e.g., [35,36]. In our case, we documented the response of the roe deer population shortly after the fallow deer was introduced. The aim of the study was to assess the effect of the fallow deer introduction on the local roe deer population. First, we used two population indices, i.e., population density and productivity. We assumed that if the roe is negatively affected by competition with the fallow deer, the two indices will decrease after the fallow deer is introduced. We compared those with analogue indices but calculated for a nearby population (assumed to be affected by similar environmental conditions but unaffected by the fallow deer). Finally, we also assumed that the roe deer, as a competitively weaker species, will avoid areas that are highly penetrated by the fallow deer, i.e., its spatial distribution will be negatively affected by the presence of the fallow deer. We also supposed that this effect would be weaker in areas where the fallow deer is present for a long time. We also tried to assess income and spending connected to the fallow deer's presence to robustly assess the economic results of its introduction.

2. Materials and Methods

2.1. Study Area

This study was conducted in central Poland, a region that is affected by the mild oceanic climate of Western Europe and the harsh and dry continental climate of Eastern Europe and Asia. The growing season lasts for ca 210 days; the total precipitation equals ca 600 mm per year; and the mean ambient temperature ranges from −4 °C in January to +18 °C in July. A mosaic of fields and forests dominates the landscape of this part of Poland. The percentage share of forest cover is the lowest in Poland and does not exceed 20%.

The first study area was over 1000 ha in Głuchów forest (51°45′11.8″ N, 20°06′24.1″ E). In January 2008, fallow deer started to be introduced there in order to increase the profitability of game management. The forest is surrounded by open fields, woodlots, and the Rawka River valley (a nature reserve); fish ponds border the forest to the south. The main tree species is Scots pine (*Pinus sylvestris*) with an admixture of oaks (*Quercus* spp.). Black alder (*Alnus glutinosa*) dominates part of the forest.

The second (and the reference to Głuchów) study area (Strzelna) was located near Rogów village (51°49′17.98″ N, 19°53′54.15″ E) in a smaller (710 ha in total) and fragmented forest, covered by Scots pine and oaks with an admixture of larch (*Larix decidua*) ca 15 km from the first study area. A busy, national road and a railway cut through the forested area. Surrounding arable fields and meadows mostly belong to private owners. The fallow deer was absent from this area.

Both study areas belong to Warsaw University of Life Sciences and are managed by the Experimental Forest Station in Rogów. The two aforementioned study sites were located in central parts of two hunting grounds (their areas being 5917 ha and 4575 ha, respectively) also managed by the university. The two study sites were separated by arable lands, sparse built-up areas, and a national road. In the first study area (Głuchów), fencing of young forest plantations was very rare prior to fallow deer introduction. In the second study site (Strzelna), most of deciduous plantations were fenced due to the high density of roe deer. In both study sites, chemical methods of forest protection against browsing were applied.

The third area was Spała Forest, located approximately 15 km south of the Głuchów forest. It is a big, compact forest complex to which the fallow deer was introduced before World War II, so the roe deer and the fallow deer have coexisted for decades. It is a part of Pilicka Forest, which is a forest complex of around 50,000 ha managed by the State Forests. It is pine-dominated and crosscut by the Pilica River. Our research was conducted in the northern part of the complex (Spała Forest Distrct) north of the Spała village (51°32′28″ N 20°08′17″ E).

2.2. Data Collection

2.2.1. Estimation of Fallow Deer Population Development after Its Introduction to the Głuchów Forest

At first, all released to the Głuchów forest (the first study area) animals had highly visible earrings, and their flight initiation distance was relatively short. Thus, during the first three years after the introduction (until 2011), their total abundance was assessed directly at feeding stations or in places where they tended to aggregate (such as meadows and food plots). When the population grew and most of the individuals were not marked, such a census was no longer possible. Density estimation by drive counts (see Section 2.2.2) was also not reliable as fallow deer tended to aggregate (thus either no individuals or the whole herd could be recorded). Therefore, we applied winter counts, aiming to assess an index showing a relative number of fallow deer for year-to-year comparisons. The counts were conducted in winter (December–February 2011/2012–2021/2022), on snowy and frosty days when fallow deer no longer stayed in open fields and congregated in mid-forest meadows, food plots, feeding stations, etc. All such known places were monitored by camera traps and additionally by direct observations. Data were captured by 12 camera traps (Reconyx HyperFire: PC90, PC800, PC850, PC900, RECONYX, Inc., Holmen, WI, USA), which were permanently located in the forest (see Section 2.2.3), and additional eight camera traps set in places where fallow deer were observed by local foresters or where their tracks/feeding signs were found. Simultaneously, direct observations were conducted by two or three people. Different spots in the forest where fallow deer aggregated were visited one by one during on the same day. Data from the camera traps and from direct observations (collected on the same day) were combined. The number of animals in a herd and their individual characteristics were taken into account. Finally, we pooled these observations to estimate a minimal number of fallow deer in the study site.

2.2.2. Estimation of Roe Deer Density Prior to and after Fallow Deer Introduction

The drive counts [46] were used to estimate roe deer density in the first study site (Głuchów forest) and in the second study area (Strzelna, no fallow deer, a reference area). The obtained results can be biased by a range of factors and the error produced can be high, e.g., [47]. Thus, we did not study year-to-year variation but compared data from longer periods of at least 10 years (see below). The drive counts were performed prior to fallow deer introduction (1997–2007, n = 10) and after introduction (2008–2018, n = 15). For the second study area, we used comparable data (1997–2007, n = 10 and 2008–2018, n = 11). The drive counts were performed by students of biology and forestry (WULS). In the first study area, they were performed every year in late October or at the beginning of November; in the second study area, they were carried out in early March (data from the Department of Forest Zoology and Wildlife Management, WULS). This difference was due to students' schedule. Nevertheless, we did not compare data from the two study sites but compared data for each study site between the two periods. Animals were counted in selected plots (i.e., one plot consisted of one or two forest compartments). We censused ca 30% of the whole forest complex. The plots were selected in terms of stand age, species composition, and the distance from the forest edge. Observers stood 50–100 m apart (to maintain visual contact) around each rectangle plot. Around one hundred students were simultaneously involved. Next, the observers along three sides of the plot stayed in place, while those along the fourth side walked inward and through the entire area, rousing the animals from the plot. All observers recorded animals passing through the line of observers. Each person counted only animals passing by on their left side to avoid duplications [48]. On the basis of the number of observed roe deer within the plots and the total area of the plots, the density of animals per 100 ha was calculated.

2.2.3. Estimation of Roe Deer Productivity Prior to and after Fallow Deer Introduction

The number of offspring of roe deer was estimated in the first study site (Głuchów forest) and in the second study area (Strzelna, no fallow deer, a reference area). The counts were made prior to fallow deer introduction (2001–2007) and after introduction (2008–2018). Data from the same periods were compared in the second area, Strzelna. In both study areas, we estimated the number of roe deer offspring mostly on the basis of direct observations collected during extensive field work or by local foresters. In the second study period, apart from direct observations, data from camera traps were added. In the first study site, 12 camera traps were set in permanent spots to monitor ungulates [49]. In the second study site, data from camera traps that were used to monitor dens of carnivores were used [50]. We collected data between May (beginning of giving birth time) and the first days of October (when hunting season for females and fawns starts). Finally, we calculated the mean number of juveniles per female. We took under consideration only females with at least one (visible) young.

2.2.4. Spatial Relationship between Roe Deer and Fallow Deer

We used the faecal standing crop method [51] to show the relationship between fallow deer and roe deer distribution, i.e., that areas highly penetrated by fallow deer are potentially avoided by roe deer [52]. In Poland, leaves that fall down in October/November cover the feces [52], and disappearance of faces is lowest in winter [53]. Thus, we performed the counts in late March and the beginning of April (i.e., the period after the snow melt but before the vegetation started)., i.e., pellets that accumulated through late autumn–winter were counted. Because feces of both species may be misidentified [54], we applied a metric procedure to tell them apart [55]. The counts were performed in the two study areas in which the fallow deer was present. In 2021 and 2022, this was the first study area (Głuchów forest, where the fallow deer was recently introduced), and in 2021, this was the third study area (Spała forest, where the fallow deer had been present for decades). The counts were made along 2 m wide transects placed randomly within the forests. The number of pellet groups was recorded per each 80 m long section of a transect. The total length of transects equaled 54.1 km (676 80-m long stretches) in Głuchów, and 35.1 km (439 stretches) in Spała.

Additionally, behavioral observations of interactions between roe and fallow deer were conducted in the Głuchów forest. The whole description of the method and obtained results are shown in Appendix B.

2.2.5. Hunting Bag and Financial Data

Data on the hunting bag of fallow deer (Głuchów forest—the first study area) and roe deer (Głuchów and Strzelna forests), trophy mass and carcasses mass of fallow deer, costs of forest protection against cervids (fencing and repellents), and compensations paid to farmers for damage in which fallow deer were involved were provided by the Forest Experimental Station in Rogów. Data on the hunting bag of roe deer and fallow deer in the first and second study area refer to the total area of the hunting grounds. Data on 15-year spending (2008–2022) on supplementary feed (beets, carrot, corn, hay, apples, salt and others food types) for fallow deer was impossible to evaluate precisely. Thus, we roughly estimated a minimum value. We also added the costs of wooden feeding stations that were built before fallow deer were introduced. It was impossible to even roughly assess the costs of damage done by fallow deer in the forest. Finally, we only took into consideration the costs of fencing young forest plantations from the last 5 years (2016–2020), when population of fallow deer was relatively big. Moreover, before fallow deer introduction, this method was applied only accidentally. Robust final revenue and spending statistics were calculated according to current (2023) EUR price.

Additionally, data on death cases of fallow deer in the Głuchów forest were collected (2008–2022), i.e., any remnants found were recorded, and foresters and game managers were interviewed (Appendix A).

2.3. Statistical Analysis

The Mann–Kendall trend test was used to assess whether the time series presenting the abundance of fallow deer had a monotonic upward trend. We separately compared the density of roe deer in the two periods (before and after fallow deer introduction or analogue periods) and in the two areas, Głuchów (where fallow deer was introduced) and Strzelna (where fallow deer was absent. Although hunting bag might affect roe deer density, we skipped hunting bag in the analysis, because there was no correlation between these two variables (Appendix C). Nevertheless, we compared mean hunting bag values between the two periods. The Kruskal–Wallis test was used, as data did not follow normal distribution (Shapiro–Wilk test, $p < 0.05$). To compare roe deer productivity before and after fallow deer introduction in the two research areas (Głuchów and Strzelna), we used the Scheirer–Ray–Hare test, as data did not follow normal distribution (Shapiro–Wilk test, $p < 0.05$).

We used generalized linear models to explore the influence of the fallow deer on roe deer. We used the number of roe deer pellet groups as a response variable, and as exploratory variables, we used the number of fallow deer pellet groups in the area (Głuchów, Spała) and the interaction between number of fallow deer pellet groups and area. The half-normal plot with a simulated envelope ('hnp' package; [56]) was used to check models with Poisson, quasi-Poisson, and negative binomial distribution. A model with negative binomial distribution was chosen as the best one. Analyses were performed using R [57].

3. Results

3.1. Fallow Deer Population Development in Głuchów Forest

A total of 22 fallow deer were released in Głuchów forest (first study area), the first 10 individuals in 2008 and the next 10 in 2009. Additionally, one female and a calf were introduced in 2011. In the following years, the population grew steadily, and after 14 years, the minimum population abundance (in winter) was estimated at 69 individuals

(Tau–Kendall trend test, S = 72, $p < 0.001$, Figure 1), which amounted to a population density of 6–7 individuals/100 ha of forest area. Between 2013 and the end of February 2022, 39 individuals were legally harvested. Over 16 years (up to summer 2023), we also documented 31 other cases of fallow deer deaths (Appendix A).

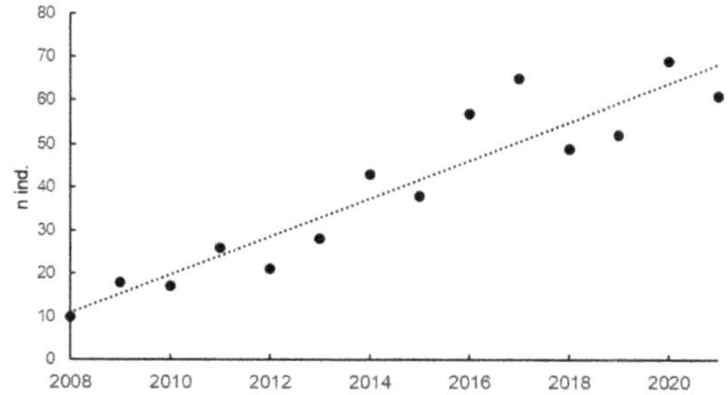

Figure 1. An increase in the estimated number (the minimum population abundance) of fallow deer (in winter) in the study area after the introduction of 22 individuals in the years 2008–2011.

3.2. Changes in Roe Deer Density after Fallow Deer Introduction

Mean roe density before fallow deer introduction into the Głuchów forest (first study area) was 17.6 ind./100 ha and dropped after the introduction of fallow deer to 10.5 ind./100 ha (Kruskal–Wallis test = 9.0069, df = 1, $p < 0.005$). In the Strzelna forest (fallow deer absent, reference area), roe deer density was initially much higher, at 32.2 ind./100 ha, and did not change when compared to the second study period (Kruskal–Wallis test = 0.079494, df = 1, $p > 0.05$) (Figure 2). At the same time, a significant increase in the hunting bag of roe deer was recorded at both study sites. In the first study area, the hunting bag rose by 91%, i.e., from an average of 14.5 (2000–2007) to 27.7 (2008–2019) (Mann-Whitney test, Z = −2106, $p < 0.05$). However, in the second area, the reference area, the hunting bag rose by 127%, i.e., from 23.6 (2000–2007) to 53.6 (2008–2019) (Mann-Whitney test, Z = −2.518, $p < 0.05$).

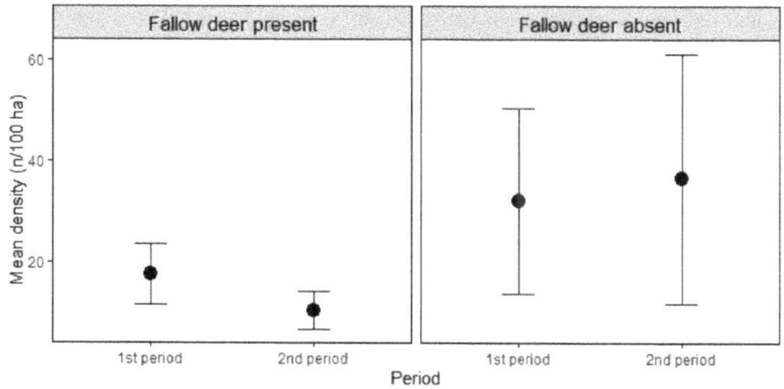

Figure 2. Changes in the mean (±SD) density of the roe deer before (1997–2007) and after (2008–2018) fallow deer introduction into Głuchów forest (first study area, fallow deer present) and comparison of roe deer density between the same study periods in Strzelna forest (reference, fallow deer absent) study area. In the area where the fallow deer was present, 10 and 15 estimations were made in the first and second period, respectively. In the area where the fallow deer was absent, 10 and 11 estimations were made.

3.3. Roe Deer Population Productivity before and after Fallow Deer Introduction

In both study sites, roe deer productivity in the first period (before 2008, i.e., pre-fallow deer introduction) was similar and accounted 1.6 juv./females. However, in the first study area, the productivity dropped to 1.4, while in the reference study area, it slightly increased to 1.75 (Figure 3). The results of Scheirer–Ray–Hare test showed that area and interaction between area and period influenced the productivity of the roe deer (Table 1).

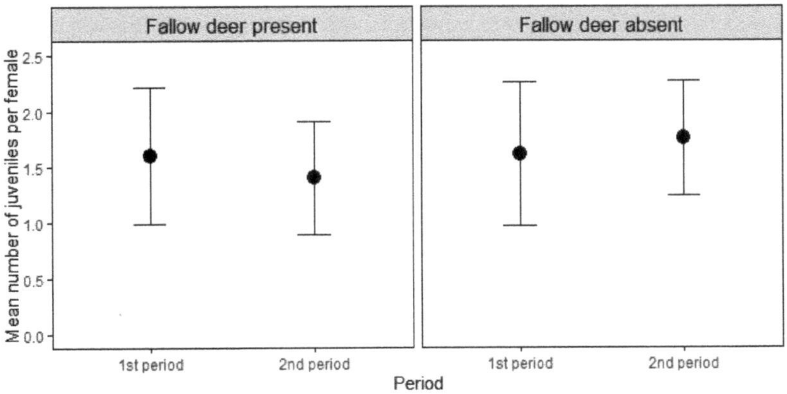

Figure 3. Changes in the mean (±SD) productivity of roe deer before (2001–2007) and after (2008–2018) fallow deer introduction into Głuchów forest (first study area, fallow deer present, n = 91) and comparison of roe deer density between the same study periods in Strzelna forest (reference, fallow deer absent, n = 106). Only females that had at least one calf were counted (i.e., the min number was 1).

Table 1. Results of Scheirer–Ray–Hare test for the variables, which influenced the productivity of the roe deer in two areas (Głuchów forest = fallow deer present and Strzelna forest = fallow deer absent) in the two periods (2001–2007 and 2008–2018, i.e., before and after introduction of fallow deer in the first area, and in analogue study periods in the second area).

Variables	H	Df	p Value
Area	14.1373	1	<0.001
Period	0.0578	1	>0.05
Area ∗ Period	10.6360	1	<0.01

3.4. *Spatial Relationship between Roe Deer and Fallow Deer*

Amongst three built models, the one with negative binomial distribution was selected as the optimal model. The model supported the hypothesis that the presence of fallow deer and area of the research influenced roe deer presence negatively, i.e., the number of pellet groups of roe deer decreased with the number of fallow deer feces and was lower in Spała forest (where fallow deer had been present for many years). The probability of roe deer presence was the lowest when both variables (presence of fallow deer in Spała forest) were combined (Table 2).

Table 2. Model output for the probability of roe deer presence in relation to fallow deer presence (expressed as the number of feces/pellet groups of each deer species recorded within a certain 80 m stretch of a transect).

Variables	Estimate	Std. Error	z Value	p Value
Intercept (roe deer, Głuchów)	0.30438	0.05616	5.420	<0.0001
Fallow deer	−0.10178	0.04362	−2.333	<0.05
Location_Spała	−0.84147	0.10231	−8.2255	<0.0001
Fallow deer ∗ Location_Spała	−1.09219	0.41819	−2.612	<0.01

3.5. Economical Effect of Fallow Deer Introduction

The cost of purchase of fallow deer that were introduced to Głuchów forest (the first study area) was 5350 EUR. To the best of our knowledge and according to the official prices we assessed, the minimum cost of fallow deer feeding is at least EUR 1000. Compensation paid to farmers in the years 2008–2022 was at least EUR 1000. Spending on forest protection (fencing) related to fallow deer presence was around EUR 1500/year (adding up to EUR 7500 over five years). To summarize, all spending (assessed via the minimum level, cost of fallow deer introduction, and the cost of maintaining of their population) reached EUR 14,850. Between 2013 and February 2022, hunters shot 39 individuals, which gave 1040 kg of venison (and generated an income of EUR 2080). Trophies of 10 males produced revenue of EUR 1201. The heaviest trophy was 2.3 kg (the min 1.1 kg) (including the skull and excepting the jaw). Altogether, the total revenue (EUR 3281) did not reimburse the costs attributed to fallow deer.

4. Discussion

In our study, we showed that introduction of the fallow deer affected the roe deer population, which led to a decrease in the population density and productivity of this smaller and competitively weaker cervid. Interactions between the two species were antagonistic, and the fallow deer was always dominant (Appendix B). We also showed that in both areas, if the fallow deer had been introduced recently or had been present for decades, it similarly negatively affected space use by the roe deer. Finally, we pointed to the fact that the introduction of fallow deer did not increase the profitability of the hunting ground, which is often the reason for the introduction of this species.

The population of fallow deer developed very quickly after its introduction to Głuchów forest. The real number of fallow deer is expected to be even higher than that recorded by direct observations. Yet, our main goal was not to estimate its abundance precisely but to show how it grew over time. Moreover, some animals have migrated to the surrounding private forests, the nearby area of fish ponds, and the surrounding hunting grounds and could therefore not be registered [58]. Indeed, in 2022, the first fallow deer was spotted in the vicinity of Rogów (Strzelna, our reference area, from which fallow were absent). This was also the reason why the study finished at that time. This quick development of the fallow deer population in our case is in line with its sharp increase in Europe, which was higher than that recorded for native cervids [18].

We showed a sharp decrease in roe deer density in the area where the fallow deer had recently been introduced. This was in contrast to the nearby population, where the fallow deer was absent and roe deer grew in number. The roe deer is considered a selective feeder [59,60], being susceptible to competition from other cervids [52,61]. It was shown that the competition may affect its population trends more than weather conditions [34]. It is smaller than the fallow deer and relies on highly nutritional vegetation for survival and reproduction (review in: ref. [62]). In turn, the fallow deer is thought to be a strong competitor with the roe deer and other cervids (review in: [63]). Its great dietary flexibility in food sources [15,62,63] and habitat [15,64] use makes it a competitive species with native cervids. This alien species also shows high behavioral plasticity, shown through, e.g., modified aggregation or mating patterns [65,66].

Competition for food may be the main reason for the observed decrease in the productivity of the roe deer in the area where the fallow deer was introduced. For the roe deer, food availability (rather than fat reserves) is crucial for reproduction and reproductive success (review in [67]). In spring, the last months of pregnancy, highly nutritional forb-dominated feed is typical for the roe deer, while in the case of the fallow deer, a higher proportion of hemicellulose in the ingested food was noted [68]. Interference was particularly frequent at the latest stages of pregnancy, births, and early maternal care and was greater in solitary roe deer, i.e., females [62].

The abundance of the roe deer pellet groups was significantly lower in places (i.e., on the stretches) where the fallow deer was present, and this was true of both study sites (the one where fallow deer had recently been introduced and the one where it had been present for decades). It was shown that inter-specific competition from the fallow deer may influence the spatial behavior of roe deer [36]. Roe deer avoided areas where the local density of fallow deer was highest, and the roe deer occupancy was negatively affected by local abundance of fallow deer [34,62]. Densities of roe deer were great where fallow deer were rare and low where fallow deer were abundant [35]. It needs to be kept in mind that the abundance of pellet groups may not be a function of the number of individuals or of time spent in that place because defecation rates may depend on activity. Yet, pellet group counts can help to estimate relative habitat use and have been used for several decades in various contexts (review in [35]).

Our field experiment was disturbed by a significant increase in the hunting bag of roe deer in both study sites. This may be treated as an unsuspected result of the cost of fallow deer introduction and the rising costs of protecting forests against damage done by the fallow deer and game in general. There has been strong pressure to reduce populations of native ungulates in Poland, i.e., roe deer, red deer or moose *Alces alces*, due to the increasing problem of damage [69–72]. Ironically, new alien ungulate populations (fallow deer and mouflon *Ovis aries musimon*) were established all over the country [20,73] at the same time. Productivity of roe deer population is density-dependent [74,75]. Even though no correlation between hunting bag and population density was found, increased productivity might be an effect of stronger hunting pressure [76]. Nevertheless, increased hunting pressure was observed in the two study sites, while productivity increased only where the fallow deer was absent.

The fallow deer was introduced to our study site in order to increase game management profitability. However, 14 years after introduction, not even the costs of the introduced animals could not be reimbursed, and the financial results were roughly minus EUR 11,569. This deficit is unlikely to be reduced as the costs of forest protection and crop damage compensation have risen while venison prices have been quite stable. Additionally, trophy hunting is not of essential economic value as antlers are of rather low quality. It needs to be kept in mind that our financial estimates are rough due to data scarcity. Nevertheless, we counted costs cautiously, and they are under- rather than overestimated.

To conclude, the introduction of the fallow deer was definitely successful and its population grew quickly. Yet, the economic effect, assessed for the hunting ground, was far from satisfactory, which waives the argument that the introduction of this new species may bring some profits. At the same time, the fallow deer impacted the population of the native cervid, the roe deer. Consequently, the number of roe deer decreased, which we assumed was due to competition affecting productivity parameters. Pellet group analysis showed that the roe deer avoided spots heavily exploited by the fallow deer. This effect was shown regardless of the time that had passed from the introduction. This shows that the fallow deer should be considered an invasive species threatening local biodiversity. We believe that these days, with the range of field data supporting the competitive role of the fallow deer alongside native cervids (review in: [63]), this alien species should not be further introduced. At the same time, the harvest rate of the roe deer, as a competitively weaker species being affected by the alien species, should be kept low in areas where it must compete with fallow deer. The key point is to use reliable, science-based data to monitor the population of cervids. As has been shown for many alien and unquestionably invasive species, after they establish a new population and grow in number, their eradication is mostly impossible [7,77]. Thus, we claim that new populations of alien cervids must not be established.

Author Contributions: Conceptualization: J.G.; Data curation: J.G., D.K.-G.; Formal analysis: D.K.-G. and K.D.J.; Funding acquisition: J.G.; Investigation: J.G., D.K.-G. and K.D.J.; Methodology: J.G.; Project administration: J.G.; Visualization: D.K.-G. and K.D.J.; Writing—original draft: J.G., D.K.-G. and K.D.J.; and Writing—review and editing: J.G., D.K.-G. and K.D.J. All authors have read and agreed to the published version of the manuscript.

Funding: The research was partially financed by the Forests Research Institute, within the framework of the research topic entitled "Spatial, numerical and behavioural interactions between the fallow deer *Dama dama* and the roe deer *Capreolus capreolus*" (260123).

Data Availability Statement: The raw data supporting the conclusions of this article will be made available by the authors on request.

Acknowledgments: We are grateful to all who delivered data used in this paper, especially Dariusz Dałkowski and other employees of Experimental Forest Station in Rogów. We would like to thank two anonymous reviewers, whose suggestions helped us to review and improve the manuscript.

Conflicts of Interest: The authors declare no conflicts of interest.

Appendix A

Table A1. Causes of deaths of fallow deer in Głuchów forest (the first study area) in the years 2008–2022.

Cause of Death	n of Cases
Harvested by hunters	39
Killed by stray dogs	6
Died due to starvation *	3
Trapped by a baller twine **	2
Killed by car	2
Drowned	1
Poached	1
Unidentified reasons	16

* during the 2012/2013 winter season, which was characterized by very deep snow cover and low temperature [78]; ** antlers were entangled in baler twine (such 'wig' may be deadly as an animal can get trapped on branches and break their neck or die due to starvation).

Appendix B. Behavioral Interactions between Fallow and Roe Deer

Starting from the introduction of fallow deer into the Głuchów forest (the first study area), we recorded all interactions between fallow and roe deer (2008–2022) throughout the whole year. This was mostly carried out on the basis of direct observations conducted in the whole forest complex and in spots where fallow deer aggregated. Later, data from camera traps set to monitor fallow deer abundance and to monitor ungulates (see methods) were also included. We assumed that contact occurred between the two species when the animals stayed within 50 m of each other (Ferretti et al. 2011b). We defined three types of antagonistic interactions:

1. Direct attack of a fallow deer male on roe deer; roe deer is chased away.
2. Roe deer interrupts its previous activity and moves away when fallow deer approaches.
3. Roe deer waiting in the vicinity of a feeding place, young forest plantation, or margins of forest meadow until the fallow deer is gone.

Between 2008 and 2022, we observed 56 interactions between fallow and roe deer, and all of them were antagonistic: there were 6 direct attacks of fallow deer males on roe deer, 32 cases of roe deer escape when the fallow deer approached, and 18 cases when roe deer waited at a margin of a meadow or feeding station until the fallow deer left. Most of those observations were made at feeding stations and forest meadows, where fallow deer aggregated. We never observed reverse situations, e.g., fallow deer being attacked by or escaping from a roe deer. We never registered fallow and roe deer feeding simultaneously in the same place.

Appendix C. Correlation between Roe Deer Density and Hunting Bag in Two Research Areas

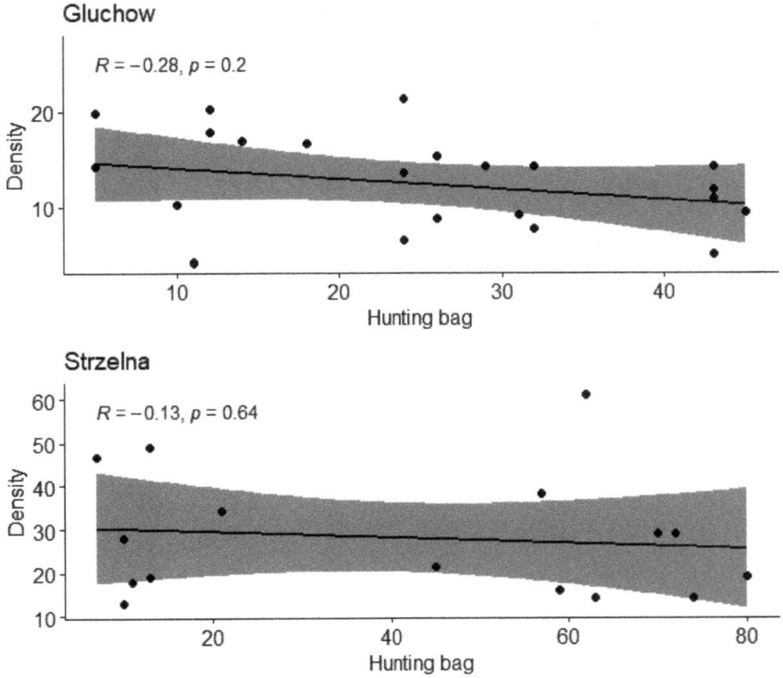

Figure A1. Correlation between roe deer density and hunting bag in two research areas (Głuchów—where the fallow deer had been introduced and Strzelna—where the fallow deer was absent).

References

1. Craik, C. Long-term effects of North American Mink *Mustela vison* on seabirds in western Scotland. *Bird Study* **1997**, *44*, 303–309. [CrossRef]
2. Simberloff, D. Eradication of island invasives: Practical actions and results achieved. *Trends Ecol. Evol.* **2001**, *16*, 273–274. [CrossRef]
3. Bellard, C.; Cassey, P.; Blackburn, T.M. Alien species as a driver of recent extinctions. *Biol. Lett.* **2016**, *12*, 20150623. [CrossRef] [PubMed]
4. Wauters, L.A.; Lurz, P.W.W.; Santicchia, F.; Romeo, C.; Ferrari, N.; Martinoli, A.; Gurnell, J. Interactions between native and invasive species: A systematic review of the red squirrel-gray squirrel paradigm. *Front. Ecol. Evol.* **2023**, *11*, 1083008. [CrossRef]
5. Spear, D.; Chown, S. Non-indigenous ungulates as a threat to biodiversity. *J. Zool.* **2009**, *279*, 1–17. [CrossRef]
6. Volery, L.; Jatavallabhula, D.; Scillitani, L.; Bertolino, S.; Bacher, S. Ranking alien species based on their risks of causing environmental impacts: A global assessment of alien ungulates. *Glob. Chang. Biol.* **2021**, *27*, 1003–1016. [CrossRef] [PubMed]
7. Genovesi, P.; Bacher, S.; Kobelt, M.; Pascal, M.; Scalera, R. Alien Mammals of Europe. In *Handbook of Alien Species in Europe, Invading Nature*; Springer Series in Invasion Ecology; Springer: Dordrecht, The Netherlands, 2009; Volume 3, pp. 119–128.
8. Chapman, D.; Chapman, N. *Fallow Deer. Their History, Distribution and Biology*; Terence Dalton Ltd.: Lavenham, Suffolk, UK, 1975.
9. Dzięciołowski, R. *Daniel*; Warsaw University of Life Science: Warsaw, Poland, 1994.
10. Masseti, M.; Cavallaro, A.; Pecchioli, E.; Vernesi, C. Artificial occurrence of the fallow deer, *Dama dama dama* (L, 1758), on the island of Rhodes (Greece): Insight from mtDNA Analysis. *Hum. Evol.* **2005**, *21*, 167–175. [CrossRef]
11. Heidemann, G. Damwild, *Cervus dama* Linné, 1758, in Kleinasien. Bestand un Schutz. *Säugetierkundliche Mitteilungen* **1976**, *24*, 124–132.
12. Marinis, A.D.; Masseti, M. Mammalian fauna of the Temessos National Park, Turkey. *ZooKeys* **2009**, *31*, 221–228. [CrossRef]
13. Baker, K.H.; Miller, H.; Doherty, S.; Gray, H.W.I.; Daujat, J.; Çakırlar, C.; Spassov, N.; Trantalidou, K.; Madgwick, R.; Lamb, A.L.; et al. The 10,000-year biocultural history of fallow deer and its implications for conservation policy. *Proc. Natl. Acad. Sci. USA* **2024**, *121*, e2310051121. [CrossRef]
14. Chapman, N.G.; Chapman, D.I. The distribution of fallow deer: A worldwide review. *Mamm. Rev.* **1980**, *10*, 61–138. [CrossRef]

15. Esattore, B.; Saggiomo, L.; Sensi, M.; Vanessa, F.; Cherin, M. Tell me what you eat and I'll tell you. . .where you live: An updated review of the worldwide distribution and foraging ecology of the fallow deer (*Dama dama*). *Mamm. Biol.* **2022**, *102*, 321–338. [CrossRef]
16. Chankaya, C.; Dokora, A.; Muchenje, V.; Hoffman, L. The fallow deer (*Dama* spp); endangered of not? *Zool. Gart* **2016**, *85*, 160–172. [CrossRef]
17. Apollonio, M. Dama dama (Linnaeus, 1758). In *The Atlas of European Mammals*; Mitchell-Jones, A.J., Amori, G., Bogdanowicz, W., Krystufek, B., Reijnders, P.J.H., Spitzenberger, F., Stubbe, M., Thissen, J.B.M., Vohralík, V., Zima, J., Eds.; T & AD Poyser Natural History: London, UK, 1999; pp. 1–484.
18. Bijl, H.; Csányi, S. Fallow Deer (*Dama dama*). Population and Harvest Changes in Europe since the Early 1980s. *Sustainability* **2022**, *14*, 12198. [CrossRef]
19. Daszkiewicz, T.; Hnatyk, N.; Dąbrowski, D.; Janiszewski, P.; Gugolek, A.; Kubiak, D.; Śmiecińska, K.; Winarski, R.; Koba-Kowalczyk, M. A comparison of the quality of the *Longisissimus lumborum* muscle from wild and farm raised fallow deer (*Dama dama* L). *Small Rumin. Res.* **2015**, *129*, 77–83. [CrossRef]
20. Solarz, W. Dama dama (Linnaeus, 1758). In *Gatunki Obce w Faunie Polski*; Głowaciński, Z., Okarma, H., Pawłowski, J., Solarz, W., Eds.; Instytut Ochrony Przyrody PAN: Kraków, Poland, 2012.
21. Kamieniarz, R.; Rzońca, Z. Metody oraz skala introdukcji, reintrodukcji i wsiedlania zwierzyny w Polsce. In *Łowiectwo w Zrównoważonej Gospodarce Łowieckiej, Zimowa Szkoła Leśna*; Gil, W., Ed.; Instytut Badawczy Leśnictwa: Sękocin Stary, Poland, 2015; pp. 91–103.
22. *Statistical Yearbook of Forestry*; Central Statistical Office: Warsaw, Poland, 2022.
23. Kamieniarz, R.; Jakubowski, M.; Dyderski, M.K.; Górecki, G.; Nasiadka, P.; Okarma, H.; Pudełko, M.; Skubis, J.; Tomek, A.; Wajdzik, M.; et al. Application of the tyraliera counting method to the large-scale inventory of red deer *Cervus elaphus* in the northern part of Western Pomerania, Poland. *Ann. For. Res.* **2023**, *66*, 33–34. [CrossRef]
24. Jamrozy, G. Carnivores, even-toed ungulates, lagomorphs and large rodents in Polish national parks. *Ann. Zool. Fenn.* **2008**, *45*, 299–307. [CrossRef]
25. Jamrozy, G. (Ed.) *Ssaki Polskich Parków Narodowych: Drapieżne, Kopytne, Zajęczaki i duże Gryzonie*; University of Agriculture: Kraków, Poland, 2014.
26. Olszewski, A. Using camera-trapping as a non-invasive technique the study of wildlife in Kampinos National Park—The first results. *Stud. Mater. CEPL* **2016**, *49*, 42–49.
27. Solarz, W. Fallow Deer. In Atlas of Polish Mammals. Available online: https://wwwiopkrakowpl/Ssaki/gatunek/124 (accessed on 13 May 2024).
28. Pucek, Z.; Raczyński, J. (Eds.) *Atlas Rozmieszczenia Ssaków w Polsce*; PWN: Warsaw, Poland, 1983.
29. Bombik, E.; Kondracki, S. Znaczenie hodowli zwierząt wolno żyjących (cz II). *Przegl. Hod.* **2007**, *75*, 18–21.
30. Borys, B.; Bogdaszewska, Z.; Bogdaszewski, M. Dynamiczny wzrost fermowej hodowli jeleni i danieli w Polsce. *Wiad. Zootech.* **2012**, *1*, 33–44.
31. European Union. Regulation (EU) no 1143/2014 of the European parliament and of the Council of 22 October 2014 on the prevention and management of the introduction and spread of invasive alien species. *Off. J. Eur. Union* **2014**, *L317*, 35–55.
32. Langbein, J.; Thirgood, S.J. Variation in Mating Systems of Fallow Deer (*Dama dama*) in Relation to Ecology. *Ethology* **1989**, *83*, 195–214. [CrossRef]
33. Obidziński, A.; Kiełtyk, P.; Borkowski, J.; Bolibok, L.; Remuszko, K. Autumn-winter diet overlap of fallow, red, and roe deer in forest ecosystems, Southern Poland. *Cent. Eur. J. Biol.* **2013**, *8*, 8–17. [CrossRef]
34. Ferretti, F.; Fattorini, N. Competitor densities, habitat, and weather: Effects on interspecific interactions between wild deer species. *Integr. Zool.* **2020**, *16*, 670–684. [CrossRef] [PubMed]
35. Ferretti, F.; Bertoldi, G.; Sforzi, A.; Fattori, L. Roe and fallow deer: Are they compatible neighbours? *Eur. J. Wildl. Res.* **2011**, *57*, 775–783. [CrossRef]
36. Focardi, S.; Aragno, P.; Montanaro, P.; Riga, F. Inter-specific competition from fallow deer *Dama dama* reduces habitat quality for the Italian roe deer *Capreolus capreolus italicus*. *Ecography* **2006**, *29*, 407–417. [CrossRef]
37. Bartoš, L.; Vaňková, D.; Šiler, J.; Losos, S. Fallow deer tactic to compete over food with red deer. *Aggress. Behav.* **1996**, *22*, 375–385. [CrossRef]
38. Dróżdż, J. Studies on helminths and helminthiases in Cervidae. III. Historical formation of helminthofauna in Cervidae. *Acta Parasitol. Pol.* **1967**, *14*, 287–300.
39. Moriarty, A. The liberation, distribution, abundance and management of wild deer in Australia. *Wildl. Res.* **2004**, *31*, 291–299. [CrossRef]
40. Dolman, P.M.; Wäber, K. Ecosystem and competition impacts of introduced deer. *Wildl. Res.* **2008**, *35*, 202–214. [CrossRef]
41. Novillo, A.; Ojeda, R.A. The exotic mammals of Argentina. *Biol. Invasions* **2008**, *10*, 1333–1344. [CrossRef]
42. Relva, M.A.; Nuñez, M.A.; Simberloff, D. Introduced deer reduce native plant cover and facilitate invasion of non-native tree species: Evidence for invasional meltdown. *Biol. Invasions* **2010**, *12*, 303–311. [CrossRef]
43. Martin, T.G.; Arcese, P.; Scheeder, N. Browsing down our natural heritage: Deer impacts on vegetation structure and songbird populations across an island archipelago. *Biol. Conserv.* **2011**, *144*, 459–469. [CrossRef]

44. Tanentzap, A.J.; Kirby, K.J.; Goldberg, E. Slow responses of ecosystems to reductions of deer (Cervidae) populations and strategies for achieving recovery. *For. Ecol. Manag.* **2012**, *264*, 159–166. [CrossRef]
45. Wright, D.M.; Andrew, J.; Tanentzap, A.J.; Flores, O.; Husheer, S.W.; Duncan, R.P.; Wiser, S.K.; Coomes, D.A. Impacts of culling and exclusion of browsers on vegetation recovery across New Zealand forests. *Biol. Conserv.* **2012**, *153*, 64–71. [CrossRef]
46. Borkowski, J.; Palmer, S.; Borowski, Z. Drive counts as a method of estimating ungulate density in forests: Mission impossible? *Acta Theriol.* **2011**, *56*, 239–253. [CrossRef] [PubMed]
47. Marcon, A.; Battocchio, D.; Apollonio, M.; Grignolio, S. Assessing precision and requirements of three methods to estimate roe deer density. *PLoS ONE* **2019**, *14*, e0222349. [CrossRef] [PubMed]
48. Gryz, J.; Krauze-Gryz, D. Why Did Brown Hare *Lepus europaeus* Disappear from Some Areas in Central Poland? *Diversity* **2022**, *14*, 465. [CrossRef]
49. Gryz, J.; Krauze-Gryz, D. Occurrence of moose *Alces alces* L in the Rogów Forest (central Poland). *For. Res. Pap.* **2022**, *82*, 179–186. [CrossRef]
50. Krauze-Gryz, D.; Gryz, J. Den dwelling carnivores in central Poland: Long term trends in abundance and productivity. *Diversity* **2022**, *15*, 32. [CrossRef]
51. Mayle, B.A.; Peace, A.J.; Gill, M.A. *How Many Deer? A Field Guide to Estimating Deer Population Size*; Forestry Commission: Dry Branch, GA, USA, 1999; Volume 18.
52. Borkowski, J.; Banul, R.; Jurkiewicz-Azab, J.; Hołdyński, C.; Święczkowska, J.; Nasiadko, M.; Dariusz Załuski, D. There is only one winner: The negative impact of red deer density on roe deer numbers and distribution in the Słowiński National Park and its vicinity. *Ecol. Evol.* **2021**, *11*, 6889–6899. [CrossRef] [PubMed]
53. Aulak, W.; Babińska-Werka, J. Estimation of roe deer density based on the abundance and rate of disappearance of their faeces from forest. *Acta Theriol.* **1990**, *35*, 111–120. [CrossRef]
54. Spitzer, R.; Churski, M.; Felton, A.; Heurich, M.; Kuijper, D.P.J.; Landman, M.; Rodriguez, E.; Singh, N.J.; Taberlet, P.; van Beeck Calkoen, S.; et al. Doubting dung: eDNA reveals high rates of misidentification in diverse European ungulate communities. *Eur. J. Wildl. Res.* **2019**, *65*, 28. [CrossRef]
55. Gryz, J.; Krauze-Gryz, D.; Klich, D. Alien or Native? How to Distinguish Feces of Fallow and Roe Deer Using Central Poland as a Case Study. *Animals* **2022**, *12*, 290. [CrossRef] [PubMed]
56. Moral, R.A.; Hinde, J.; Demétrio, C.G.B. Half-Normal Plots and Overdispersed Models in R: The hnp Package. *J. Stat. Softw.* **2017**, *81*, 1–23. [CrossRef]
57. R Core Team. *R: A Language and Environment for Statistical Computing*; R Foundation for Statistical Computing: Vienna, Austria, 2023; Available online: https://wwwR-projectorg (accessed on 10 September 2023).
58. Forest Data Bank. Hunting Reports. Available online: https://wwwbdllasygovpl/portal/ (accessed on 8 September 2023).
59. Hofmann, R.R. Digestive physiology of the deer—Their morphophysiological specialisation and adaptation. *R. Soc. N. Z. Bull.* **1985**, *22*, 393–407.
60. Latham, J. Interspecific interactions of ungulates in European forests: An overview. *For. Ecol. Manag.* **1999**, *120*, 13–21. [CrossRef]
61. Latham, J.; Staines, B.W.; Gorman, M.L. Correlations of red (*Cervus elaphus*) and roe (*Capreolus capreolus*) deer densities in Scottish forests with environmental variables. *J. Zool.* **2009**, *242*, 681–704. [CrossRef]
62. Ferretti, F.; Sforzi, A.; Lovari, S. Behavioural interference between ungulate species: Roe are not on velvet with fallow deer. *Behav. Ecol. Soc.* **2011**, *65*, 875–887. [CrossRef]
63. Ferretti, F.; Lovari, S. Introducing Aliens: Problems Associated with Invasive Exotics. In *Behaviour and Management of European Ungulates*, 1st ed.; Putman, R., Apollonio, M., Eds.; Whittles Publishing: Dunbeath, UK, 2014.
64. Thirgood, S.J. The effects of sex, season and habitat availability on patterns of habitat use by fallow deer (*Dama dama*). *J. Zool.* **1995**, *235*, 645–659. [CrossRef]
65. Apollonio, M.; Focardi, S.; Toso, S.; Nacci, L. Habitat selection and group formation pattern of fallow deer *Dama dama* in a submediterranean environment. *Ecography* **1998**, *21*, 225–234. [CrossRef]
66. Thirgood, S.J. Alternative Mating Strategies and Reproductive Success in Fallow Deer. *Behaviour* **1991**, *116*, 1–10. [CrossRef]
67. Ferretti, F.; Sforzi, A.; Lovari, S. Intolerance amongst deer species at feeding: Roe deer are uneasy banqueters. *Behav. Proc.* **2008**, *78*, 487–491. [CrossRef] [PubMed]
68. Rautiainen, H.; Bergvall, U.A.; Felton, A.M.; Tigabu, M.; Kjellander, P. Nutritional niche separation between native roe deer and the nonnative fallow deer—A test of interspecific competition. *Mamm. Res.* **2021**, *66*, 443–455. [CrossRef]
69. Balik, B.; Moskalik, T.; Sadowski, J.; Zastocki, D. Selected aspects of forest protection from game. *Stud. Mater. CEPL* **2016**, *46*, 181–191.
70. Wawrzyniak, P. Dynamika liczebności, jej wpływ na środowisko bytowania, a konieczność zarządzania populacja łosia *Alces alces* w Polsce. In *Zarządzanie Populacjami Zwierząt*; Polski Związek Łowiecki, Łowiec Polski: Warsaw, Poland, 2016; pp. 17–27.
71. Wawrzyniak, P. Funkcjonowanie Populacji łosia (Alces alces), a Skala Konfliktów z Działalnością Człowieka w Warunkach Polski Północno-Wschodniej. 2017. Available online: https://wwwuwmedupl/wbz/sites/default/files/uploads/WBZ/Dokumenty/Katedry/KHZFiL/Konferencja24-11-2017/2_wawrzyniak_p_uwm_2017_lospdf (accessed on 8 September 2023).
72. Bobek, B.; Błaszczyk, J.; Frąckowiak, W.; Furtek, J.; Morow, K.; Wojciuch-Poskonka, M.; Wyrobek, K. *Łoś Alces Alces-Biologia, Ochrona i Gospodarka Łowiecka*; Międzynarodowy Instytut Ekologii: Kraków, Poland, 2021.

73. Nasiadka, P.; Wajdzik, M.; Skubis, J. A comprehensive over 100 years history of mouflon (*Ovis musimon*) in Poland: From the promising beginning in 1902 to questionable future in 2014. A case study of wildlife management history. *Appl. Ecol. Environ. Res.* **2021**, *19*, 993–1017. [CrossRef]
74. Vincent, J.; EBideau, E.; Hewison, A.; Angibault, J. The influence of increasing density on body weight, kid production, home range and winter grouping in roe deer (*Capreolus capreolus*). *J. Zool.* **1995**, *236*, 371–382. [CrossRef]
75. Andersen, R.; Linnell, J.D.C. Irruptive potential in roe deer: Density-dependent effects on body mass and fertility. *J. Wildl. Manag.* **2000**, *64*, 698–706. [CrossRef]
76. Clutton Brock, T.H.A.; Illius, W.; Wilson, K.; Grenfell, B.T.; Maccoll, A.D.C.; Albon, S.D. Stability and Instability in Ungulate Populations: An Empirical Analysis. *Am. Nat.* **1997**, *149*, 195–219. [CrossRef]
77. Genovesi, P. Eradications of invasive alien species in Europe: A review. *Biol. Invasions* **2005**, *7*, 127–133. [CrossRef]
78. Gryz, J.; Chojnacka-Ożga, L.; Krauze-Gryz, D. Long-term stability of tawny owl (*Strix aluco*) population despite varying environmental conditions—A case study from central Poland. *Pol. J. Ecol.* **2019**, *67*, 75–83. [CrossRef]

forests

MDPI

Article

Forest Environmental Conditions Shape Carcass Mass but Not Antler Investment of Red Deer Stags (*Cervus elaphus* L.)—Study from Western European Populations

Jacek Skubis [1,*], Grzegorz Górecki [1], Emilia Pers-Kamczyc [2] and Jacek Kamczyc [1]

[1] Department of Game Management and Forest Protection, Poznan University of Life Sciences, Wojska Polskiego 28, 60-637 Poznań, Poland; grzegorz.gorecki@up.poznan.pl (G.G.); jacek.kamczyc@up.poznan.pl (J.K.)

[2] Institute of Dendrology, Polish Academy of Sciences, Parkowa 5, 62-035 Kornik, Poland; epk@man.poznan.pl

* Correspondence: jacek.skubis@up.poznan.pl

Abstract: Population stability depends on environmental conditions and their changes, as well as the availability of energy resources. Animals allocate their energy to maintenance, growth, reproduction, and energy storage; therefore, trade-offs are expected between life history traits. Access to abundant resources is expected to manifest itself in the investment of male individuals in sexually selected traits, such as carcass mass and antler size. The study aimed to analyze environmental climate conditions on the carcass and antler mass, as well as on antler form in red deer (*Cervus elaphus* L.) populations. We analyzed the carcasses and the antler masses and forms of 550 red deer stags from three populations in Central–Western Europe that differ in climate conditions that were hunter-harvested between the 2017 and 2021 hunting seasons. Our data indicated that carcass mass was shaped by the location of the population, stag age, precipitation, and temperature, as well as the number of frost days from January to the harvest date. Antler mass and antler investment depended on stag age but not climatic factors. Regular antler forms were more often observed in the harsh environmental conditions. Our observation confirms that resource trade-off is related to carcass mass of red deer.

Keywords: *Cervus elaphus*; carcass mass; antlers; mass; antler investment; antler form; Central Europe; climatic conditions

1. Introduction

According to resource allocation theory [1], animals must use energy for various activities, like reproduction or survival [2]. Trade-offs are therefore expected between life history traits [3]. In large herbivores, variations in key resources can result in spatial patterns on both the geographic and intra-population scales [4,5]. Access to abundant resources is expected to manifest itself in the investment of male individuals in sexually selected traits, such as body mass and antler size [5]. Therefore, it can be expected that spatial variations in key resources, as well as their availability during periods of limited availability, may translate into traits such as carcass size or antler weight and form. Antlers are a secondary sexual trait that, apart from in reindeer, characterizes only male cervids. Antler size and weight, like other sexual characteristics, are related to fighting ability, access to females, and position in the dominance hierarchy [6]. Antlers form can be characterized using two measurement system—CIC [7] and Safari Club International (SCI) [8]—that in detail measure different characteristics of the antlers. Measuring the size and quality of red stag trophies is crucial for hunters, conservationists, and wildlife managers. Moreover, the antler's mass indicates not only the condition of the population, but also the position of a particular individual in the herd. Recent studies have also included analyses of antler form and weight [9,10]; however, they do not cover the impacts of climatic and environmental conditions on carcass and antler mass, as well as antler investment and frequency of antler form.

We used European red deer (*Cervus elaphus* L.) as a model species because it is a typical forest animal of the order Artiodactyla and the family Cervidae, crucial to the functioning of the forest ecosystem. This species is widely distributed in North and South America, North Africa, Asia, Europe, and New Zealand [11–14]. However, they are also observed on agricultural land, such as grasslands that are frequently eaten and damaged. Those areas can provide better forage but also resting zones; therefore, they can improve individual condition [15]. As one of the most important forest species, *Cervus elaphus* L. has been subject to significant impacts from human activities, such as selective hunting, translocations, and habitat fragmentation [16,17], which may have locally resulted in high levels of genetic drift, inbreeding, and the blurring of natural patterns of genetic structure [16]. Phylogeographically, according to mitochondrial DNA sequences there are three genetic lineages of red deer in Europe: (1) the eastern populations from the Balkans and the Romanian Carpathians, reaching north to Austria, Hungary, and the Czech Republic from Sardinia and North Africa; (2) the Sardinian and North African population; and (3) the western population from Iberia and the British Isles to Scandinavia, Germany, and Western Poland [17]. However differences between populations are still discussed [11,18,19].

The red deer populations in Poland vary in density as well as the size of individuals. For instance, the highest densities per 1000 ha of forest area occur in the southwest and north, whereas the lowest is in Central Poland. In addition, historical data have indicated that the weight of red deer in Poland has varied. The lightest deer have been found in Southwestern Poland and the heaviest in the northeastern regions of the country and the Carpathians [10], although the size of individuals has varied both in Europe [17] and in Poland [20]; the causes of this fact are not clear and may depend on several factors. It is assumed that the carcass size of red deer is affected by the interaction of several factors, such as the food availability expressed by the proportion of the broadleaved habitat, stress caused by human activity (including selective hunting), natural selection caused by predators (e.g., wolves), and climatic factors (including snow cover) [10]; however, the effects of climatic conditions on red deer carcass size have not been broadly investigated. Moreover, antler mass can indicate the reproductive success of male individuals, and since antlers are costly to produce, their growth can be affected by environmental conditions [21], but there is little knowledge about the impact of environmental conditions on antler investment and antler form.

We hypothesize that environmental conditions shape body mass and antler size, but can also affect antler investment and regularity of antler form. Taking into account the above criteria, the purpose of our study was to analyze the relationship between carcass mass and antler mass in three different populations of red deer living under conditions of varying access to environmental resources and influenced by different climatic conditions. The present study may help to obtain knowledge regarding the reproduction of this wild species and its relationship with the environment.

2. Materials and Methods

2.1. Analysis of Red Deer Features

We analyzed carcass and antler mass of hunter-harvested red deer (*Cervus elaphus* L.) collected within five hunting seasons (2017–2022) from three locations in Poland (Figure 1). Moreover, the stag hunting season in Poland lasted from the 21st of August to the 28th of February; therefore, only individuals harvested within this period could be analyzed. In Poland, red deer are harvested according to the rules established for the entire country. Stags in their second year of life can be harvested only if their spikes are less than 30 cm; stags in third year (2nd head) with 4 tins on each beam, inclusive; and stags aged 4 and 5 (3rd and 4th head) up to five tins on each beam. In the range of 6–10 year old stags (5th to 9th head); all individuals without the crown can be harvested. The oldest stags, from 11 years old (10th head), can be harvested regardless of the form of antlers. In the case of does, it is recommended to harvest individuals with the weakest condition first. Moreover, according to the Polish legal rules of stag hunting and harvesting, only the following stags can be harvested: stags from 1 to 5 years old without the presence of a crown on antlers; between 5 and 10 years old, only stags with the presence of a crown on only one antler, but always individuals characterized by below-average antler and carcass quality from the populations. These restrictions are not obligatory when older individuals are harvested. However, although such regulations apply throughout Poland and have not changed during the harvest period, it can be assumed that all data obtained are influenced by the selection rules applicable at a given time, allowing them to be compared and treated as random data. No animals were killed specifically for the study.

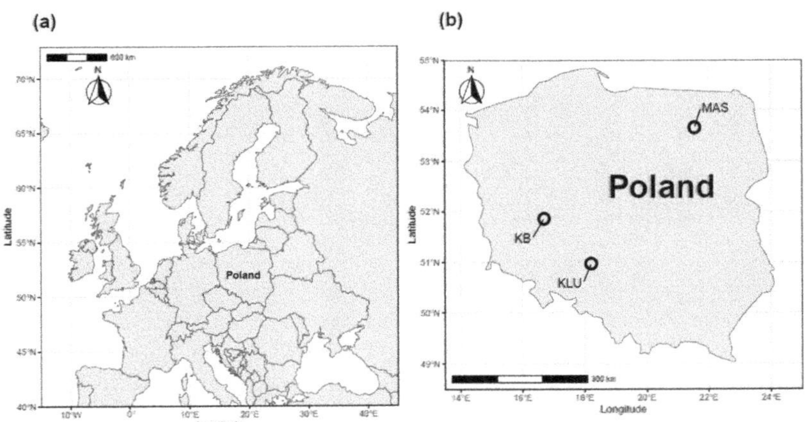

Figure 1. Locations of harvested populations in Europe (**a**) and in Poland (**b**). Abbreviations indicate analyzed populations: MAS—Maskulińskie; KB—Karczma Borowa; KLU—Kluczbork.

The number of hunted stags in each location depended on the population size and structure connected to the total forest area. However, the following data have been collected for each individual: region of harvesting, the date of the harvest, stags age, carcass mass (kg), antler mass (kg), and the form of the antlers (number of tines).

To determine the age of the animals, the dentition features and the degrees of their clashes were analyzed [9]. The red deer stags were divided into five classes: 1st class—stags 2 years old; 2nd class—stags 3–5 years old; 3rd class—stags 6–8 years old; 4th class—stags 9–11 years old; and 5th class—12 years and older. All red deer individuals younger than 2 years were excluded from the analysis.

To determine body mass, we used stag carcass mass (±1 kg) that was measured without head and antlers, whereas antler mass (±0.01 kg) was measured with skull without mandible. We used a ratio of antler mass over body mass, which, according to Mattioli [5], can quantify antler investment in terms of efforts to build conspicuous secondary sexual traits.

Antler forms were classified into regular (R) and non-regular (NR) forms (Figure 2). The regular form indicated antlers with the same number of tines on each branch [22]. Non-regular forms indicated a higher number of tines on one branch. The forms of the antlers were classified as follows: none (n), spike (SP), 6R, 6NR, 8R, 8NR, 10R, 10NR, 12R, 12NR, 14R, 14NR, 16R, 16NR, 18R, 18NR, 20NR, 22NR, fork (F), and bizarre (B). The number of tines on less developed branches was not determined, and such antlers were classified as non-regular.

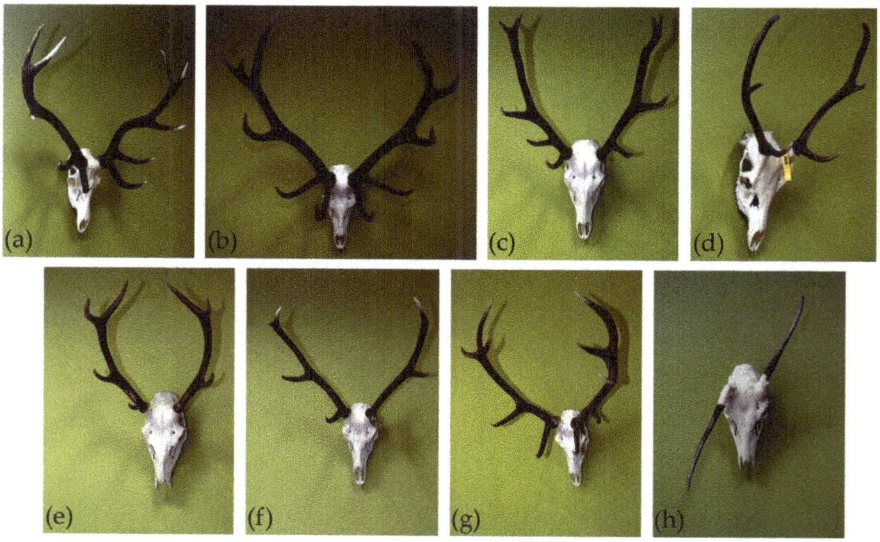

Figure 2. Different forms of antlers: (**a**) stag with antler form 10 R; (**b**) stag with antler form 12 R; (**c**) stag with antler form 10 NR; (**d**) stag with antler form 6 R; (**e**) stag with antler form 8 R; (**f**) stag with antler form 8 NR; (**g**) stag with antler form 10 R; and (**h**) spike with broken pedicle. R—regular, NR—non-regular antler form.

Therefore, for each harvested individual, the following data were collected: day of harvest, location, carcass and antler mass, and type of antler form.

2.2. Study Sites

Studies were conducted in three different forest districts—Karczma Borowa, Kluczbork, and Maskulińskie—which represent varied climatic (e.g., temperature, precipitation, snow cover) and habitat conditions (forest type) (Table 1).

We classified the forests as follows:

1. Karczma Borowa (KB)—Central–Western Poland. The climate of this region is transitional between oceanic and continental, and milder than in the eastern parts of Poland [23];
2. Kluczbork (KLU)—mild weather conditions prevail in these areas, especially in the western part (located on the Oder River). This location has the longest growing season in Poland and very favorable amounts of precipitation, especially for agriculture [24];

3. Maskulińskie (MAS)—the climate is harsher here than in other locations; the presence of large lakes leads to high precipitation (677 mm) and relative humidity (80%) [25].

Table 1. Characteristics of the forest of Red Deer (*Cervus elaphus* L.) populations [23–25].

Parameter	Location of Population		
	KB	**KLU**	**MAS**
Geographical coordinates	51°84′–51°42′ N 16°23′–16°56′ E	50°51′–51°10′ N 17°54′–18°27′ E	53°47′–53°50′ N 21°33′–21°48′ E
Total area (ha)	12,698.24	19,032.07	28,329.85
Forest area (ha)	12,211.61	18,484.85	26,065.35
Percentage of broadleaved forests (%)	48.9	61.3	10.0
Vegetation period (days)	210	220	200–205
Mean forest age (years)	58	60	83
Grazing forest plots (ha)	9.0	36.8	31.0
The density of red deer per 1000 ha	21	33	20
Optimal density of red deer per 1000 ha	25–30	25–30	25–30

2.3. Climatic Data Collection

We assumed that resource allocation including antler building by red deer stags could be affected by main climatic factors, as severe climatic conditions may interfere with the ability of individuals to store carcass reserves, creating a trade-off between growth, survival, and subsequent reproduction [26].

Among selecting climate factors, we focused on the factors during which antler growth and development occur, as well as changes in individual red deer stag condition before and after the deer mating period.

Each month within the hunting season was characterized by mean monthly precipitation (mm/month), monthly average daily temperatures (°C), and the mean number of frost days (days with temperatures below zero) as the main climatic factors affecting populations to determine resource allocation. These climatic characteristics were calculated from January to the harvest date for carcass mass analysis from the last hunting season.

We downloaded a dataset from the Climatic Research Unit (University of East Anglia) and Met Office (CRU TS Version 4.06 database [27]). The climatic data for the red deer populations were obtained from the database using the 'raster' package [28].

2.4. Statistical Analysis

We started the analysis by using all predictors and using the Akaike Information Criterion (AIC) to choose the most parsimonious model to analyze overdispersion. We used population location as a random factor (1 | location), whereas precipitation, temperature, and the number of frost days (January–harvest date) were fixed factors. We used the lmer4 package for mixed models [29]. Results were treated as significant when $p < 0.05$. Maps were provided using the 'ggmap', 'sf', and 'ggspatial' packages. Plots were produced based on the ggplot package. Statistical analyses were conducted using R 4.2.2 software [30]; available online: 21 June 2023).

We performed ANOVA followed by Tukey test to describe the effect of localization of population on analyzed features of carcass and antlers mass and antler investment. We also used the chi-square test to compare the frequency of age classes of hunted stags and the frequency of antler forms between populations. Data were presented as mean values with standard deviation (SD). Data were described as statistically significant when $p < 0.05$.

3. Results

In five hunting seasons (2017–2021), 550 red deer stags were hunted between late August and February, and they were hunted as follows: 343 stags in the KLU Forest District, 118 stags in the MAS, and 89 stags in the KB Forest District. The total number of hunted stags varied over the years (2017: 123; 2018: 95; 2019: 141; 2020: 140; 2021: 51) and differs over the months (in total: September: 170; October: 46; November: 95; December: 77; January: 117; February: 45).

The proportion of stags from different age classes differed between populations and ranged as follows: in the 1st class (none in KB), from 0.29% in KLU to 2.54% in MAS; in the 2nd class, from 14.61% in KB to 38.14% in MAS; in the 3rd class, from 35.59% in MAS to 45.77% in KLU; in the 4th class, from 18.95% in KLU to 32.58% in KB; and in the 5th class, from 0.29% in KLU to 8.99% in KB (χ^2 (2, 550) = 46.8032, $p < 0.001$) (Figure 3, Table A1).

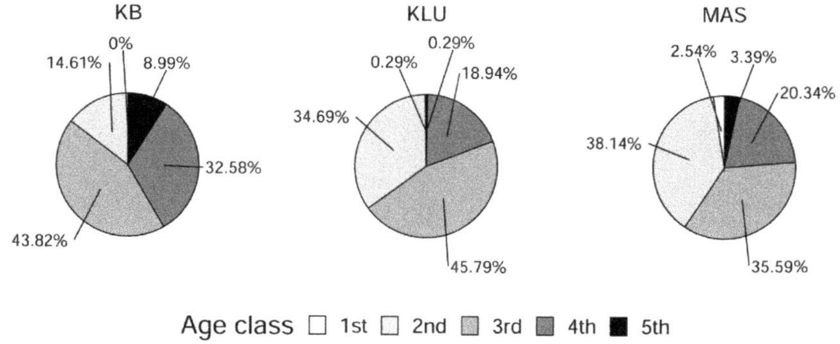

Figure 3. Proportion of different age classes of Red Deer stags harvested by hunters in three populations in Poland between 2017–2021. KB—Karczma Borowa Forest; KLU—Kluczbork Forest and MAS—Maskulinskie Forest.

Carcass mass was shaped by the location of the population, stag age, mean precipitation, mean temperature, and mean number of frost days from January to the harvest date (AIC = 4592.06, $R^2_m = 0.364$, $R^2_c = 0.588$, Table 2). Stags from KB had the highest mean carcass mass and differed from other populations (Table 2).

Table 2. Predictors for carcass and antler mass, and antler investment of red deer (*Cervus elaphus* L.) stags of three populations in Poland in 2017–2021. CI—confidence interval, *p*—probability.

Predictor	Carcass Mass			Antler Mass			Antler Investment		
	Estimate	CI	*p*	Estimate	CI	*p*	Estimate	CI	*p*
(Intercept)	−135.39	−197.69–−73.08	<0.001	−0.58	−4.30–3.15	0.760	1.48	−2.76–5.72	0.494
Stag age	3.54	3.00–4.08	<0.001	0.53	0.50–0.57	<0.001	0.34	0.31–0.36	<0.001
Minimal temperature	21.33	16.01–26.66	<0.001	−0.32	−0.94–0.30	0.313	−0.02	−0.38–0.35	0.3968
Mean temperature	-	-	-	0.31	−0.09–0.70	0.132	-	-	-
Frost days	9.45	6.19–12.71	<0.001	0.01	−0.21–0.23	0.953	0.05	−0.14–0.24	0.610
Precipitation	0.45	0.28–0.62	<0.001	-	-	-	0.00	−0.02–0.01	0.368
Random effects: *n* = 3		$R^2_m = 0.364$ $R^2_c = 0.588$			$R^2_m = 0.556$ $R^2_c = 0.741$			$R^2_m = 0.546$ $R^2_c = 0.650$	

The mean carcass mass varied over the months. The highest carcass mass in the pooled data was noted in September (136.1 ± 12.0 kg), and the lowest in February (98.6 ± 16.9 kg); however, these values differed between the locations. The reduction reached 28.23% in KLU (from September to February), 26.65% in KB (from September to January), and 22.2% in MAS (from October to February) (Figure 4a). The reductions in carcass mass during autumn and winter differed between populations. It was also revealed that the changes in carcass mass varied between stag age classes. The reduction in carcass mass was the lowest in the 4th class (16.52%), whereas the highest was in the 5th class (27.2%). The carcass mass increases also differed between the populations. Our data revealed higher carcass mass increases from the 2nd to 5th age classes in KLU (+40.0%, from 96 to 135 kg), while the lowest increase was in KB (16.5%, from 127 to 147 kg) (Figure 4b).

Antler mass (AIC = 1564.7, R^2_m = 0.556, R^2_c = 0.741), as well as antler investment (AIC = 1223.7, R^2_m = 0.546, R^2_c = 0.650), were affected by stag age (Table 2). The lowest mean values were recorded in KLU, and the highest mean values in KB (Table 3). The increases in antler mass from the 2nd to 5th age classes differed between the populations. The lowest increase was recorded in KLU (+107%); it was slightly higher in KB (+120%), and the highest in MAS (+125%) (Figure 4c).

Table 3. Mean values of carcass and antler mass, and antler to carcass mass ratio of red deer (*Cervus elaphus* L.) stags of three populations in Poland in 2017–2021. Data were presented as mean and standard deviation. Values with different letters indicate statistically significant differences between populations from different locations.

Parameter	Locations			
	Karczma Borowa	Kluczbork	Maskulinskie	*p*
Mean carcass mass (kg)	138.7 ± 21.1 a	112.8 ± 20.2 b	112.1 ± 20.8 b	0.001
Mean antler mass (kg)	6.39 ± 1.95 a	3.71 ± 1.47 b	4.04 ± 1.95 b	0.001
Antler investment (%)	4.59 ± 1.17 a	3.22 ± 0.97 b	3.49 ± 1.37 b	0.001

The antlers of the analyzed stags were classified into 20 types of antler forms. Antler forms characterized as SP, F, and B were observed in 5.27% of stags (29/550; 3, 2, and 24, respectively), whereas 44.9% (247/550) of stags possessed a regular form of antlers, while 49.82% (274/550) of them had a non-regular form of antlers. In total, from all 20 types of antler forms the most frequently observed were 12 NR (15.2%), 10 R (13.8%), and 10 NR (13.4%). Two of them were also frequently observed within populations; however, their participation differs between populations. According to share of antler forms, the three most frequent forms in KB were 14 NR (22.47%), 10 R (15.73%), and both 12 N (11.24%) and 12 NR (11.24%); 10 NR (18.08%), 12 NR (17.49%), and 10 R (13.12%) in KLU; and 8 R (16.10%), 10 R (14.41%), and 12 NR (11.86%) in MAS (Table A1).

When only stags with R and NR antler forms were considered, the highest frequency of stags with R form was observed in MAS (60.36%, 67/111), compared to KB (42.86%, 36/84) and KLU (44.17%, 144/326) (χ^2 (2, n = 521) = 9.5395, $p < 0.01$). We observed that stags with NR antler forms had higher antler mass (4.54 kg) compared to regular form counterparts (3.92 kg; $p < 0.0001$), and these was statistically significant for two populations: KB and KLU; however, similar trend was observed in MAS (Table 4). Moreover, there were no significant differences in mean carcass mass between R and NR stags ($p > 0.05$); however, populations differed between each other. NR stags were characterized by statistically significant higher index of antler investment in KB and KLU, but not in MAS (Table 4).

According to climatic conditions, we have recorded higher mean monthly precipitation in KLU (29.8–89.1 mm) during the whole year, which exceeded the values recorded from KB (25.2–75.6 mm) and MAS (25.5–74.8 mm). The precipitation in KB and MAS reached similar values during the year. The mean monthly temperature was the lowest in MAS (−3.64–+19.78) during the whole year. Temperature was similar between KB (−0.7–+20.9) and KLU (−0.9–+20.1). Similarly, the number of days with frost during the year was higher in MAS (up to 117 days in a year) when compared to KLU (106 days) and KB (up to 96 days) (Figure 5a–c).

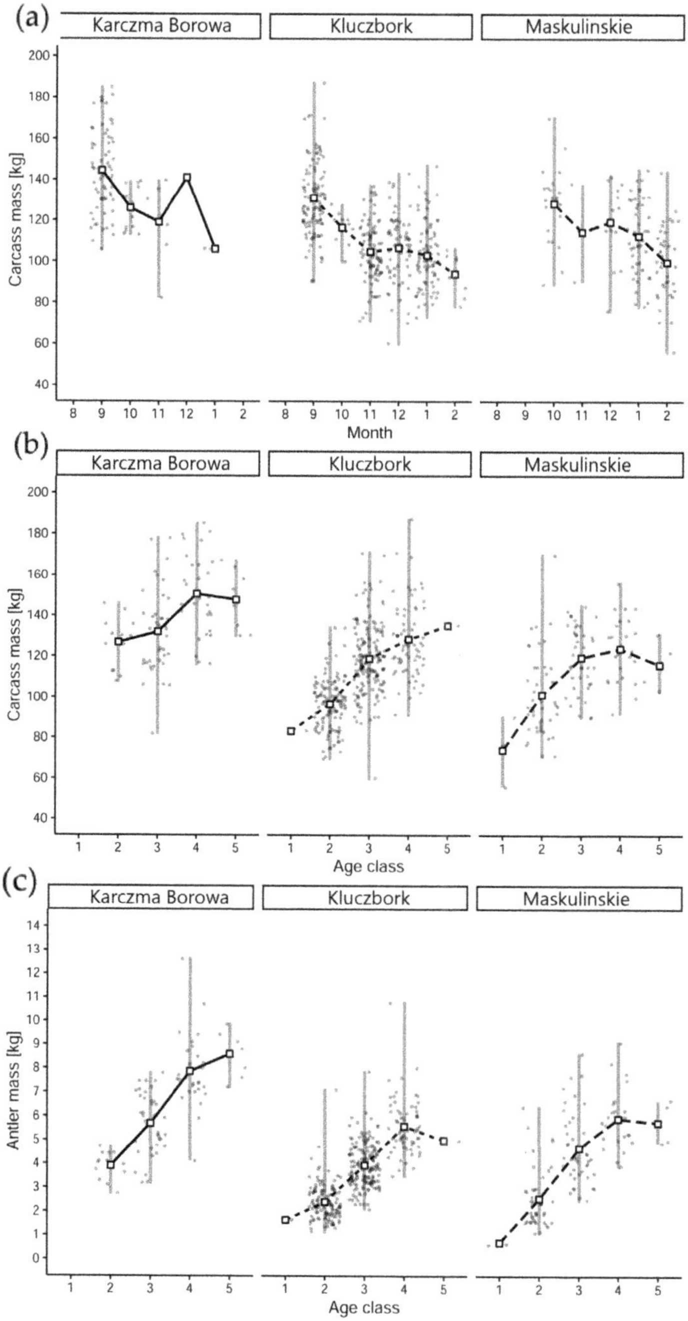

Figure 4. Mean carcass mass (kg) within (**a**) months, mean carcass mass (**b**) as well as mean antler mass (kg) (**c**) in age classes of red deer (*Cervus elaphus* L.) stags of three populations in Poland in 2017–2021 (August–February). The points represent the data for individual stag, the vertical lines represent the range of the value, and the white square indicates the mean value of each month or age class.

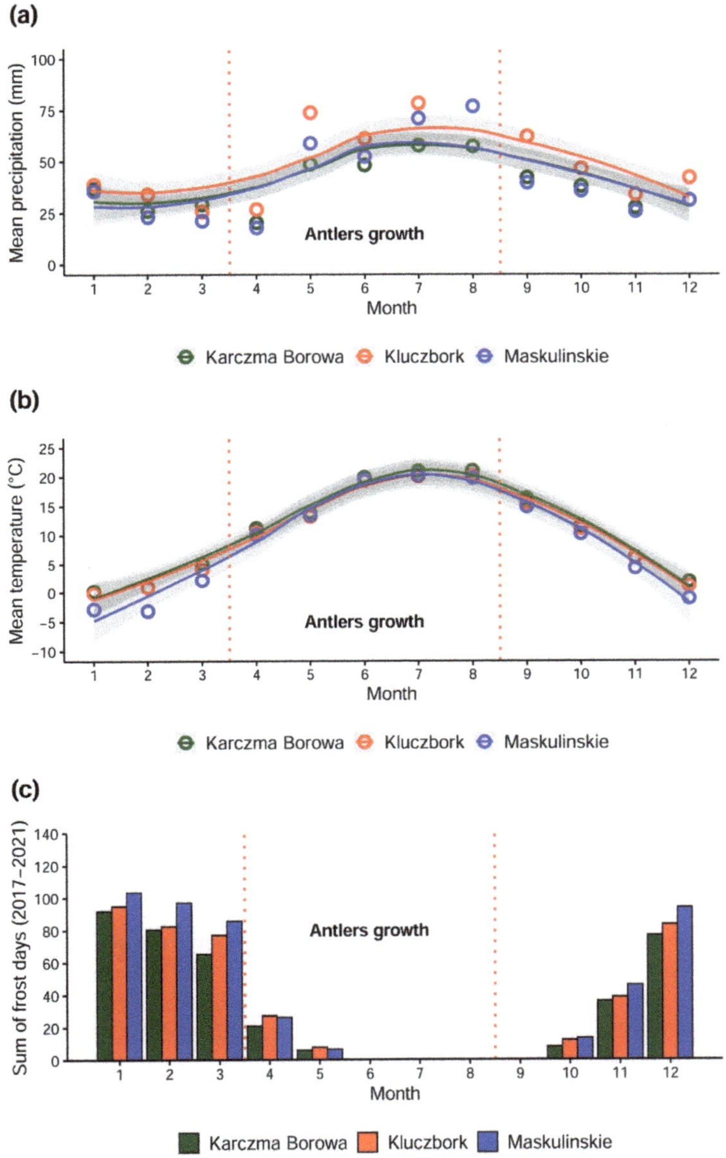

Figure 5. Characteristics of mean environmental conditions of three populations of red deer in 2017–2021 described by (**a**) mean precipitation (mm), (**b**) mean temperature (°C), and (**c**) sum of frost days.

Table 4. Mean values of carcass and antler mass, as well as antler investment of red deer (*Cervus elaphus* L.) stags of three populations in Poland in 2017–2021 in relation to antler form (R—regular and NR—non-regular). Data were presented as mean and standard deviation. Values with different letters indicate statistically significant differences between populations from different locations (a,b) and groups with different antler's form (x,y).

Parameter		Locations			
		Karczma Borowa	Kluczbork	Maskulinskie	*p*
Mean carcass mass (kg)	R	134.05 ± 22.38 a	110.84 ± 21.55 b	110.34 ± 21.93 b	<0.001
	NR	139.67 ± 18.94 a	115.05 ± 18.97 b	117.21 ± 16.79 b	<0.001
	p	0.218	0.062	0.081	
Mean antler mass (kg)	R	5.68 ± 1.79 ya	3.51 ± 1.45 yb	3.84 ± 2.00 b	<0.001
	NR	6.78 ± 1.93 xa	3.96 ± 1.46 xb	4.50 ± 1.64 b	<0.001
	p	0.009	0.007	0.069	
Antler investment (%)	R	4.21 ± 1.03 ya	3.09 ± 0.91 yb	3.35 ± 1.42 b	<0.001
	NR	4.85 ± 1.25 xa	3.40 ± 1.00 xb	3.79 ± 1.14 b	<0.001
	p	0.0145	0.0046	0.089	

4. Discussion

The reproductive success of red deer stags depends on its success in defending harems or mating territories, as red deer is a highly polygynous species [31]. Therefore, sexual selection can promote the evolution of secondary sexual traits, such as body and antler mass, thus a stag's dominance is largely determined by both traits. However, sexually selected traits are costly and can be better afforded by individuals living in better conditions or with better resource availability [32]. Our data indicates that environmental conditions such as precipitation, mean temperature, and mean number of frost days affect mean carcass mass, but not mean antler mass or antler investment. In our studies, those two features were affected by stag age. Moreover, we observed that mean antler mass, as well as antler investment, were higher in stags characterized by the presence of non-regular antler form compared to regular counterparts. Regular antler form was more frequently observed in populations characterized by a greater scale of environmental stress. Furthermore, stags from harsh conditions lose carcass mass more slowly during autumn and winter time, compared to their counterparts.

Our observation pointed out that the highest stag carcass mass (over 138 kg on average) was recorded in the population in Karczma Borowa, which is located in the southwestern part of Poland, which is characterized by higher mean monthly temperature but also the lowest number of days with frost during the year. Therefore, costs related to thermoregulation were lower compared to red deer living in Maskulinskie, where the highest number of frost days as well as the lowest mean monthly temperature were observed together with lower carcass mass. Our data are in line with others [9,33–35]. Carcass and antler mass vary between population localization, which may be caused by habitat conditions, the genetic characteristics of individual populations, varying population densities and climatic conditions, among other factors [9]. Our data indicate that carcass mass is shaped by climatic conditions, such as precipitation and temperature (minimal and maximal), but those environmental factors had no effect on antler mass and antler investment. Analyzed climatic factors, through their influence on plant phenology [36], shape the quality and digestibility of food. For instance, delayed lignification resulting from a cool onset of vegetation (May–June) leads to longer access to high-quality forage and, consequently, greater weights [36,37].

As an example, a positive relationship between temperature and calf mass has been shown in red deer populations in Highland [38]. A similar relationship has been shown by Borkowski [38] in calves in the Raciborskie Forests. Our results are also consistent with those of previous reports on stag carcass mass in Poland [20,39–41]. Interestingly, the carcass mass of the stags from the population in Kluczbork was about 10% higher than the results reported from the areas of Wielkopolska [34] and Opole [9], which may suggest a slight recent increase in the carcass mass of harvested stags. We also revealed varying increases in stag carcass mass among the age classes. The highest total increase in stag carcass mass from the youngest to the oldest individuals was observed in the population in Kluczbork (a cumulative increase of 40%), and the lowest was observed in the population in Karczma Borowa (of 17%). In addition, the rates of weight increase varied based on age class and were the highest in individuals representing younger age classes. These higher increases are consistent with a study by Czyżyk et al. [42], who reported that carcass mass can increase by as much as 25% between 2 and 3 years and then more slowly in older age classes. In our study, the highest stag carcass mass was recorded in the population in Kluczbork (on average, about 139 kg), which is similar to the results obtained by Bobek et al. [33] in the area of Southeastern Poland, where bull carcass mass has ranged from 108 to 148 kg [41]. In addition, Wajdzik [9] has reported an average stag carcass mass of about 110 kg in the Opole area; this is lower than our results for the nearest population in Kluczbork. Therefore, a new question arose: When did the full physical development of the red deer stags occur in the populations studied? The answer is between the ages of 9 and 11 (the 4th age class in our study), which has been confirmed by the data from Karczma Borowa and Maskulińskie. Our results have also revealed the smallest range of mass variation in older individuals, i.e., over 12 years of age in the populations from Karczma Borowa and Maskulińskie. According to Zalewski and Szczepanski [43], in populations in Northeastern Poland, the full physical development of red deer stags occurs between 6 and 7 years of age. Our study indicated a loss of carcass mass during the post-horn period in all populations studied. The lowest mass loss occurred in the Maskulińskie population (a decrease of about 22.6%) and the highest in the Kluczbork (a decrease of about 27.8%) within a winter period. The results obtained are consistent with those presented by Nasiadka et al. [10], who reported a range of carcass mass loss between 20 and 35% over that period. The observed pronounced decrease in the carcass mass of stags is a result of the high energy costs of the rigor they underwent.

Beside the importance of climatic conditions, the carcass mass may be also shaped by the forage and food, which is connected with the forest habitat type. It has been also proved that the proportional abundance of broadleaved forest in the total cover area is crucial for the condition of individual red deer [9]. In our forests, the total forest area ranges from ~12 to ~28 thousands of ha, which means that we had large forest areas where abundant populations may occur. Further, the proportional abundance of broadleaved forest was highest in Kluczbork (~60%) and Karczma Borowa (~50%) and lowest in Maskulińskie (~10%) forests. Previous studies of Wajdzik et al. [9] proved that less afforested areas (forest area < 40%) with higher proportion of fields were characterized by higher red deer carcass mass in all age classes when compared to stags living in forest areas (forest area > 40%). Similar observation was also done by Mysterud et al. [15]. Our studies were conducted in typical forest areas (forest area > 40%); therefore, among environmental factors, only the proportional abundance of broadleaved species may be important. Moreover, all analyzed populations had similar densities. Earlier start of the growing season has an impact on sooner access to high-quality forage, which affects the condition of individuals (carcass mass) and can manifest itself in better antler quality. In our study, the growing season was similar between locations and ranged from 200 to 210 days. This difference may be crucial in the case of particularly harsh winter conditions (when it is difficult for animals to access food for a longer period of time), which could manifest itself in lower carcass mass in the population from Maskulinskie. The observed effect of environmental conditions on the increase in carcass mass of stags may reflect selection related to access to does and thus to reproduction success and the transmission and maintenance of the genetic pool in the population [44].

We observed that antler mass depends on stag age but is not affected by environmental conditions such as temperature or precipitation. The effect of stag age on antler development, including antler mass, has been repeatedly documented by other researchers [42,43,45]. We observed that the average antler mass of the stags in Karczma Borowa was higher in each age class compared with those in the other populations. On the other hand, the highest increase in antler mass occurred at up to 9 to 11 years of age (age class 4), which is consistent with the results presented by Łabudzki [34] from the Greater Poland area and Zalewski [46] from the Warmia and Mazury areas. Our results revealed the largest antler mass to be in Karczma Borowa (8.60–9.40 kg) in 11- to 13-year-old stags. In other populations, we obtained lower values, i.e., in Kluczbork (5.8–6.3 kg) for ages 10 and 11, and in Maskulińskie (6.3–6.7 kg) for ages 11–15. Our results are in line with those of Zalewski [46], who noted the culmination of antler mass at 6.3 kg at 9–12 years of age in the Warmia and Mazury regions. On the other hand, Łabudzki [34] showed climaxes that occurred at 11 (5.13 kg) and 12 (5.04 kg) years of age. Similar results from the Pisz Forest area were obtained by Żurkowski [47], who noted that antler mass climaxed at the age of 7–10 years and obtained values from 4.1 to 4.7 kg. The highest antler mass was for stags at age 11 (5.6 kg). Similar correlations were recorded for the population in Bory Tucholskie, where the highest average antler mass (5.9 kg) was recorded for stags aged 11 and older [48]. The development of antler mass in all the presented populations did not differ from the results presented by other authors.

Antler mass of stags was age-related, but not related to climatic factors. Both antler mass, as well as antler investment, were stag age dependent. Red deer from Karczma Borowa had heavier antlers, and antler investment reached higher values than other populations. However, Karczma Borowa and Kluczbork were characterized by lower frequency (~40%) of regular antler form compared to Maskulinskie, where more than 60% of all harvested stags had R antler form. Red deer that had NR antler forms were characterized by higher index of antler development and higher antler mass, which can be connected to fluctuating asymmetry theory. According to that theory, in stress conditions, red deer stags could invest resources into production of regular, symmetrical antler form that further underlines the importance of antler shape in reproductive success [49]. Moreover, cost related to secondary sexual traits, e.g., antler development, can be modulated not only by production cost, but also by benefits resulting from their presence. Antler size is also dependent on the level of intra-sexual competition; thus, red deer males living in female groups produced smaller antlers compared to red deer living in male groups [50]. It was also observed that, each year, approximately 6% of red deer are injured in a fight [51]. Therefore, higher antler mass as well as better quality, described by antler investment and symmetrical form, could be quantitative markers of stag condition. We observed that, in harsh conditions, red deer had lower carcass mass and antler mass, and although antler investment was decreased, red deer more often had symmetrical antler form. On the other hand, red deer living in better environmental conditions with potentially more available resources and lower cost of thermoregulation had higher carcass mass, antler mass, and antler investment, but more often individuals produced non-regular antler form. Thus, sexually selected traits are expected to represent the quality of the individual bearing them [52], and more symmetrical antler form results in lower probability of antler damage during fighting [53], as well as possibly higher reproductive success; therefore, red deer living in harsh environmental conditions can be described by better individual condition compared to individuals living in better conditions. Further, red deer living in better environmental conditions with higher resource availability can invest more in antler mass;

thus, it can compensate for the presence of non-regular antler form. Moreover, as others have pointed out, antler building (antler investment) depends on age but is also affected by food availability, red deer density, and climate [5,54–56]. In our study, density of red deer population was highest in Kluczbork, but not in Karczma Borowa where we observed the highest antler mass. Mateos et al. [49] found that environmental variability is less important in individuals at the prime age. The development of antlers is a product of individual potential, functional needs, and the availability of building substances, which mainly come from food intake. The availability of agricultural land determines increased carcass mass [57], as well as carcass and antler mass [9]. In Southwestern Poland, stags with higher antler mass were recorded, which is consistent with the results obtained from the Kluczbork area in our study. Similar relationships have been reported in populations in northern Italy, in areas characterized by large areas of agricultural land [5]. These relationships can be explained by increased environmental diversity [33] and a long ecotone that fills the needs of this species [9,58,59]. However, our outcomes similar to other studies conducted on wild animals, could also undergo hunter-selection pressure.

5. Conclusions

Environmental conditions shape individuals' features related to secondary sexual traits. Our data indicated that individual quality of red deer stags described by carcass mass is controlled by environmental factors (precipitation, mean minimal temperature, and number of days with frost), but those factors had no impact on antler mass and antler investment. Moreover, harsh environmental conditions can promote the presence of regular antler form, and these can also be controlled by genetic selection or adaptation. Our observation confirms that resource trade-off is related to secondary sexual traits of red deer.

Author Contributions: Conceptualization—J.S.; methods—J.S., G.G. and J.K.; investigation—J.S., G.G. and J.K.; formal analysis—J.S., G.G., J.K. and E.P.-K.; writing—original draft preparation—J.S., G.G. and J.K.; writing—review and editing—J.S., G.G., J.K. and E.P.-K. All authors have read and agreed to the published version of the manuscript.

Funding: The publication was co-financed by the Polish Minister of Science and Higher Education as part of the Strategy of the Poznan University of Life Sciences for 2024–2026 in the field of improving scientific research and development work in priority research areas. Partial support also came from the Department of Game Management and Forest Protection, Poznan University of Life Sciences, and statutory research of the Institute of Dendrology, Polish Academy of Sciences.

Institutional Review Board Statement: Approval from the ethical committee for animal experiments is required in Poland for studies or experimentation on animals. According to Article 2, point 2, subpoint 1 of the Act of 15 January 2015, on the protection of animals used for scientific or educational purposes (https://dziennikustaw.gov.pl/D2015000026601.pdf, accessed on 7 November 2024), we did not study animals. Our research involved post mortem examination of bones (jaw) of red deer (*Cervus elaphus* L.) hunted by hunters under the Hunting Law (https://isap.sejm.gov.pl/isap.nsf/download.xsp/WDU20230001082/T/D20231082L.pdf, accessed on 7 September 2024) and the Regulation of the Minister of Climate and Environment on the list of game species (https://isap.sejm.gov.pl/isap.nsf/download.xsp/WDU20050450433/O/D20050433.pdf, accessed on 7 November 2024).

Data Availability Statement: Data is contained within the article. The original contributions presented in the study are included in the article, further inquiries can be directed to the corresponding author.

Acknowledgments: We would like to thank the following Forest Districts: Karczma Borowa, Kluczbork, and Maskulińskie for providing the data.

Conflicts of Interest: The authors declare no conflicts of interest.

Appendix A

All data were presented within the manuscript.

Table A1. Numbers of Red Deer classified according to different forms of antlers within different age classes that were hunted-harvested in three populations—Karczma Borowa (KB), Kluczbork (KLU), Maskulińskie (MAS)—in the years 2017–2021. Each number represents the maximum tines for all beams. NR—irregular form; R—regular form, SP—spike form; F—fork, B—bizarre form.

Antler Form	KB					KLU						MAS						Total
	2nd	3rd	4th	5th	Total	1st	2nd	3rd	4th	5th	Total	1st	2nd	3rd	4th	5th	Total	
SP	-	-	-	-	-	-	-	-	-	-	-	3	-	-	-	-	3	3
F	-	-	-	-	-	-	2	-	-	-	2	-	-	-	-	-	-	2
6NR	-	-	-	-	-	-	2	-	-	-	2	-	-	-	-	-	-	2
6R	-	-	-	-	-	-	33	1	-	-	34	-	11	-	-	-	11	45
8NR	3	-	-	-	3	-	35	5	1	-	41	-	4	2	-	-	6	50
8R	2	3	-	-	5	-	20	22	1	-	43	-	14	5	-	-	19	67
10NR	1	3	-	-	4	-	9	48	4	-	62	-	6	2	-	-	8	73
10R	5	8	1	-	14	-	5	32	7	1	45	-	8	7	2	-	17	76
12NR	-	8	2	-	10	-	3	36	21	-	60	-	1	8	4	1	14	84
12R	1	6	3	-	10	-	1	4	8	-	13	-	-	7	1	-	8	31
14NR	1	6	9	4	20	-	1	2	6	-	9	-	1	3	7	1	12	41
14R	-	1	3	1	5	-	-	1	8	-	9	-	-	3	5	2	10	24
16NR	-	1	3	-	4	-	-	-	7	-	7	-	-	1	-	-	1	12
16R	-	-	-	1	1	-	-	-	-	-	-	-	-	1	-	-	1	2
18NR	-	-	1	2	3	-	-	-	1	-	1	-	-	-	1	-	1	5
18R	-	-	1	-	1	-	-	-	-	-	-	-	-	1	-	-	1	2
20NR	-	1	2	-	3	-	-	-	1	-	1	-	-	-	1	-	1	5
22NR	-	-	-	-	-	-	-	-	-	-	-	-	-	-	1	-	1	1
24NR	-	-	1	-	1	-	-	-	-	-	-	-	-	-	-	-	-	1
B	-	2	3	-	5	1	8	6	-	-	15	-	-	2	2	-	4	24
Total	13	39	29	8	89	1	119	157	65	1	343	3	45	42	24	4	118	550

References

1. Levins, R. Evolution in Changing Environments: Some Theoretical Explorations. In *Evolution in Changing Environments*; Princeton University Press: Princeton, NJ, USA, 2020; ISBN 978-0-691-20941-8.
2. Moyes, K.; Morgan, B.J.T.; Morris, A.; Morris, S.J.; Clutton-Brock, T.H.; Coulson, T. Exploring Individual Quality in a Wild Population of Red Deer. *J. Anim. Ecol.* **2009**, *78*, 406–413. [CrossRef]
3. Stearns, S.C. Trade-Offs in Life History Evolution. *Funct. Ecol.* **1989**, *3*, 259–268. [CrossRef]
4. Kavčić, K.; Corlatti, L.; Safner, T.; Budak, N.; Šprem, N. Contrasting Patterns of Sexually Selected Traits in Mediterranean and Continental Populations of European Mouflon. *Ecol. Evol.* **2020**, *10*, 2085–2092. [CrossRef]
5. Mattioli, S.; Ferretti, F.; Nicoloso, S.; Corlatti, L. Spatial Variation in Antler Investment of Apennine Red Deer. *Ecol. Evol.* **2021**, *11*, 7850–7864. [CrossRef]
6. Gómez, J.A.; Ceacero, F.; Landete-Castillejos, T.; Gaspar-López, E.; García, A.J.; Gallego, L. Factors Affecting Antler Investment in Iberian Red Deer. *Anim. Prod. Sci.* **2012**, *52*, 867–873. [CrossRef]
7. Zalewski, D.; Szczepański, W. Contribution of Particular Features of Red Deer (*Cervus elaphus* L.) Antlers in the CIC Score. *Sylwan* **2004**, *10*, 30–37.
8. Schwabland, J.; Barnhart, L.; Paz, G.; Aguilar, G. *Official Measurer's Manual*; Safari Club International: Washington, DC, USA, 2019.
9. Wajdzik, M.; Hink, K.; Szyjka, K.; Nasiadka, P.; Skubis, J. Wpływ lesistości na jakość osobniczą samców jelenia szlachetnego (*Cervus elaphus*) na terenie opolszczyzny. *Acta Sci. Pol. Silvarum Colendarum Ratio Et Ind. Lignaria* **2018**, *17*, 69–77. [CrossRef]
10. Nasiadka, P.; Kowalski, M.; Sadowski, B.; Wojtan, R.; Merta Nasiadka, D.P.; Merta, D. Charakterystyka masy tuszy i poroży byków jeleni (*Cervus elaphus* L.) bytujących w Lasach Spalskich i Lasach Gostyńsko-Włocławskich w Polsce centralnej. *Sylwan* **2016**, *160*, 777–788.
11. Skog, A.; Zachos, F.E.; Rueness, E.K.; Feulner, P.G.D.; Mysterud, A.; Langvatn, R.; Lorenzini, R.; Hmwe, S.S.; Lehoczky, I.; Hartl, G.B.; et al. Phylogeography of Red Deer (*Cervus elaphus*) in Europe. *J. Biogeogr.* **2009**, *36*, 66–77. [CrossRef]
12. Jiang, Z.; Jiang, J.; Wang, Y.; Zhang, E.; Zhang, Y.; Li, L.; Xie, F.; Cai, B.; Zheng, G. Red List of China's Vertebrates. *Biodivers. Sci.* **2016**, *24*, 500–551. [CrossRef]
13. Flueck, W.T. Exotic Deer in Southern Latin America: What Do We Know about Impacts on Native Deer and on Ecosystems? *Biol. Invasions* **2010**, *12*, 1909–1922. [CrossRef]

14. Forsyth, D.M.; Richardson, S.J.; Menchenton, K. Foliar Fibre Predicts Diet Selection by Invasive Red Deer *Cervus elaphus scoticus* in a Temperate New Zealand Forest. *Funct. Ecol.* **2005**, *19*, 495–504. [CrossRef]
15. Mysterud, A.; Langvatn, R.; Meisingset, E.L.; Rivrud, I.M. Agricultural Grasslands Buffer Density Effects in Red Deer Populations. *J. Wildl. Manag.* **2023**, *87*, e22357. [CrossRef]
16. Zachos, F.E.; Hajji, G.M.; Hmwe, S.S.; Hartl, G.B.; Lorenzini, R.; Mattioli, S. Population Viability Analysis and Genetic Diversity of the Endangered Red Deer *Cervus elaphus* Population from Mesola, Italy. *Wildl. Biol.* **2009**, *15*, 175–186. [CrossRef] [PubMed]
17. Niedziałkowska, M.; Jedrzejewska, B.; Honnen, A.C.; Otto, T.; Sidorovich, V.E.; Perzanowski, K.; Skog, A.; Hartl, G.B.; Borowik, T.; Bunevich, A.N.; et al. Molecular Biogeography of Red Deer *Cervus elaphus* from Eastern Europe: Insights from Mitochondrial DNA Sequences. *Acta Theriol* **2011**, *56*, 1–12. [CrossRef]
18. Meiri, M.; Kosintsev, P.; Conroy, K.; Meiri, S.; Barnes, I.; Lister, A. Subspecies Dynamics in Space and Time: A Study of the Red Deer Complex Using Ancient and Modern DNA and Morphology. *J. Biogeogr.* **2018**, *45*, 367–380. [CrossRef]
19. Ludt, C.J.; Schroeder, W.; Rottmann, O.; Kuehn, R. Mitochondrial DNA Phylogeography of Red Deer (*Cervus elaphus*). *Mol. Phylogenet Evol.* **2004**, *31*, 1064–1083. [CrossRef] [PubMed]
20. Brewczynski, P. Ciezar ciala i jakosc poroza jelenia *Cervus elaphus* L. w osrodkach hodowli zwierzyny na terenie Regionalnej Dyrekcji Lasow Panstwowych w Krosnie (Karpaty). *Sylwan* **2002**, *146*, 63–75.
21. Mysterud, A.; Meisingset, E.; Langvatn, R.; Yoccoz, N.G.; Stenseth, N.C. Climate-Dependent Allocation of Resources to Secondary Sexual Traits in Red Deer. *Oikos* **2005**, *111*, 245–252. [CrossRef]
22. Okarma, H.; Tomek, A. *Łowiectwo*; Wydawnictwo Edukacyjno-Naukowe H2O: Bibice, Poland, 2008; ISBN 978-83-927737-0-2.
23. Office of Forest Management and Forest Surveying. *Forest Management Plan for Karczma Borowa Forest District*; Poznan State Forest Holding: Poznan, Poland, 2015.
24. Office of Forest Management and Forest Surveying. *Forest Management Plan for Kluczbork Forest District*; Krakow State Forest Holding: Kraków, Poland, 2018.
25. Office of Forest Management and Forest Surveying. *Forest Management Plan for Maskulinskie Forest District*; Białystok State Forest Holding: Białystok, Poland, 2014.
26. Ayotte, P.; Simard, A.M. Cote Reproductive Plasticity of Female White-Tailed Deer at High Density and under Harsh Climatic Conditions. *Oecologia* **2019**, *189*, 661–667. [CrossRef]
27. Harris, I.; Osborn, T.J.; Jones, P.; Lister, D. Version 4 of the CRU TS Monthly High-Resolution Gridded Multivariate Climate Dataset. *Sci. Data* **2020**, *7*, 109. [CrossRef] [PubMed]
28. Hijmans, R.J.; van Etten, J. Raster: Geographic Analysis and Modeling with Raster Data 2012. Available online: http://raster.r-forge.r-project.org/ (accessed on 20 December 2024).
29. Bates, D.; Mächler, M.; Bolker, B.; Walker, S. Fitting Linear Mixed-Effects Models Using Lme4. *J. Stat. Softw.* **2015**, *67*, 1–48. [CrossRef]
30. R Core Team, R. *A Language and Environment for Statistical Computing*; R Foundation for Statistical Computing: Vienna, Austria, 2024.
31. Carranza, J.; Alvarez, F.; Redondo, T. Territoriality as a Mating Strategy in Red Deer. *Anim. Behav.* **1990**, *40*, 79–88. [CrossRef]
32. Andersson, M.; Iwasa, Y. Sexual Selection. *Trends Ecol. Evol.* **1996**, *11*, 53–58. [CrossRef]
33. Bobek, B.; Perzanowski, K.; Kosobucka, M. *Jeleń*; Świat: Warszawa, Poland, 1992.
34. Łabudzki, L. Charakterystyka Wybranych Cech Biometrycznych Jelenia Szlachetnego (*Cervus elaphus* L.) w Wielkopolsce [Characteristics of Selected Biometric Traits of the Red Deer (*Cervus elaphus* L.) Dwelling in the Region of Wielkopolska]. *Roczniki Akademii Rolniczej w Poznaniu* **1993**, *241*, 1–59.
35. Dzięciołowski, R. Jeleń Szlachetny. In *Łowiecki Podręcznik Selekcjonera*; Dziedzic, R., Ed.; Oficyna Wydawnicza Forest: Warszawa, Poland, 2011; pp. 15–21.
36. Langvatn, R.; Albon, S.D.; Burkey, T.; Clutton-Brock, T.H. Climate, Plant Phenology and Variation in Age of First Reproduction in a Temperate Herbivore. *J. Anim. Ecol.* **1996**, *65*, 653. [CrossRef]
37. Sykut, M.; Pawelczyk, S.; Borowik, T.; Pokorny, B.; Flajšman, K.; Hunink, T.; Niedzialkowska, M. Environmental Factors Shaping Stable Isotope Signatures of Modern Red Deer (*Cervus elaphus*) Inhabiting Various Habitats. *PLoS ONE* **2021**, *16*, e0255398. [CrossRef] [PubMed]
38. Borkowski, J. *Wpływ Środowiska Leśnego Na Populacje Jeleniowatych w Warunkach Restytucji Lasu Po Wielkoobszarowym Pożarze*; Instytut Badawczy Leśnictwa: Sękocin Stary, Poland, 2009; Volume 13.
39. Dzięciołowski, R. Relations between the Age and Size of Red Deer in Poland. *Acta Theriol* **1970**, *15*, 253–268. [CrossRef]
40. Tomek, A. *Właściwości i Struktura Populacji Jelenia [Cervus elaphus L.] w Lasach Krynickich [Karpaty]*; Uniwersytet Rolniczy im. Hugona Kołłątaja w Krakowie: Kraków, Poland, 2002.
41. Janiszewski, P.; Szczepański, W. Charakterystyka Masy Tusz Byków, Łań i Cieląt Jelenia Szlachetnego (*Cervus elaphus* L.) Pozyskanych w Okresie Jesienno-Zimowym. *Sylwan* **2004**, *1*, 33–38.

42. Czyżyk, P.; Żurkowski, M.; Ciepluch, Z.; Struziński, T.; Czajka, W. Parametry Populacyjne Jelenia Szlachetnego (*Cervus elaphus* L.) w Leśnym Kompleksie Promocyjnym "Lasy Mazurskie" Część I. Ocena Masy Poroża i Masy Tuszy Byków Pozyskanych w Wyniku Odstrzałów Selekcyjnych. *Sylwan* **2007**, *9*, 41–50.
43. Zalewski, D.; Szczepański, W. Charakterystyka populacji jelenia szlachetnego (*Cervus elaphus* L.) na Warmii i Mazurach. *Sylwan* **2004**, *3*, 35–39.
44. Post, E.; Langvatn, R.; Forchhammer, M.C.; Stenseth, N.C. Environmental Variation Shapes Sexual Dimorphism in Red Deer. *Proc. Natl. Acad. Sci. USA* **1999**, *96*, 4467–4471. [CrossRef] [PubMed]
45. Dzięgielewski, S. *Jeleń (Red Deer)*; Państwowe Wydawnictwo Rolnicze i Leśne: Warszawa, Poland, 1970.
46. Zalewski, D. *An Analysis of Selected Phenotypic Traits of Red Deer (Cervus elaphus L.) Males and an Evaluation of Red Deer Population Management in the Hunting Areas in the Province of Warmia and Mazury (NE Poland)*; Uniwersytet Warmińsko-Mazurski w Olszynie: Olsztyn, Poland, 2010.
47. Żurkowski, M.L.; Chartanowicz, W.; Żurkowski, M.W. Charakterystyka Jelenia Szlachetnego (*Cervus elaphus* L.) w Puszczy Piskiej. *Sylwan* **2000**, *11*, 55–63.
48. Janiszewski, P.; Gugołek, A.; Hanzal, V.; Kowalska, H. Quality of Red Deer (*Cervus elaphus*) Stags Harvested in the Hunting Grounds of the Tuchola Forest. *Acta Sci. Pol. Silvarum Colendarum Ratio Ind. Lignaria* **2007**, *3*, 41–49.
49. Mateos, C.; Alarcos, S.; Carranza, J.; Sánchez-Prieto, C.B.; Valencia, J. Fluctuating Asymmetry of Red Deer Antlers Negatively Relates to Individual Condition and Proximity to Prime Age. *Anim. Behav.* **2008**, *75*, 1629–1640. [CrossRef]
50. Carranza, J.; Pérez-Barbería, J.; Mateos, C.; Alarcos, S.; Torres-Porras, J.; Pérez-González, J.; Sánchez-Prieto, C.B.; Valencia, J.; Castillo, L.; de la Peña, E.; et al. Social Environment Modulates Investment in Sex Trait versus Lifespan: Red Deer Produce Bigger Antlers When Facing More Rivalry. *Sci. Rep.* **2020**, *10*, 9234. [CrossRef] [PubMed]
51. Clutton-Brock, T.H.; Albon, S.D.; Gibson, R.M.; Guinness, F.E. The Logical Stag: Adaptive Aspects of Fighting in Red Deer (*Cervus elaphus* L.). *Anim. Behav.* **1979**, *27*, 211–225. [CrossRef]
52. Andersson, M. *Sexual Selection*; Princeton University Press: New York, NY, USA, 1994.
53. Jennings, D.J.; Gammell, M.P. Fluctuating Asymmetry of Fallow Deer Antlers Is Associated with Weapon Damage but Not Tactical Behaviour during Fights. *Symmetry* **2023**, *15*, 829. [CrossRef]
54. Peláez, M.; Perea, R.; Díaz, M.; San Miguel, A.; Rodríguez-Vigal, C.; Côté, S.D. Use of Cast Antlers to Assess Antler Size Variation in Red Deer Populations: Effects of Mast Seeding, Climate and Population Features in Mediterranean Environments. *J. Zool.* **2018**, *306*, 8–15. [CrossRef]
55. Brown, R.D. Nutrition and Antler Development. In *Horns, Pronghorns, and Antlers*; Bubenik, G.A., Bubenik, A.B., Eds.; Springer: New York, NY, USA, 1990; pp. 426–441.
56. Smolko, P.; Garaj, P.; Lebocký, T.; Bútora, Ľ.; Pataky, T.; Jaňáková, Z.; Babic, M.; Veselovská, A.; Kubala, J.; Kropil, R. Soil Nutrients and Deer Density Affect Antler Size of the Carpathian Red Deer. *Mamm. Biol.* **2022**, *102*, 119–130. [CrossRef]
57. Mysterud, A.; Yoccoz, N.G.; Stenseth, N.C.; Langvatn, R. Effects of Age, Sex and Density on Body Weight of Norwegian Red Deer: Evidence of Density-Dependent Senescence. *Proc. R. Soc. B Biol. Sci.* **2001**, *268*, 911–919. [CrossRef]
58. Geist, V. *Front Cover Image for Deer of the World: Their Evolution, Behaviour, and Ecology Deer of the World: Their Evolution, Behaviour, and Ecology*; Stackpole Books: Mechanicsburg, PA, USA, 1998.
59. Mitchell, B.; Staines, B.W.; Welch, D. *Ecology of Red Deer: A Research Review Relevant to Their Management in Scotland*; Institute of Terrestrial Ecology Natural Environment Research Council: Cambridge, UK, 1977.

forests

MDPI

Communication

Monitoring Wildlife Using Camera Traps: Effect of Bait Type in Pine Plantations and Natural Hardwood Stands

Todd Fredericksen * and Wylie Martin

School of Natural Sciences and Mathematics, Ferrum College, Ferrum, VA 24088, USA
* Correspondence: tfredericksen@ferrum.edu

Abstract: Camera traps across from cages baited with either sardines or suet were installed in forests on the campus of Ferrum College in Virginia, USA, during the Fall and Spring seasons over two years. The objective of this study was to determine the vertebrate wildlife abundance and species composition in natural hardwood forests with mixed pine species compared to pine plantations. We found that the forest type and bait preference differed among the species by season. The relative abundance of natural foods and the need for winter thermal cover may explain the capture success in this study.

Keywords: wildlife habitat use; camera trapping; wildlife monitoring; Virginia; forests

1. Introduction

Camera traps are motion-triggered devices that are useful and cost-effective for short- and long-term passive monitoring of wildlife and allow for estimates of species detection, relative population abundance, and activity patterns [1,2]. The placement of camera traps is important for camera captures, and includes choosing the correct habitat for the target species, locating traps along travel corridors and log features, and the season of wildlife activity [3,4].

Using baits or lures could increase the capture efficiency of animals, but could also include bias, especially for estimates of the population size [5]. Studies have shown that different types of baits have varying success in attracting wildlife to camera traps. For example, [2] Randler [2] found that different baits, such as tuna and mustelid mix, affected the detection and the number of images taken by camera traps. Ferreira-Rodríguez and Pombal [6] tested four attractants (chicken, canned sardines, peanut butter, and cat urine) and found that fish and meat baits had the best performance.

The forest type may also influence wildlife habitat use and camera traps may allow for the quantification of species habitat use to inform natural history studies and management [7]. Deciduous (hardwood) forests typically have a better habitat structure for wildlife than pine plantations for many species [8]. Deciduous forests typically also offer more food resources, particularly hard mast provided by tree species such as oaks (*Quercus*), hickories (*Carya*), and beech (*Fagus*) [9]. These are important food resources for species such as White-tailed Deer (*Odocoileus virginianus*), White-footed Mice (*Peromyscus leucopus*) and Gray Squirrels (*Sciurus carolinensis*), among others.

Pine plantations provide pine seeds, which are food sources for some species, but because they are monocultures, they offer less structure and food diversity for wildlife. There is a lack of comparative studies in the literature comparing natural hardwood forests and pine plantations using camera traps and whether the type of bait may influence the captures of wildlife in these forest types.

In this study, we used baited camera traps to determine wildlife use of pine plantations compared to natural mixed deciduous–pine forests at multiple sites at a property in southwestern Virginia over four sampling periods from 2022 to 2024. During the last two sampling periods, we also compared habitat use with cages baited with two different types of bait (canned sardines and suet cakes).

2. Materials and Methods

The study was conducted at four sites on the campus of Ferrum College in Franklin County, Virginia, USA, during the Fall and Spring seasons from 2002 to 2024. At each site, we placed two camera traps in stands dominated by deciduous forest with some pine trees (hereafter referred to as "hardwood") stands and also an adjoining pine plantation (hereafter referred to as "pine") stands. Hardwood stands were mature (>80 years old) mixed pine–hardwood forests with the canopy dominated by Oak (*Quercus* spp.), Red Maple (*Acer rubrum*), Tuliptree (*Liriodendron tulipifera*), and White Pine (*Pinus strobus*) tree species. All hardwood stands were selectively logged approximately 30 years ago. Pine stands were either Loblolly Pine (*Pinus taeda*) stands approximately 60 years old or White Pine stands approximately 30 years old. We selected sampling sites in different parts of the Ferrum College campus where there were adjacent hardwood and pine stands.

All camera traps in this study were STC-PC12 V2 Stealth cameras (Stealth Cam, LLC, Irving, TX, USA). They were mounted on trees using a strap at a height of 1.5 m. Across from the camera traps at the same height and at a distance of 5–10 m were 12 × 12 cm metal cages affixed to a tree with elastic cords that contained either a partially opened can of sardines in soybean oil or a high-energy suet cake (C&S products Inc., Fort Dodge, IA, USA) (Figures 1 and 2). Suet cakes are typically used for attracting songbirds, but can also attract mammals. Traps were set to take two photos with a delay time of 30 s between camera captures. We recorded one capture of an animal once per photo session. Repeated captures that appeared to be same animal were not counted unless there was a 30 min delay in a photo capture. We did not differentiate between animals upon appearance and acknowledge that captures may include repeated captures of the same animal even if they included the 30 min new capture counting delay. In some cases, captures included more than one animal of the same species. In this case, all animals were counted.

We installed camera traps during two seasons (Fall and Spring) in adjoining hardwood and pine stands in two sampling periods: 2022–2023 and 2023–2024. Fall sampling occurred between August and November and Spring sampling occurred between February and April. In both Fall and Spring sampling seasons, we sampled two different pairs of pine and hardwood stands with camera sampling periods in each stand lasting 14 days. Cameras were checked for battery status and photos were downloaded every 7 days. We also replaced bait each 7 days.

In the 2022–2023 sampling period, we only baited traps with canned sardines, but in the 2023–2024 sampling period, we baited one trap with sardines and the other with suet. To reduce the possibility of attracting wildlife to both traps in the same location, we separated traps with different bait types by at least 20 m.

Because we did not uniformly trap all bait types, forest types, and seasons, we mostly used descriptive statistics in this paper for overall results that were summed over all camera traps per forest type and bait. We also used chi-square contingency tests for the various treatments for overall captures and individual species where there appeared to be a possible statistically significant differences between forest types or bait types.

Figure 1. Camera photo of a Black Bear at sardine bait in a White Pine plantation.

Figure 2. Camera photo of a Raccoon at suet bait in hardwood forest.

3. Results

Comparing only the habitat use of pine and hardwood stands during the first year of the study, we recorded six identifiable species in the hardwood stands and only three identifiable species in the pine stands in the Fall of 2022 (Table 1). There were more than 3× the number of captures in the hardwood stands ($p < 0.001$), mainly due to a very large number of captures of White-tailed Deer (*Odocoileus virginianus*) ($p < 0.001$). We captured five photos of Black Bear (*Ursus americanus*) in the hardwood stands and none in the pine stands. There were numerous captures of Raccoon (*Procyon lotor*) in both forest types. Only Southern Flying Squirrel (*Glaucomys volans*) had more camera captures in the pine plantations than hardwood forests.

Table 1. Total camera photos of wildlife species in the Fall of 2022 in deciduous forests (hardwood) and pine plantations. Photos were considered unique if an individual animal of a species was photographed at least 30 min later than the first photo.

Fall 2022	**Forest Type**	
Species	**Hardwood**	**Pine Plantation**
Black Bear (*Ursus americanus*)	5	0
Gray Squirrel (*Sciurus carolinensis*)	1	0
House Cat (*Felis catus*)	1	0
Raccoon (*Procyon lotor*)	16	15
Southern Flying Squirrel (*Glaucomys volans*)	3	6
Virginia Opossum (*Didelphis virginiana*)	1	0
White-tailed Deer (*Odocoileus virginianus*)	45	2
Unidentified	5	2
TOTAL	**77**	**25**

In the Spring of 2023, there were seven identifiable species of wildlife with camera captures in hardwood stands and five in pine plantations (Table 2). Summed over all the species, there were nearly 1.5× more camera captures of wildlife in the pine plantations than the hardwood forests, but this difference was not significant ($p = 0.19$). This trend was mostly due to a larger number of captures of Raccoon ($p = 0.003$) and White-tailed Deer ($p = 0.09$) in the pine plantations compared to hardwood forests. Unlike in the Fall season, there was a larger number of captures of Southern Flying Squirrel in the hardwood forests ($p = 0.03$).

Table 2. Total camera photos of wildlife species in the Spring of 2023 in deciduous forests (hardwood) and pine plantations. Photos were considered unique if an individual animal of a species was photographed at least 30 min later than the first photo.

Spring 2023	**Forest Type**	
Species	**Hardwood**	**Pine Plantation**
Gray Squirrel (*Sciurus carolinensis*)	15	16
Pileated Woodpecker (*Dryocopus pileatus*)	1	0
Raccoon (*Procyon lotor*)	5	27
Southern Flying Squirrel (*Glaucomys volans*)	13	2
Virginia Opossum (*Didelphis virginiana*)	8	10
White-footed Mouse (*Peromyscus leucopus*)	2	0
White-tailed Deer (*Odocoileus virginianus*)	3	12
Unidentified	2	2
TOTAL	**49**	**69**

Unlike in the Fall of 2023, there were more species captures in the pine plantations (7) than in the hardwood forests (3) (Table 3). The number of captures was also higher in the pine plantations, mainly due to a larger number of White-tailed Deer and Raccoons ($p = 0.09$). Only Gray Squirrel (*Sciurus carolinensis*) had higher captures in the hardwood stands. Although not statistically significant, captures were higher for the cameras baited with suet than those with sardines, particularly for White-tailed Deer and Raccoons. Only Gray Squirrel and Black Bear had higher captures for the cameras baited with sardines. Gray Fox (*Urocyon cinereoargenteus*), Virginia Opossum (*Didelphis virginiana*), and Tufted Titmouse (*Baeolophus bicolor*) were only captured at sites baited with suet.

Table 3. Total camera photos of wildlife species in the Fall of 2023 in deciduous forests (hardwood) and pine plantations with contrasting bait types (sardines and suet). Photos were considered unique if an individual animal of a species was photographed at least 30-min later than the first photo.

Fall 2023	**Forest Type**			
	Hardwood		**Pine Plantation**	
Species	**Sardines**	**Suet**	**Sardines**	**Suet**
Black Bear (*Ursus americanus*)	0	0	2	1
Gray Squirrel (*Sciurus carolinensis*)	3	0	1	1
Gray Fox (*Urocyon cinereoargenteus*)	0	0	0	1
Raccoon (*Procyon lotor*)	10	12	17	20
Tufted Titmouse (*Baeolophus bicolor*)	0	0	0	1
Virginia Opossum (*Didelphis virginiana*)	0	0	0	2
White-tailed Deer (*Odocoileus virginianus*)	2	8	5	16
Unidentified	0	0	2	0
TOTAL	**15**	**20**	**27**	**42**

In the Spring of 2024, there were 11 species captured in the pine plantations and only 6 species in the hardwood stands (Table 4). There were larger numbers of captures of species in pine stands than in the hardwood stands ($p < 0.001$), particularly for pine stand cameras baited with sardines, driven mostly by captures of Raccoons ($p < 0.001$) and Opossums ($p = 0.03$). Songbirds, in general, were only captured at the sites baited with suet, most of these in pine plantations. In the hardwood stands, there was an equal number of captures for the cameras baited with sardines and suet. White-tailed Deer had more captures in the hardwood stands ($p = 0.05$).

Table 4. Total camera photos of wildlife species in the Spring of 2024 in deciduous forests (hardwood) and pine plantations with contrasting bait types (sardines and suet). Photos were considered unique if an individual animal of a species was photographed at least 30 min later than the first photo.

Spring 2024	**Forest Type**			
	Hardwood		**Pine Plantation**	
Species	**Sardines**	**Suet**	**Sardines**	**Suet**
American Crow (*Corvus brachyrhynchos*)	2	0	0	0
Coyote (*Canis latrans*)	0	0	1	0
Gray Squirrel (*Sciurus carolinensis*)	0	0	3	1
Gray Fox (*Urocyon cinereoargenteus*)	0	0	1	0
Hairy Woodpecker (*Leuconotopicus villosus*)	0	2	0	0

Table 4. *Cont.*

Spring 2024	Forest Type			
	Hardwood		Pine Plantation	
Species	**Sardines**	**Suet**	**Sardines**	**Suet**
Hermit Thrush (*Catharus guttatus*)	0	0	0	2
Raccoon (*Procyon lotor*)	4	7	29	18
Southern Flying Squirrel (*Glaucomys volans*)	2	0	4	0
Tufted Titmouse (*Baeolophus bicolor*)	0	0	0	4
Virginia Opossum (*Didelphis virginiana*)	0	5	13	6
White-breasted Nuthatch (*Sitta carolinensis*)	0	0	0	2
White-footed Mouse (*Peromyscus leucopus*)	0	0	0	3
White-tailed Deer (*Odocoileus virginianus*)	10	3	1	2
Unidentified	0	0	0	1
TOTAL	**18**	**17**	**52**	**39**

4. Discussion

The number of captures among species differed by season and bait type (bait was only used in the second year of sampling). The most common species captured were White-tailed Deer and Raccoons, and to a lesser extent Virginia Opossums, Gray Squirrels, and Southern Flying Squirrels.

White-tailed Deer are very abundant in the area and the photos of this species varied greatly between seasons. As herbivores, they were unlikely attracted to the bait, but perhaps were captured more due to other factors, such as the hard mast availability or winter thermal cover [9,10]. The large number of captures of White-tailed Deer in the hardwood stands in the Fall of 2022 may be due to an abundance of acorn nuts (*Quercus* hard mast) in that season during that year. Interestingly, deer captures were much lower in the following Spring and following Fall and tended to be higher in the pine plantations. Then, they were captured more frequently in the hardwood stands in the Spring of 2024.

Raccoons were commonly captured in all years and seasons, but in the Fall seasons, they were equally abundant in the hardwood and pine plantations. The camera captures indicated a strong preference, however, for the pine plantations in the Spring seasons. Raccoons are opportunistic omnivores [11], but perhaps were more attracted to bait during the Spring seasons when there was the most food scarcity and their preference for pine stands may also have been related to the greater thermal winter cover provided there. They were particularly attracted to the sardines compared to the suet in the second year, perhaps because they often locate food by olfactory sensing [11]. Virginia Opossums are also opportunist omnivores [11] and their captures showed a similar trend to those of Raccoons.

The two squirrel species captured in this study included the Eastern Gray Squirrel and the Southern Flying Squirrel. Gray Squirrels are diurnal and Flying Squirrels are nocturnal. Both rely heavily on hard masts, particularly in the Fall, but also rely on acorns in the Winter and Spring by scatter-hoarding or caching food [11]. Their captures indicate a general preference for hardwood stands, but Southern Flying Squirrels also were more abundant in some seasons in the pine plantations, contrary to a previous live capture study in the forests at these sites [6].

Black Bears were only captured on camera in the Fall, which was expected because this species feeds heavily in the Fall and hibernates in Virginia from November to April.

The few captures indicated a preference for the hardwood stands in the Fall of 2022 and for the pine plantations in the fall of 2023.

Few bird species were captured during the first year, when only sardines were used as bait. The camera captures of songbirds, although not in large numbers, became more numerous at the suet-baited cameras in year 2. All of the captures occurred in the Spring of 2024 in the pine plantations, which could be explained by the food scarcity at that time of year and the need for more thermal cover provided by the pine plantations [12].

One surprising result of this study was the larger number of wildlife photos recorded in the pine plantations compared to the hardwood forests. Hardwood forests tend to have more food and habitat structure for wildlife [8,9]. It should be noted, however, that most of the species in this study have relatively large home ranges and their occurrence in pine plantations in close proximity to hardwood stands may not be unexpected, particularly in spring, when there is more thermal cover in pine plantations.

5. Conclusions

Like other studies, this study indicates the usefulness of camera trapping for detecting and monitoring wildlife species and determining their habitat preferences. It also showed that the type of bait provided in front of the camera traps can attract different species of wildlife and may vary in capture success during different seasons. The relative abundance of habitat structure and natural foods during certain seasons may also play a strong role in camera capture success.

Author Contributions: Both authors contributed equally to field work and analysis. T.F. prepared original draft and W.M. reviewed and edited the final draft. All authors have read and agreed to the published version of the manuscript.

Funding: Ferrum College provided funding for equipment and supplies.

Data Availability Statement: The original contributions presented in this study are included in the article. Further inquiries can be directed to the corresponding author.

Acknowledgments: We thank Bailey Peek and Ethan Fragoso for their help in this study.

Conflicts of Interest: The authors declare no conflicts of interest.

References

1. Oliver, R.Y.; Iannarilli, F.; Ahumada, J.; Fegraus, E.; Flores, N.; Kays, R.; Birch, T.; Ranipeta, A.; Rogan, M.S.; Sica, Y.V.; et al. Camera trapping expands the view into global biodiversity and its change. Philos. *Trans. R. Soc. Lond. B Biol. Sci.* **2023**, *378*, 20220232. [CrossRef] [PubMed]
2. Randler, C.; Katzmaier, T.; Kalb, J.; Kalb, N.; Gottschalk, T.K. Baiting/luring improves detection probability and species identification—A case study of mustelids with camera traps. *Animals* **2020**, *10*, 2178. [CrossRef] [PubMed]
3. Kolowski, J.M.; Forrester, T.D. Camera trap placement and the potential for bias due to trails and other features. *PLoS ONE* **2017**, *12*, e0186679. [CrossRef] [PubMed]
4. O'Connor, K.M.; Nathan, L.R.; Liberati, M.R.; Tingley, M.W.; Vokoun, J.C.; Rittenhouse, T.A.G. Camera trap arrays improve detection probability of wildlife: Investigating study design considerations using an empirical dataset. *PLoS ONE* **2017**, *12*, e0175684. [CrossRef] [PubMed]
5. Stewart, F.E.; Volpe, J.P.; Fisher, J.T. The debate about bait: A red herring in wildlife research. *J. Wildl. Manag.* **2019**, *83*, 985–992. [CrossRef]
6. Fredericksen, T.S.; Banton, J.; Davis, A. Habitat selection of the Southern Flying Squirrel in the Virginia Piedmont. *Banisteria* **2014**, *44*, 3–6.
7. Trolliet, F.; Vermeulen, C.; Huynen, M.C.; Hambuckers, A. Use of camera traps for wildlife studies: A review. *Biotechnol. Agron. Société Environ.* **2014**, *18*, 446–454.
8. Hunter, M.L., Jr. (Ed.) *Maintaining Biodiversity in Forest Ecosystems*; Cambridge University Press: Cambridge, UK, 1999.
9. McShea, W.J.; Schwede, G. Variable acorn crops: Responses of white-tailed deer and other mast consumers. *J. Mammal.* **1993**, *74*, 999–1006. [CrossRef]

10. DelGiudice, G.D.; Fieberg, J.R.; Sampson, B.A. A long-term assessment of the variability in winter use of dense conifer cover by female white-tailed deer. *PLoS ONE* **2013**, *8*, e65368. [CrossRef] [PubMed]
11. Lindzey, D.W. *The Mammals of Virginia*, 2nd ed.; The McDonald & Woodward Publishing Company: Newark, OH, USA, 2021.
12. Morrison, M.L.; With, K.A.; Timossi, I.C. The structure of a forest bird community during winter and summer. *Wilson Bulletin* **1986**, *98*, 214–230.

forests

MDPI

Article

Habitat Composition and Preference by the Malabar Slender Loris (*Loris lydekkerianus malabaricus*) in the Western Ghats, India

Smitha D. Gnanaolivu [1,2,*], Joseph J. Erinjery [3], Marco Campera [4,*] and Mewa Singh [2]

[1] Department of Advanced Zoology and Biotechnology, Loyola College, University of Madras, 24, 9, Nelson Manickam Rd, Nungambakkam, Chennai 600034, India

[2] Biopsychology Laboratory, University of Mysore, Mysore 570006, India; msingh@psychology.uni-mysore.ac.in

[3] Department of Zoology, Kannur University, Mananthavady Campus, Mananthavady 670645, India; josephjerinjery@kannuruniv.ac.in

[4] Sustainable Agroforestry Research Group, School of Biological and Medical Sciences, Oxford Brookes University, Oxford OX3 0BP, UK

* Correspondence: simmy.smitha@gmail.com (S.D.G.); mcampera@brookes.ac.uk (M.C.)

Abstract: Habitat degradation poses a critical threat to the Malabar slender loris (*Loris lydekkerianus malabaricus*), yet little is known about its microhabitat requirements in intact forest. In Aralam Wildlife Sanctuary, we combined nocturnal trail surveys (337 loris sightings) with plotless sampling of 2830 trees (86 species from 35 families) to characterize both vegetation structure and loris presence. Our results show that lorises occur almost exclusively in mildly degraded wet evergreen and secondary moist deciduous subcanopies, where understory trees and climber networks provide continuous pathways. Individuals are most often encountered at heights of 5–15 m—ascending into higher strata as the night progresses—reflecting a balance between foraging access and predator avoidance. Substrate analysis revealed strong preferences for twigs ≤ 1 cm (36.98%) and small branches 2–5 cm in diameter, oriented obliquely to minimize energetic costs and maintain stability during slow, deliberate arboreal locomotion. Day-sleeping sites were overwhelmingly located within dense tangles of lianas on large-girth trees, where intertwined stems and thorny undergrowth offer concealment from both mammalian and avian predators. Vegetation surveys documented a near-equal mix of evergreen (50.6%) and deciduous (49.4%) species—including 26 endemics (18 restricted to the Western Ghats)—with *Aporosa cardiosperma* emerging as the most abundant riparian pioneer, suggesting both ecological resilience and potential simplification in fragmented patches. Complementing field observations, our recent habitat-suitability modeling in Aralam indicates that broad-scale climatic and anthropogenic factors—precipitation patterns, elevation, and proximity to roads—are the strongest predictors of loris occupancy, underscoring the interplay between landscape-level processes and microhabitat structure. Together, these findings highlight the imperative of multi-strata forest restoration—planting insect-hosting native trees, maintaining continuous canopy and climber networks, and integrating small "mini-forest" modules—to recreate the structural complexity vital for slender loris conservation and the broader resilience of Western Ghats biodiversity.

Keywords: prosimian; microhabitat selection; arboreal locomotion; sleep-site concealment; habitat restoration; terrestriality; mini-forest reforestation; insect-hosting tree species

1. Introduction

The examination of how nonhuman primates use space is a central concern in socio-ecological research, particularly because patterns of spatial behavior are intimately linked to a species' ability to survive, thrive, and reproduce within a given landscape [1,2]. The choices that individuals or groups make in selecting and occupying specific microhabitats within their home ranges are shaped by a variety of ecological factors—ranging from food availability to shelter, predator presence, and microclimatic conditions. Numerous studies have demonstrated that primates often exhibit a strong preference for areas of intact, minimally disturbed primary forest or mature secondary forest. These habitats are typically characterized by greater structural complexity, including well-developed canopy layers, and tend to offer more reliable and abundant food resources, both of which are crucial for the daily energetic demands and long-term reproductive success of primate populations [3–5].

However, across the globe, the integrity of such habitats is increasingly compromised by anthropogenic pressures, particularly habitat fragmentation. Among the various forms of disturbance that threaten primate survival, fragmentation remains one of the most pervasive and ecologically consequential [6]. Fragmentation disrupts the continuity of forest cover, leading to a patchy landscape in which the size, shape, and isolation of habitat fragments can drastically affect species richness, alter species interactions—including predator–prey dynamics—and reduce overall habitat suitability [7–9]. In addition to these biological consequences, fragmentation also changes the physical environment: temperature, humidity, and light penetration are all subject to alteration along forest edges and within smaller patches, which in turn affects biotic interactions in complex and often unpredictable ways [10]. Moreover, most forest fragments, especially in heavily settled regions, are not only small and isolated but also suffer from ongoing degradation due to illegal logging, livestock grazing, invasive species, and other forms of human activity [11]. For conservation efforts to be successful, particularly in fragmented and degraded landscapes, it is imperative to restore habitats in ways that are informed by a clear understanding of a species' ecological requirements under more pristine conditions. Unfortunately, for many primate species and subspecies, particularly those that are nocturnal, cryptic, or understudied, such baseline information from intact habitats remains unavailable or insufficiently documented.

This gap in knowledge is especially critical in the context of the Western Ghats of India, a global biodiversity hotspot that harbors an exceptionally rich assemblage of flora and fauna, including several endemic primates. Despite its ecological significance, the region faces intense conservation challenges due to widespread habitat fragmentation, high human population density, and competing demands on land use [12]. While large tracts of forest in the Western Ghats have been formally brought under a protected area network, these protected zones are interspersed with, and often surrounded by, highly fragmented and human-impacted landscapes. Restoration of these degraded fragments is both ecologically urgent and logistically complex, and demands a grounded understanding of how focal species interact with forest structure in relatively undisturbed settings. Against this backdrop, the present study focuses on the habitat structure and habitat use of the slender loris, a small, nocturnal, and arboreal primate that occurs in the Western Ghats, specifically within relatively intact rainforest environments.

The genus *Loris* comprises two species: the Red Slender Loris (*Loris tardigradus* Linnaeus, 1758) and the Grey Slender Loris (*Loris lydekkerianus* Cabrera, 1908), both of which are distributed in southern India and Sri Lanka. Of the two, *L. lydekkerianus* has a broader geographic distribution [13]. Within India, *L. lydekkerianus* is further subdivided into two subspecies: the Mysore Slender Loris (*L. l. lydekkerianus*) and the Malabar Slender Loris

(*L. l. malabaricus*), the latter being endemic to the southwestern slopes of the Western Ghats [14,15]. In India, the slender loris is accorded the highest level of protection at the national level, under Schedule I, Part I of the Wildlife (Protection) Act, 1972, while internationally, IUCN lists the species as 'Near Threatened' as they are significantly declining because of widespread habitat loss and hunting through much of its range [16]. Although these animals are relatively widespread in terms of their historic range, recent decades have witnessed significant declines in their populations due to forest fragmentation, habitat degradation, and direct anthropogenic threats such as hunting and trapping. As a result, many populations today are restricted to isolated patches of remnant forest, with little connectivity between them [16]. Conservation strategies and long-term management plans remain largely undeveloped, in part because of a critical lack of data on the species' fine-scale habitat preferences within natural, undisturbed forest ecosystems [16,17].

Loris l. malabaricus is distributed throughout the western slopes of the Western Ghats, south of the Tapti River [18]. The species has been reported from a variety of habitat types, including undisturbed rainforests, rainforest scrub at higher elevations, moist deciduous forests, degraded rainforest edges, and even within human-modified landscapes such as cardamom plantations [19,20]. Despite its apparent ecological plasticity, the species is strictly arboreal and highly dependent on canopy connectivity for its locomotion, foraging, and predator avoidance. [21–23]. In highly fragmented habitats, slender lorises are occasionally forced to descend to the ground—a behavior that exposes them to increased risk and suggests a compromised habitat structure [17]. They are assumed to have very slow life history parameters and a low rate of reproduction due to their consumption of toxic insects [24]. Lorises employ obscure and noiseless movements, immobility, inconspicuous retreat, or rapid flight upward as modes of defence [15].

Yet, despite its ecological importance and conservation urgency, research on *L. l. malabaricus* remains limited, sporadic, and geographically uneven. In order to better understand the ecological needs of this species and to inform restoration efforts in fragmented landscapes, we conducted the present study in Aralam Wildlife Sanctuary—one of the few remaining areas known to support a relatively high abundance of slender lorises [25]. Our primary aim was to characterize the structure of habitats preferred by *L. l. malabaricus* within a largely intact rainforest setting. This is especially critical given recent predictive models suggesting that suitable habitat for the species may decline by as much as 52% by the year 2070, primarily due to ongoing land-use changes and climate-related shifts in vegetation patterns [26].

A recent study conducted in the same sanctuary identified basal area, tree species richness, and certain forms of human disturbance—such as branch lopping and tree felling—as positive determinants of loris occupancy, suggesting that the species may tolerate, or even prefer, mildly disturbed secondary forest [14]. While previous observations indicated that lorises were less likely to be active in areas with dense climber cover, we hypothesized that such tangled vegetation might serve as preferred sleeping sites due to the concealment it offers from predators and environmental exposure. Given the loris' arboreal lifestyle and its inability to leap, we also expected the species to favor microhabitats with slender twigs and small branches that facilitate cautious, deliberate locomotion. Additionally, we predicted that lorises would prefer supports that are oblique in orientation—rather than strictly vertical or horizontal—as such structures likely offer a more energetically efficient and biomechanically feasible path through the canopy. Through this study, we aim to contribute much-needed ecological insights into the habitat preferences of *L. l. malabaricus* in a relatively undisturbed rainforest context, thereby informing future conservation and habitat restoration efforts across the Western Ghats.

2. Materials and Methods

2.1. Study Site

We conducted the present study in Aralam Wildlife Sanctuary in the south Indian state of Kerala (Figure 1). Spread over 55 km^2, the sanctuary is situated in the western slopes of the Western Ghats range. Lying between 11°59′ N and 11°54′ N and 75°47′ E and 75°57′ E, the elevation of the site varies from 50 m to 1145 m. The vegetation consists of moist deciduous forest, semi-evergreen forest, evergreen forest, and plantations [14,27]. The temperature varies from 21 °C to 40 °C at the foothills and 8 °C to 25 °C at high altitudes, and the annual rainfall in the region is about 3000 mm [14,27]. A tribal settlement forms a fringe around the study site, with the Aralam Farm on one side and shares boundaries with three townships/human habitation on the other. The Valapattanam river creates a natural boundary line on the township area side [14,27].

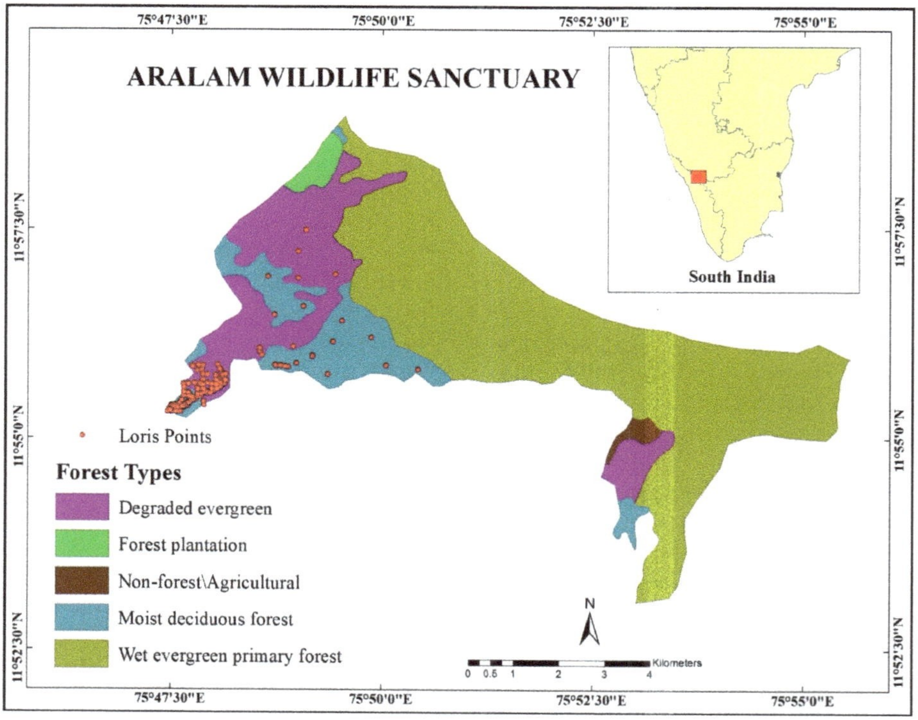

Figure 1. Map of Aralam Wildlife Sanctuary showing the occurrence of lorises.

2.2. Observation of Slender Loris Habitat Use

SDG carried out behavioral observations of slender lorises over a period of two years from March 2014 to March 2016, for a total of 1560 h. Due to a lack of continuous visibility under the dense forest conditions, SDG recorded the behavior of the animal at the time of the first contact following the method used by Charles-Dominique and Bearder [28], Gamage [21], and Nekaris [29,30].

SDG conducted the study between 19.00 h and 6.00 h on foot with a walking speed averaging 1 km/h [25,31]. SDG carefully scanned the trees to find lorises during the walks. SDG spotted the animal by its unique orange-red tapetal reflection to light emitted by reduced intensity (~200 lumens) Petzel headlamps covered with red cellophane sheets.

SDG marked the loris presence trees (trees in which lorises were spotted) using a broad satin ribbon to facilitate vegetation sampling during the day.

Lorises were detected via eyesight, and a range finder (Bushnell Medalist), was subsequently used to estimate their height on the tree. On spotting a loris, SDG recorded the data pertaining to the age class of the individual (adult male, adult female, juvenile, or infant) [32], the tree species the loris was spotted on, the height at which the loris was spotted and the tree height, the part of the tree used, clumped or single status of the tree, the size and the angular orientation (vertical, horizontal, oblique) of the substratum, and the behavioral activity of the loris. SDG also recorded the information on the presence or absence of climbers, species of climbers, and the angle of orientation of the climbers [29,32]. SDG collected the feeding ecology data only when we spotted a loris feeding or foraging for food, leading to the catching of prey and its consumption. The data included the tree species, the height at which SDG spotted the loris, and the food item it was feeding on.

2.3. Vegetation Sampling

SDG employed a plotless sampling technique [33] around 337 loris presence trees to ascertain the characteristics of the habitat used by the loris. We defined a tree as one with a girth > 10 cm. SDG divided the region around the loris presence tree into four quadrats by placing two sticks perpendicular to each other. In each quadrat, SDG measured the distance between the nearest tree (N) and the loris encountered tree (MP), and the nearest neighbor of the tree (NN) in the same direction. SDG collected all data within the area, including the four quadrats [33,34]. This area is being referred to as a plot in our study.

For classification and nomenclature of trees, we followed Sasidharan [35,36]. Data related to tree density included species and family names, distance from the previous tree (MP-N) and to the nearest neighbor (N-NN), tree height measured with a range finder, and girth (circumference) of the trunk at breast height (CBH).

As lorises cannot jump beyond 0.3 m [37], continuity of arboreal substrate is important for the slender loris' locomotion [30,34]. We defined the arboreal connectivity as the connectivity between one tree and another through the intersection between leaves or branches (leaf connectivity) or with the help of climbers (climbers' connectivity). We visually measured the percentage of arboreal connectivity (leaf connectivity and climbers' connectivity) between the trees and the canopy cover using the Braun-Blanquet cover-abundance scale [34,38,39]. We examined the canopy strata of the trees based on the spread of their leaves and branches. We classified them as canopy, subcanopy, and bushes. SDG also observed lorises feeding off the ground; hence, we also measured the percentage of ground cover using the Braun-Blanquet cover-abundance scale, relative density of saplings with CBH > 10 cm, and herbaceous vegetation between MP-N and N-NN trees [34,38,39].

SDG recorded the presence or absence of climbers on each tree and the species of these climbers. Previous studies reported that Lorisiforms used dense tangles or holes as their sleeping sites [40]. So, we identified the potential sleeping sites by looking for dense climbers' tangles, densely tangled branches forming a nest or holes on trees inside the plot or next to the plot. We recorded the presence or absence of sleeping sites, including climber tangles or branch tangles and tree holes, the tree species they were found on, and the name of the climbers they were associated with [34,41].

2.4. Data Analysis

We used the shortest distance between MP and N among the four quadrats to calculate the tree density using the T-square method [33,34]. The equation used was $D = m^2/(2.828\, \Sigma_i x_i \times \Sigma_i z_i)$, where D is the tree density (trees/ha), m is the number of loris presence trees (MP), x_i is the distance from MP to the closest neighbor (N), and z_i is the distance to the nearest neighbor (NN). We also carried out a test of random distribution using the equation $t = \{\Sigma_i [x_i^2/(x_i^2 + z_i^2)/2] - m/^2\}\sqrt{12/m}$. If t is greater than +1.96, the distribution is significantly more regular than a random distribution; if it is less than −1.96, it is significantly clumped [33,34]. CBH was divided by π to give the diameter at the breast height (DBH). We calculated the basal area of the trees using the equation $CBH^2/4\pi$ in m^2/ha.

We used chi-square tests to compare leaf and climbers' connectivity, frequency of lorises at various heights, the size of the substrates used by the lorises, orientation of substrates and climbers, and time spent by lorises at various places. We used the Kruskal–Wallis Analysis of Variance test for temporal distribution at different time periods, and across seasons.

2.5. Ethical Note

The study was non-invasive and followed the guidelines of best practices for field Primatology from the International Primatological Society. We followed recommendations for behavioral observations on nocturnal primates, such as the use of red light and appropriate distances maintained during observations. The Principal Chief Conservator of Forests and Chief Wildlife Warden, Forest Headquarters, Vazhuthacaud, Thiruvananthapuram—695014 (Permit No. WL10-17697/2012)—approved the research protocol and adhered to the legal requirements of the Kerala Forest Department. The authors have no conflicts of interest to declare.

3. Results

3.1. Habitat Structure

3.1.1. Tree Species

The 2830 trees recorded during the vegetation study represented 86 species belonging to 35 families, with 50.57% of them being evergreen trees and 49.43% being deciduous trees. Of these, 26 species were endemic to India, among which 18 were endemic to the Western Ghats. Eight species were endemic to both India and Sri Lanka. The most abundant tree species recorded was *Aporosa cardiosperma* (n = 289; 10.21%). The list of floristic composition is presented in Table A1.

3.1.2. Density, Distribution, Height, Girth, and Basal Area of Trees

The tree density in the study area was 1521.04 ± 94.73 trees/ha. The tree with the highest density was *Dysoxylum malabaricum* with 9855.26 trees/ha. The "t" value for the test of random distribution was greater than +1.96 for three tree species (+4.04 for dead trees, +3.65 for *Artocarpus hirsutus*, and 2.03 for *Gmelina arborea*), suggesting a regular distribution, and the "t" value was less than −1.96 for five tree species, viz., *Terminalia crenulata* (−2.07), *Dalbergia lanceolaria subsp. paniculata* (−2.48), *Butea monosperma* (−2.34), *Shorea roxburghii* (−2.25), and *Lagerstroemia speciosa* subsp. *speciosa* (−2.03), suggesting significant clumping. The "t" value of the remaining 45 tree species fell between −1.96 and +1.96. While certain species showed significant deviations—either clumping or regularity—the majority of species did not deviate significantly from randomness; therefore, the overall tree distribution in the loris habitat can be interpreted as random.

The average height of the trees in the study area was 11.36 ± 0.12 m (n = 2813, range –0.30 to 67.96 m), with the largest tree being *Aporosa cardiosperma* with a CBH of 220 cm. Only 3 species (0.85%) out of 86 species had an average height of <5 m, viz., *Terminalia catappa* (3.33 ± 0.33 m), *Ixora polyantha* (4.99 ± 0.58 m), and *Agrostistachys borneensis* (5 ± 0.0 m). The average CBH was 51.71 ± 0.98 cm with a minimum of 10 cm and a maximum of 451 cm. The average DBH was 16.46 ± 0.31 cm. The average basal area was

430.41 ± 20.08 m^2/ha (range 7.95 to 16,186.13). The tree with the largest basal area was *Actinodaphne maderaspatana*. The average basal area of the three most common families was 195.28 ± 19.47 m^2/ha (n = 468; *Phyllanthaceae*), 856.62 ± 56.19 m^2/ha (n = 260; *Fabaceae*), and 215.29 ± 32.80 m^2/ha (n = 255; *Dipterocarpaceae*).

3.1.3. Canopy and Ground Cover

The study area had 40.70% of canopy tree species, 48.84% of subcanopy tree species, and 10.47% of bushy tree species. We recorded the canopy cover and the ground cover between the nearest neighbor trees 317 times, using the Braun-Blanquet scale. The canopy cover was high (50%–75%) in 43.63%, followed by moderate (25%–50%) in 24.52%, and very high (>75%) in 22.29% of the plots. Only 8.92% and 0.64% of the plots recorded low (5%–25%) and very low (<5%) canopy cover, respectively. The ground cover was low (5%–25%) in 50.79% of the plots, high (50%–75%) in 22.40% of plots, moderate (25%–50%) in 17.67%, very high (>75%) in 6.62%, and very low (<5%) in 2.52% of the plots.

3.1.4. Climbers

We identified a total of 16 species of climbers belonging to 10 families in the study area. However, we spotted the lorises on only eight species of climbers. Among them, we found the frequency of *Gentum edule* was found to be the highest (21.01%), followed by *Acacia caesia* (20.38%) (Table 1). The species richness of climbers ranged from zero to seven species in each plot. About 25.87% of the plots had two species of climbers, 23.66% had only one species, 19.56% and 19.24% of plots had three and four species of climbers, respectively. About 7.57% of the plots had five species of climbers, six species had 0.95% of the plots, only one plot had seven species of climbers, and nine plots were devoid of climbers.

Table 1. Composition of climbers in the study area at Aralam Wildlife Sanctuary.

Scientific Name	Type	Distribution	% Composition
Gnetum edule	Woody lianas	India	22.16
Acacia caesia	A prickly climbing shrub	Indo-Malesia	21.50
Calycopteris floribunda	Scandent climbing shrubs	Indo-Malesia	11.75
Piper sarmentosum	Perennial herb with creeping rhizomes	India and Malesia	11.08
Derris scandens	Woody lianas	Indo-Malesia	9.75
Aspidopterys canarensis	Woody lianas	Endemic to Western Ghats	5.87
Ipomoea marginata	Extensive twiners	Paleotropics	3.60
Cissus repens	Creepers	Indo-Malesia	3.34
Ventilago maderaspatana	Climbing shrubs	Indo-Malesia	2.54
Cissus latifolia	Climbing shrubs	India and Sri Lanka	2.14
Piper nigrum	Glabrous climbers, climbing shrub	India and Sri Lanka	1.34
Bauhinia scandens	Woody Lianas	Indo-Malesia	1.20
Argyreiaelliptica sp.	Twiners	India and Sri Lanka	0.93
Piper mullesua	Woody Lianas	India	0.93
Calamus travancoricus	Very slender climbing canes	Endemic to Western Ghats	0.67
Unidentified			0.67
Mimosa diplotricha	Rambling shrubs, exotic climber	Native of Tropical America; a weed in India	0.53

3.1.5. Climber and Leaf Connectivity

About 32.80% of the plots had climbers' connectivity between < two trees, 23.25% between three and four trees, 17.83% between five and six trees, and 26.11% between seven and eight trees. About 46.50% of the plots had leaf connectivity between seven and eight trees, 35.35% between five and six trees, 12.10% between three and four trees, and 6.05% between ≤two trees. The average climbers' connectivity between trees was 4.09 ± 0.14 (range = 0–8), and the leaf connectivity was 5.94 ± 0.10 (range = 1–8). Relatively, the leaf connectivity between trees represented by percentages of connectivity (χ^2 = 58.39, df = 3, p <0.001) was more than the climbers' connectivity (χ^2 = 6.21, df = 3, p = 0.102).

3.2. Habitat Use

3.2.1. Tree Use by Lorises

We collected a total of 337 data points from the opportunistic sampling, i.e., the point at the moment of first visual contact. Out of the 86 species of trees present in the study area, the slender lorises used 51 species belonging to 28 families. They used *Aporosa cardiosperma* (13.38%, n = 38) and *Xylia xylocarpa* (11.62%, n = 33) the most, followed by *Naringi crenulata* (9.15%, n = 26), *Holigarna arnottiana* (5.28%, n = 15), *Rotheca serrata* (4.23%, n = 12), and *Terminalia paniculata* (3.87%, n = 11). We spotted lorises in the subcanopy tree species for 51.92% of our observation duration, followed by 36.54% on canopy tree species and 11.54% in bushes (χ^2 = 3.00, df = 2, p = 0.223). The details on the tree species used by slender loris are summed up in Table A1. Slender lorises used trees with a mean height of 14.49 ± 0.44 m (n = 282, range 1.00–35.05 m) and were spotted at an average height of 9.57 ± 0.49 m (n = 337, range = 0.00–31.00 m). However, on height class analysis, the lorises mostly frequented trees of the 10–15 m height class. We most frequently encountered lorises in the <5 m height class (48.96%), followed by the 5–10 m height class (18.99%), and only on one occasion, above 30 m (Figure 2). The frequency of lorises at various heights varied significantly (χ^2 = 131.58, df = 6, p < 0.001), showing that they were mostly at a height between 5 and 15 m. We most often encountered the slender lorises on trees with a CBH between 30 and 60 cm, having an average basal area of 220.84 ± 50.49 m^2/ha (n = 91) (31.93%), followed by trees with a CBH between 60 and 90 cm having an average basal area of 418.24 ± 102.84 m^2/ha (n = 57) (20%) (Figure 3).

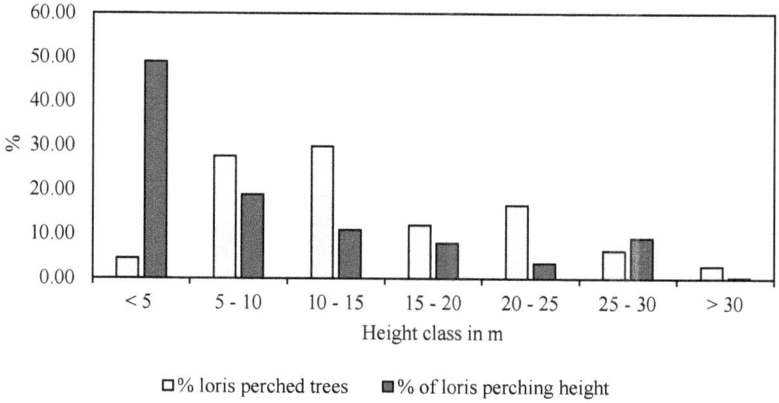

Figure 2. Height-class analysis of loris presence in trees and perching height of slender lorises.

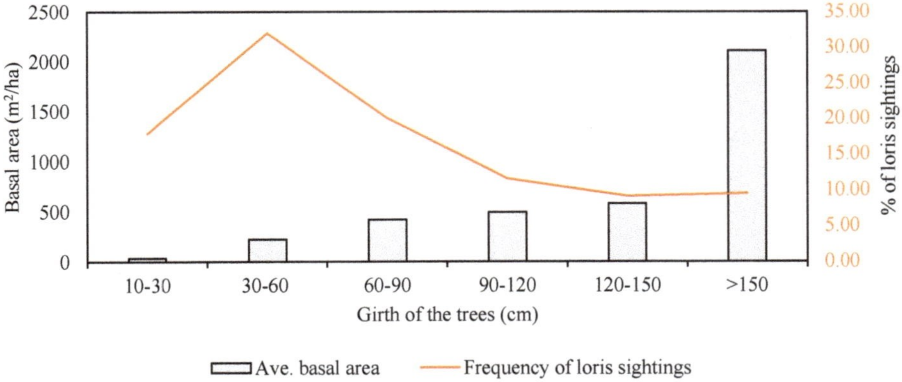

Figure 3. Frequency of the slender loris on different trees of the basal area class.

The heights used by lorises during the three seasons did not differ (Kruskal–Wallis values H = 0.08, df = 2, p = 0.962). However, we spotted lorises at greater heights as the night progressed (Kruskal-Wallis H = 17.37, df = 3, p = 0.011). The detection also decreased as the night progressed when the lorises ascended to greater heights. We sighted lorises more during early evening and midnight than during late night and early dawn (χ^2 = 88.34, df = 3, α = <0.001). Loris sightings were more during summer and monsoon than during post-monsoon (χ^2 = 11.49, df = 2, p = 0.012) (Table 2).

Table 2. Temporal distribution of slender loris sightings in different seasons.

Seasons/Time Duration	**Height at Which the Loris Was Spotted (m)**	**Dry/Summer**	**Southwest Monsoon**	**Post-Monsoon**
		February to May	**June to November**	**December to January**
Early night (19–22 h)	Average	7.35 ± 0.87	8.09 ± 1.26	7.58 ± 1.39
	Range	30–0.0	30–1.21	30–0.6
	n	68	46	25
Midnight (22–01 h)	Average	11.36 ± 1.79	11.43 ± 1.66	8.35 ± 1.57
	Range	31–0.0	30–0.0	25–0.0
	n	31	38	23
Late night (1–4 h)	Average	16.87 ± 2.82	11.17 ± 2.21	11.80 ± 1.81
	Range	30–0.0	30–0.91	25–0.0
	n	16	22	19
Early dawn (4–6 h)	Average	16.22 ± 3.12	21 ± 4.58	11.42 ± 3.17
	Range	28–2.13	30–10.00	30–2.00
	n	11	5	11

3.2.2. Substrate Size and Orientation

Lorises (n = 338) preferred twigs of ≤1 cm (36.98%), small branches of 2–5 cm (33.43%), medium sized branches of 6–10 cm (13.02%), and the large branches > 10 cm (16.57%) (χ^2 = 22.89, df = 3, p < 0.001). The orientation of these substrates (n = 330) was horizontal (27.58%), oblique (50.00%), and vertical (22.42%) (χ^2 = 19.35, df = 2, p < 0.001).

3.2.3. Time Spent at Various Substrates

The lorises were sighted 42.55% of the time on climbers at a mean height of 9.83 m ± 0.77, 2.56% on the ground, and 54.89% on trees (Figure 4). There was a significant association between the proportion of sightings on different substrates ($\chi^2 = 72.81$, df = 5, $p < 0.001$), and nonsignificant for the height at which the loris was spotted on different substrates ($\chi^2 = 6.00$, df = 5, $p = 0.306$). The loris preferred to perch on the regions of the trees associated with climbers (50.89%) rather than on trees without climbers (39.94%), and on climbers not associated with trees (9.17%) ($\chi^2 = 42.11$, df = 2, $p < 0.001$).

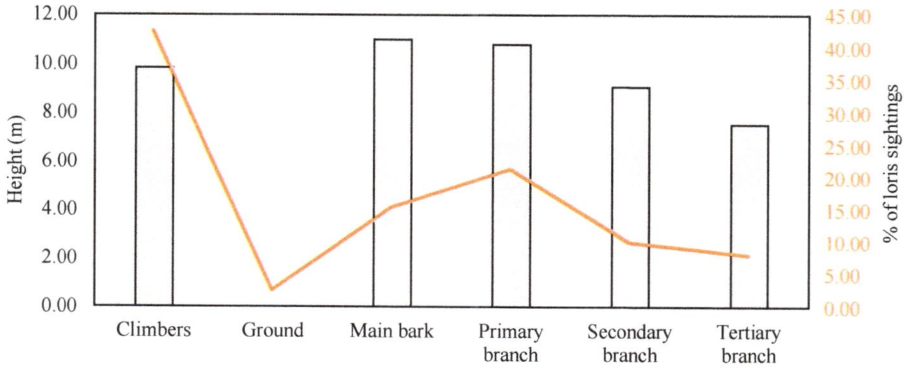

Figure 4. Percentage of sightings of slender lorises on different substrates (orange line) and relative height where the lorises were found (grey bars).

3.2.4. Climbers Used by Lorises

Acacia caesia (33.59%), *Gnetum edule* (22.14%), and *Aspidopterys canarensis* (11.45%) were the three most important climbers used by the lorises. We most frequently found the lorises (52.94%) on climbers that were intertwined with two species, followed by single species (35.29%). The intertwining of more than two species of climbers was rare, and we found lorises on them only 11.76% of the time (Table 3). The orientation of the climbers, where the slender lorises were encountered, was oblique (47.55%), horizontal (25.17%), and vertical (27.27%).

Table 3. Climbers used by the slender lorises at Aralam Wildlife Sanctuary.

Climbers the Loris Was Spotted On	Count	%
Acacia caesia	44	33.59
Gnetumedule sp.	29	22.14
Aspidopteryscanarensis sp.	15	11.45
Acacia caesia + *Gnetumedule* sp.	11	8.40
Acacia caesia + *Calycopteris floribunda*	5	3.82
Acacia caesia + *Gnetumedule* sp. + *Aspidopteryscanarensis* sp.	4	3.05
Calycopteris floribunda	4	3.05
Gnetumedule sp. + *Aspidopteryscanarensis* sp.	4	3.05
Acacia caesia + *Aspidopteryscanarensis* sp.	3	2.29
Derris scandens	3	2.29
Gnetumedule sp. + *Calycopteris floribunda*	2	1.53
Gnetumedule sp. + *Derris scandens*	2	1.53
Acacia caesia + *Gnetumedule* sp. + *Aspidopteryscanarensis* sp.	1	0.76

Table 3. *Cont.*

Climbers the Loris Was Spotted On	Count	%
Derris scandens + *Calycopteris floribunda*	1	0.76
Gnetumedule sp. + *Cissusrepens* sp.	1	0.76
Gnetumedule sp. + *Piper sarmentosum*	1	0.76
Ipomoea marginata	1	0.76

3.2.5. Vegetation Used for Sleeping and Feeding

The sleeping sites (clumped climber tangles) were found in 87.70% of the plots sampled (n = 317) on 36 species of trees. We mostly found them on *Xylia xylocarpa* (15.08%), followed by *Dillenia pentagyna* (8.54%) (Table 4).

Table 4. Top ten trees with sleeping sites of slender lorises.

Scientific Name	%
Xylia xylocarpa	15.08
Dillenia pentagyna	8.54
Artocarpus hirsutus	5.28
Syzygium cumini	5.03
Holigarna arnottiana	4.27
Terminalia paniculata	4.27
Dead Tree	4.02
Terminalia bellirica	4.02
Aporosa cardiosperma	3.77

We spotted the slender lorises feeding on 23 instances during the study period. When spotted, they were observed feeding on insects or foraging for insects. They fed on insects at an average height of 4.79 ± 0.91 m. One or more climbers (52.17%) were found most often associated with the substrate trees. However, on 30.43% of the occasions, we spotted the lorises on climbers that lacked connectivity with trees. We spotted the lorises feeding on trees without climbers on only four occasions. We observed that all the climbers, with the exception of *Acacia caesia*, were linked with either trees or other climbers. Slender lorises also used *Acacia caesia* to hold onto while feeding off the ground.

4. Discussion

4.1. Habitat Structure

The structural characteristics of the study area reflect the ecological complexity and evolutionary richness of the Western Ghats, a globally recognized biodiversity hotspot [12]. The high tree diversity and balanced composition of evergreen and deciduous species are indicative of the region's transitional climate, altitudinal gradients, and heterogeneous microhabitats [42]. The dominance of *Aporosa cardiosperma*, a riparian-endemic species, underscores both its resilience to disturbance and its potential role as a keystone resource for arboreal fauna such as lorises [43]. The notable presence of endemic species further affirms the area's biogeographic significance and aligns with floristic patterns observed in other old-growth forests across the Western Ghats [44].

The recorded tree density (1521.04 ± 94.73 trees/ha) falls within the expected range for tropical moist deciduous forests in the region (1200–1800 trees/ha) [45], suggesting that the forest retains much of its original structure. Spatial distribution patterns of dominant tree species also provide insight into forest dynamics and anthropogenic influence; the relatively uniform spacing of *Artocarpus hirsutus* and *Gmelina arborea* may reflect past selective logging or competitive exclusion mechanisms, whereas the clumped distributions of *Terminalia crenulata* and *Shorea roxburghii* could result from dispersal limitations or species-specific microhabitat preferences [46]. The overall random distribution of trees is typical of mature tropical forests, where stochastic events and ecological drift often dominate community assembly [47].

The moderate stratification in canopy height and basal area values, with emergent individuals of *A. cardiosperma*, suggests the presence of vertical niche availability, crucial for arboreal locomotion and predator avoidance among lorises [19]. The dominance of Fabaceae in basal area points to their role as nitrogen-fixing pioneers in recovering forests, whereas Dipterocarpaceae—associated with late-successional stages—signal forest maturity and successional complexity [48]. The limited development of the understory, with only 0.85% of species found below 5 m, implies a closed canopy that restricts light penetration, a characteristic feature of undisturbed Western Ghats rainforests [49].

4.2. Habitat Use

Our observations suggest that Malabar slender lorises preferentially occupy mildly degraded wet evergreen and secondary moist deciduous forests, particularly within sub-canopy layers (5–15 m height), gradually ascending higher into the canopy as the night progresses. This vertical ascent during the night may be biologically driven by the vertical stratification of insect prey, as many nocturnal insects become more abundant in higher canopy layers later in the night, especially moths and beetles, which are key components of loris diets [50]. These findings are consistent with previous studies; Kumara et al. [19] observed lorises in the 5–20 m vertical range, with a peak at 5–15 m, and similar height preferences have been recorded in *L. lydekkerianus* populations in Sri Lanka, where 79% of sightings occurred below 15 m [13]. Red slender lorises (*L. tardigradus tardigradus*) in Sri Lanka were similarly observed in the 3.5–15 m range, with average perch heights around 8.64 ± 5.00 m [22]. Nekaris and Jayewardene also documented a predominance of activity below 12 m [51], noting that higher perch levels likely reduce detectability during nocturnal surveys. Recent phylogenetic and ecological analyses further suggest vertical niche partitioning between grey and red loris taxa, with grey subspecies often selecting higher perches—likely as a predator-avoidance tactic [52]. Altogether, these data underscore the importance of preserving multi-strata forest architecture—ensuring canopy connectivity from understory to emergent layers—to maintain the full suite of microhabitats critical for nocturnal loris behavior and conservation.

The lorises in our study were primarily found in forest patches with high tree density, high basal area, and trees in the 60–90 cm girth class. These conditions suggest low-intensity disturbance, typically limited to minor activities such as firewood collection [53]. Our findings contrast with those from Sri Lanka, where *L. t. tardigradus* appears to favor highly disturbed forests and even human-dominated landscapes [22,34], and *L. l. nordicus*, a known habitat specialist, is restricted to undisturbed montane evergreen and mist forests characterized by tall, well-connected canopies [21,54]. These differences likely reflect both ecological and anthropogenic influences, while *L. t. tardigradus* has adapted to fragmented, disturbed environments due to intense historical habitat loss in Sri Lanka's wet zone, *L. l. nordicus* persists in climatically unique, high-elevation ecosystems with limited human access. In contrast, *L. l. malabaricus* appears to occupy an intermediate niche—preferring structurally complex but moderately disturbed habitats—possibly due to varying disturbance regimes, forest use patterns, and prey availability across the Western Ghats. This balance of disturbance tolerance and ecological dependency suggests both a degree of behavioral plasticity and a reliance on specific habitat features, such as vertical connectivity and microhabitat richness. These interspecific differences underscore the ecological flexibility of *L. l. malabaricus* and the need for site-specific conservation approaches.

4.3. Arboreal Locomotion and Substrate Use

The restricted terrestrial movement observed among slender lorises in this study aligns with their highly specialized arboreal morphology and behavioral ecology. These primates exhibit extreme adaptations for slow, deliberate quadrupedal locomotion and are anatomically incapable of leaping beyond approximately 0.3 m [37]. As such, they respond to canopy gaps by detouring rather than jumping, a strategy supported by anatomical studies of their limb and muscle structure. An important dimension of habitat use among primates is the degree of terrestriality and the use of forest strata [55,56]. From a socio-ecological perspective, forest strata use is primarily influenced by predation pressure [57], competition with sympatric species [58], and habitat structure and seasonality [59–62]. Descending to the ground exposes lorises to elevated predation risk from snakes, carnivores, and raptors, and disrupts their cryptic movement strategy—reinforcing the importance of maintaining uninterrupted canopy connectivity [18].

Instances of ground descent by lorises in fragmented landscapes have been well documented. In Sri Lanka, for example, populations inhabiting home gardens and agroforests show increased terrestriality due to broken canopy cover. Comparable patterns have been observed in Java, where installation of artificial canopy bridges for *Nycticebus javanicus* reduced terrestrial activity from 5.98 to 0.43 s/hour [63], and in urbanized areas where slow lorises nonetheless favored bamboo patches for movement and foraging [64]. In Kerala, similar behaviors were linked to logging and agricultural encroachment that reduced tree connectivity [17].

Our data reveal a pronounced preference for narrow substrates: twigs $\leq$1 cm diameter accounted for 36.98% of locomotor events, and small branches (2–5 cm) accounted for 33.43%. These diameters are consistent with the species' need for stable yet navigable supports, enabling effective grip and reducing energetic cost [19,65]. Early natural history observations by Petter and Hladik [66] and captive studies by Subramoniam [67] support these findings, noting lorises' avoidance of vertical supports over 2.5 cm in diameter and their inability to climb smooth trunks $\geq$10 cm, even in controlled environments.

Obliquely oriented substrates were used in 50% of locomotor activity, reflecting an energetically efficient strategy that balances mobility and stability. Inclined branches provide secure travel paths and serve as important escape routes from predators [51,52]. Biomechanical studies have shown that oblique substrates enable lorises to maintain a crouched posture with limbs positioned beneath the body, thereby reducing the muscular effort required to counteract gravity compared to vertical clinging or suspension. This orientation also facilitates continuous limb contact with the substrate, enhancing friction and reducing slip risk while minimizing energetic expenditure during slow, deliberate locomotion [68]. Additionally, oblique routes often present a gradient of mechanical support, allowing lorises to distribute body weight more efficiently across multiple limbs during movement [69]. Horizontal branches, in contrast, were generally used for behaviors requiring extended stasis—feeding, resting, or social interactions—while vertical clinging was limited to brief transitions due to its higher metabolic cost [15,29].

Slender lorises exhibit a highly selective pattern of sleep-site use. In our study, sleeping sites were located in 87.70% of surveyed plots—a markedly higher density than in most reported populations [29,41,51]. Lorises predominantly chose dense tangles of climbers on large-diameter trees, which offer concealment from diurnal predators and physical stability for daytime rest. These findings align with broader observations across lorisiforms, which show preferences for thorny understories, small entrance cavities, and well-connected canopies [40,65,70]. Larger trees often accumulate more extensive liana networks, creating microhabitats with both mechanical complexity and visual shielding [71,72]. These results reinforce the importance of maintaining structurally complex, vertically continuous forests to support both active and resting behaviors of slender lorises.

Our observations of loris foraging behavior further emphasize this dependence on habitat complexity. All feeding events involved insect prey, aligning with existing reports that 96% of loris diets consist of animal matter—primarily ants and termites [51]. Targeted foraging occurred on tree species known to support dense insect communities: *Artocarpus hirsutus* (ants, mealybugs) [73], *Garcinia gummi-gutta* (flea beetles) [74], *Lagerstroemia microcarpa* (nocturnal moths) [51], *Calophyllum inophyllum* (bark beetles) [75], and *Antidesma bunius* (hemipterans and ants) [61]. By exploiting mutualistic relationships between ants and honeydew-producing hemipterans, lorises benefit from spatially predictable prey aggregations—a strategy that requires fine-scale navigation of complex substrates [76].

Altogether, our findings underscore the central importance of preserving multi-strata forest architecture from thin, oblique twigs to dense vine tangles and large emergent trees. Such structural diversity supports the complete behavioral repertoire of the Malabar slender loris—including locomotion, foraging, and refuge use—and is essential for its conservation in both natural and human-modified landscapes.

4.4. Limitations

While this study provides valuable insights into the ecology and behavior of the Malabar slender loris (*Loris lydekkerianus malabaricus*), several limitations must be acknowledged. First, the study duration—although spanning two years—is relatively short for detecting long-term patterns or interannual variability, particularly in relation to reproductive cycles, habitat shifts, and population dynamics. Seasonal variation, especially the monsoon period, likely influenced insect availability and, by extension, loris foraging behavior; however, due to logistical constraints, fine-scale seasonal comparisons were limited.

Second, *L. l. malabaricus* is a cryptic, small-bodied nocturnal primate inhabiting dense, multilayered vegetation, making detection inherently challenging. The effectiveness of red-filtered headlamps in locating lorises relies heavily on the retro-reflective "eyeshine" produced when the animal is looking directly at the observer. If the loris is oriented away or its line of sight is obstructed, detection is virtually impossible. As such, there is a high likelihood of underestimating true encounter rates or overlooking individuals entirely.

Third, the field site is also home to large and potentially dangerous fauna—including elephants (*Elephas maximus*), tigers (*Panthera tigris*), leopards (*Panthera pardus*), and venomous snakes (e.g., *Bungarus caeruleus*, *Naja naja*, *Daboia russelii*)—which further constrained fieldwork, particularly during night surveys. Safety protocols had to be strictly followed, often reducing the duration or spatial extent of surveys.

Additional methodological limitations include the use of non-invasive visual encounter surveys without the aid of thermal imaging or radio-telemetry, which could have increased detection probabilities. Habitat data, although systematically collected, might not fully capture microhabitat use or fine-scale structural complexity due to accessibility issues in thick understory vegetation.

Taken together, these constraints suggest that while the trends reported are robust within the observed framework, future studies with longer durations, broader spatial coverage, and enhanced detection tools are essential for a more comprehensive understanding of this elusive primate's ecology.

5. Conclusions

Our study demonstrates that Malabar slender lorises exhibit finely tuned microhabitat preferences that optimize foraging efficiency and predator avoidance. These primates forage almost exclusively on insect prey—predominantly ants and termites—with their diet spanning nine insect orders, and they select twigs ≤1 cm and small branches (2–5 cm) for locomotion, exploiting oblique substrates that minimize energy expenditure, while maintaining stability [15,40,51,53]. Diurnal refuge sites are overwhelmingly located in dense liana tangles, thorny undergrowth, and small nest holes, providing concealment from mammalian and avian predators through structural complexity [50,51,70]. These behavioral specializations underscore the critical role of multi-strata forest architecture—maintaining a mosaic of canopy, subcanopy, and understory elements—to sustain the full repertoire of loris' ecological needs [29,66].

The Indian Institute of Science's "mini-forest" experiment provides a compelling model for applying these insights to habitat restoration. A 1.75 ha Deccan scrub plot planted with native Western Ghats evergreen species, structurally complex woodlands developed within 25 years, now supports a thriving Mysore slender loris (*L. l. lydekkerianus*) population [77,78]. Beyond ex situ conservation value, this mini-forest functions as a high-performance carbon sink, lowers ambient temperatures by up to 2 °C, and has driven groundwater recharge—raising local water tables by nearly 190 ft—thereby contributing to climate resilience and hydrological stability [77]. To replicate these successes across the slender loris range, restoration initiatives should prioritize multi-strata plantings of insect-hosting trees (e.g., *Artocarpus hirsutus*, *Garcinia gummi-gutta*, *Lagerstroemia microcarpa*), integrate mini-forest modules in degraded and urban fringe landscapes, and embed long-term monitoring frameworks as recommended by global restoration guidelines [79]. Our recent MaxEnt-based habitat-suitability modeling in Aralam Wildlife Sanctuary identified precipitation of the warmest quarter, precipitation of the driest month, distance from roads, and elevation as the most influential predictors of Malabar slender loris (*Loris lydekkerianus malabaricus*) distribution, emphasizing that broad-scale climatic and anthropogenic variables—rather than local canopy connectivity measures—govern habitat suitability in this landscape [26]. Collectively, these findings validate multi-strata restoration approaches—integrating insect-hosting trees, dense climbers, and varied branch architectures—as scalable conservation strategies across the slender loris range.

Supplementary Materials: The following supporting information can be downloaded at: https://www.mdpi.com/article/10.3390/f16060876/s1, Table S1: Trees, loris points, and point sampling of ribboned trees.

Author Contributions: S.D.G. contributed to conceptualization, visualization, supervision, project administration, funding acquisition, investigation, resources, data curation, validation, formal analysis, and manuscript writing—original draft preparation. J.J.E. was involved in conceptualization, validation, formal analysis, and manuscript writing. M.C. contributed to visualization, resources, and writing—review and editing. M.S. contributed to conceptualization, validation, writing—review and editing, and supervision. All authors have read and agreed to the published version of the manuscript.

Funding: This research was supported by the Women Scientist Scheme-A (WOS-A) fellowship under the Department of Science and Technology, Government of India (Grant No. SR/WOS-A/LS-89/2013) to Smitha D. Gnanaolivu.

Data Availability Statement: The original contributions presented in this study are included in the article/Supplementary Material. Further inquiries can be directed to the corresponding authors.

Acknowledgments: We thank the Forest Department of Kerala for the permission to conduct the study. We also thank the officials of Aralam Wildlife Sanctuary, Kerala, for their timely help and support in the field work, namely Madhoosudhan, Sushant, Biju, Radhakrishna, and Laxmana, for spending long hours surveying and for keeping us safe from elephants. We thank all the volunteers who helped us collect data. Mewa Singh thanks the Science and Engineering Research Board, Government of India, for the award of SERB Distinguished Fellowship (Award number: SB/S9/YSCP/SERB-DF/2018(1).

Conflicts of Interest: The authors declare no conflicts of interest. The funders had no role in the design of the study; in the collection, analyses, or interpretation of data; in the writing of the manuscript; or in the decision to publish the results.

Appendix A

Table A1. Floristic composition of the study area of Aralam Wildlife Sanctuary.

Species	All Trees	Loris Encountered	Average Height		Average DBH	Distribution	Tree Type
	Count	%	Tree (m)	Occupied (m)	(cm)		
Aporosa cardiosperma	289	13.38	9.29 ± 0.77	9.46 ± 1.36	12.77 ± 0.58	India and Sri Lanka	Evergreen
Xylia xylocarpa	226	11.62	19.02 ± 1.28	9.99 ± 1.41	29.5 ± 1.11	Indo-Malesia	Deciduous
Naringi crenulata	221	9.15	12.16 ± 1.13	7.74 ± 1.49	10.93 ± 6.46	Indo-Malesia	Deciduous
Hopea parviflora	145	2.46	11.7 ± 2.39	13.43 ± 4.56	11.93 ± 0.84	Endemic to Western Ghats	Evergreen
Antides mabunius	136	1.76	11.28 ± 2.24	14.93 ± 5.71	13.06 ± 0.91	Indo-Malesia to Australia and South China	Evergreen
Gossypium herbaceum	83	-	9.15 ± 5.07	-	10.78 ± 0.65	India, Arab, Persia, Afghanistan, Turkey, North Africa, Spain, Ukraine, China	Evergreen
Actinodaphne maderaspatana	82	2.46	8.62 ± 1.77	9.08 ± 2.84	15.5 ± 2.18	Endemic to Western Ghats	Evergreen
Antiaris toxicaria	71	-	17.29 ± 1.78	-	28.87 ± 1.79	Paleotropics	Evergreen
Terminalia paniculata	70	3.87	12.69 ± 1.87	8.73 ± 2.69	28.62 ± 2.91	India	Deciduous
Schleichera oleosa	63	2.82	22.14 ± 2.26	11.34 ± 2.91	18.3 ± 1.87	Indo-Malesia	Deciduous
Chionanthus mala-elengi subsp. *mala-elengi*	59	1.76	6.71 ± 0.67	8.41 ± 3.44	10.07 ± 0.48	Endemic to Peninsular India	Evergreen
Xanthophyllum flavescens	57	1.76	9.63 ± 1.23	11.2 ± 3.12	8.33 ± 0.51	China to Indomalaysia	Deciduous
Strychnosnux-vomica	55	0.35	21.95	9.14	7.47 ± 0.79	Indo-Malesia	Deciduous
Chionanthus albidiflorus	54	-	7.64 ± 3.95	-	8.83 ± 0.95	Indo-Malesia	Evergreen
Dead Trees	53	3.17	16.49 ± 1.69	8.41 ± 2.93	28.47 ± 1.98		
Polyalthia longifolia	53	0.35	10.06	1	6.72 ± 0.71	India and Sri Lanka	Evergreen
Stereospermum colais	50	0.7	17.83 ± 7.17	13.5 ± 8.5	19.74 ± 1.66	Indo-Malesia	Deciduous
Syzygium cumini	50	0.7	19.2 ± 9.45	15	10.25 ± 0.99	Indo-Malesia	Deciduous
Vateria indica	49	1.06	16.97 ± 7.82	13.53 ± 8.67	16.98 ± 2.16	Endemic to Western Ghats	Evergreen
Holigarna arnottiana	47	5.28	17.43 ± 1.85	12.04 ± 2.71	25.85 ± 2.96	Endemic to Western Ghats	Evergreen
Plumeria obtusa	45	2.46	11.58 ± 1.91	5.84 ± 1.61	21.55 ± 3.73	Central America, from Mexico to Panama	Deciduous
Sapindus trifoliatus	44	1.41	14.86 ± 1.82	8.65 ± 3.87	13.41 ± 1.56	South Asia	Deciduous
Artocarpus hirsutus	43	3.52	21.82 ± 0.89	16.72 ± 3.71	29.53 ± 3.43	Endemic to Western Ghats	Evergreen
Dillenia pentagyna	40	2.11	27.13 ± 2.67	10.48 ± 4.81	51.41 ± 6.1	China to Indo-Malesia	Deciduous

Table A1. *Cont.*

Species	All Trees	Loris Encountered	Average Height		Average DBH	Distribution	Tree Type
	Count	%	Tree (m)	Occupied (m)	(cm)		
Baccaurea courtallensis	39	0.35	8.53	2.74	7.19 ± 0.47	Endemic to Peninsular India	Evergreen
Myristica beddomei	38	2.82	12.03 ± 2.45	4.72 ± 2.95	12.81 ± 1.13	Endemic to Peninsular India	Evergreen
Lagerstroemia speciosa subsp. *Speciosa*	37	0.7	25.3	16.22 ± 13.79	22.07 ± 2.1	S-China (Yunnan), India,	Evergreen
Shorea roxburghii	36	0.7	7.01	4 ± 1	7.92 ± 1.16	Indo-Malesia	Deciduous
Rotheca serrata	33	4.23	10.24 ± 1.13	15.24 ± 3.21	12.26 ± 1.35	Indo-Malesia	Deciduous
Drypetes venusta	32	1.06	10.06 ± 2.14	13.45 ± 8.59	11.15 ± 1.68	Endemic to Western Ghats	Deciduous
Buchanania axillaris	30	2.11	13.56 ± 3.56	6.08 ± 2.45	13.75 ± 1.95	India and Sri Lanka, Myanmar	Deciduous
Anacolosa densiflora	26	0.7	12.04 ± 0.15	16.5 ± 13.5	17.95 ± 3.82	India	Evergreen
Madhuca longifolia	26	0.35	7.32	4.57	7.09 ± 0.69	India and Myanmar	Deciduous
Lagerstroemia microcarpa	25	0.35	22.56	18	28.01 ± 3.86	Endemic to Western Ghats	Deciduous
Clausena anisata	23	1.41	13.94 ± 0.89	9.93 ± 6.06	11.93 ± 1.53	India, Nepal, Sri Lanka, and Africa	Evergreen
Butea monosperma	20	0.7	17.53 ± 1.68	6.1 ± 6.1	19.91 ± 4.79	India, Sri Lanka and S.E. Asia.	Deciduous
Gmelina arborea	20	0.7	17.53 ± 1.37	6.86 ± 2.29	22.47 ± 4.69	Indo-Malesia	Deciduous
Terminalia bellirica	19	0.7	22.1 ± 0.15	6 ± 4	40.94 ± 5.95	Indo-Malesia	Deciduous
Vatica chinensis	18	-	11.08 ± 5.59	-	12.41 ± 1.22	India and Sri Lanka	Evergreen
Ixora polyantha	17	-	4.99 ± 2.37	-	6.13 ± 0.35	Endemic to Western Ghats	Evergreen
Tarenna monosperma	17	-	6.71 ± 4.18	-	5.82 ± 0.59	Endemic to Western Ghats	Evergreen
Lannea coromandelica	15	1.41	19.66 ± 3.55	8.75 ± 3.45	31.64 ± 5.79	Southern Asia	Deciduous
Litsea coriacea	15	1.06	10.87 ± 1.93	15.5 ± 8.23	10.12 ± 2.03	Endemic to Western Ghats	Evergreen
Knema attenuata	14	-	12.21 ± 7.15	-	8.14 ± 0.85	Endemic to Western Ghats	Deciduous
Melia azedarach	14	-	5.70 ± 1.73	-	7.55 ± 0.89	Paleotropics	Deciduous
Solanum erianthum	12	0.35	8.23	5.5	8.73 ± 0.98	South East Asia and North Australia	Deciduous
Vitex altissima	12	0.35	22.25	30	30.58 ± 4.59	India	Deciduous
Terminalia crenulata	11	1.06	25.1 ± 0.81	11.38 ± 8.33	91.56 ± 6.59	Indo-Malesia	Deciduous
Olea wightiana	11	0.7	12.19 ± 3.96	10 ± 5	10.91 ± 2.07	Endemic to Peninsular India	Deciduous
Alstonia scholaris	10	0.35	7.62	10	9.2 ± 2.41	South and South East Asia to Australia	Evergreen
Artocarpus gomezianus	10	0.35	11.89	9.14	23.78 ± 4.22	India and Sri Lanka	Deciduous
Cinnamomum keralaense	9	0.35	9.14	20	11.74 ± 2.26	Endemic to Western Ghats	Evergreen
Atalantia monophylla	8	-	11.05 ± 1.70	-	15.24 ± 2.8	Indo-Malesia	Deciduous

Table A1. *Cont.*

Species	All Trees	Loris Encountered	Average Height		Average DBH	Distribution	Tree Type
	Count	%	Tree (m)	Occupied (m)	(cm)		
Memecylon umbellatum	8	-	9.75 ± 2.96	-	11.09 ± 2.03	India and Sri Lanka	Deciduous
Polyalthia fragrans	8	-	4.23 ± 3.11	-	8.12 ± 1.93	Endemic to Western Ghats	Evergreen
Erythrina stricta	8	1.41	21.11 ± 4.17	6.69 ± 3.77	26.94 ± 3.71	India, China, Nepal, Thailand, and Vietnam	Deciduous
Scolopia crenata	8	0.35	8.53	10	7.44 ± 0.76	Indo-Malesia	Evergreen
Hopea ponga	7	-	13.44 ± 9.11	-	14.23 ± 4.51	Endemic to Western Ghats	Evergreen
Lepisanthes tetraphylla	7	-	6.14 ± 2.14	-	5.5 ± 1.37	Indo-Malesia and Africa	Evergreen
Sterculia villosa	6	-	6.14 ± 2.14	-	9.97 ± 3.43	South Asia and Myanmar	Deciduous
Dalbergia lanceolaria subsp. *paniculata*	6	1.06	9.75	5.19 ± 2.62	12.47 ± 2.19	India and Myanmar	Deciduous
Adina cordifolia	5	-	15.02 ± 4.97	-	8.94 ± 1.23	India, Myanmar, Sri Lanka, and Indo-China	Deciduous
Agrostistachys borneensis	5	-	5.00 ± 1.14	-	9.17 ± 0.23	Indo-Malaya	Evergreen
Elaeocarpus serratus	5	-	10.24 ± 6.81	-	16.81 ± 3.34	India, Nepal, Malaysia	Evergreen
Neolamarckia cadamba	5	-	8.47 ± 5.48	-	6.18 ± 1.48	Asia, the Pacific, and Australia	Deciduous
Commiphora caudata	5	0.7	21.18 ± 1.98	21.5 ± 6.5	21.84 ± 5.25	India and Sri Lanka	Deciduous
Bridelia retusa	4	-	5.87 ± 1.48	-	8.1 ± 1.12	Indo-Malaya	Deciduous
Grewia tiliifolia Vahl	4	-	15.62 ± 1.2	-	47.35 ± 13.13	Tropical Africa, India to Indo-China	Deciduous
Wrightia arborea	4	-	12.10 ± 2.9	-	27.06 ± 7.72	Indo-Malesia	Deciduous
Cinnamomum malabatrum	4	0.35	29.57	10	18.46 ± 6.24	Endemic to Western Ghats	Evergreen
Sterculia balanghas	4	0.35	10	10	5.33 ± 1.94	South Asia and Myanmar	Deciduous
Plumeria rubra	3	-	7.92 ± 2.60	-	4.35 ± 0.46	Native of Tropical America; widely naturalized elsewhere in the tropics	Deciduous
Casearia ovata	2	-	5.94 ± 1.07	-	22.76 ± 18.62	India and Sri Lanka	Evergreen
Garcinia morella	2	-	6.40 ± 0	-	6.37 ± 0	Indo-Malesia	Evergreen
Mallotus nudiflorus	2	-	13.16 ± 5.84	-	9.55 ± 1.91	Indo-Malaya	Evergreen
Terminalia alata	2	-	14.62 ± 1.2	-	6.05 ± 0.64	India	Deciduous
Terminalia catappa	2	-	3.33 ± 0.46	-	4.3 ± 0.16	Indo-Malesia	Deciduous
Azadirachta indica	1	-	16	-	21.65	Indo-Malesia	Evergreen
Calophyllum inophyllum	1	-		-	35.01	Paleotropics	Evergreen
Garcinia gummi-gutta	1	-	5.49	-	3.18	South India and Sri Lanka	Evergreen
Mangifera indica	1	-	8.53	-	10.5	Native to India and Burma.	Evergreen
Manilkara roxburghiana	1	-	5.23	-	14.01	Endemic to Western Ghats	Evergreen

Table A1. *Cont.*

Species	All Trees	Loris Encountered	Average Height		Average DBH	Distribution	Tree Type
	Count	%	Tree (m)	Occupied (m)	(cm)		
Manilkara zapota	1	-	8.23	-	53.16	India and tropical America	Evergreen
Scleropyrum pentandrum	1	-	11.28	-	6.37	India and Sri Lanka	Evergreen
Spondia spinnata	1	-	17.37	-	22.92	Indo-Malesia	Deciduous
Dysoxylum malabaricum	1	0.35	11.28	5	15.92	Endemic to Western Ghats	Evergreen
Syzygium mundagam	1	0.35	12.5	3.04	18.46	Endemic to Western Ghats	Evergreen

References

1. Overdorff, D.J. Ecological correlates to social structure in two lemur species in Madagascar. *Am. J. Phys. Anthropol.* **1996**, *100*, 487–506. [CrossRef]
2. Riley, E.P. Ranging patterns and habitat use of Sulawesi Tonkean macaques (*Macaca tonkeana*) in a human-modified habitat. *Am. J. Primatol.* **2008**, *70*, 670–679. [CrossRef] [PubMed]
3. Heiduck, S. The use of disturbed and undisturbed forest by masked titi monkeys *Callicebus personatus melanochir* is proportional to food availability. *Oryx* **2002**, *36*, 133–139. [CrossRef]
4. Li, Y. The effect of forest clear-cutting on habitat use in Sichuan snub-nosed monkey (*Rhinopithecus roxellana*) in Shennongjia Nature Reserve, China. *Primates* **2004**, *45*, 69–72. [CrossRef]
5. Campera, M.; Serra, V.; Balestri, M.; Barresi, M.; Ravaolahy, M.; Randriatafika, F.; Donati, G. Effects of habitat quality and seasonality on ranging patterns of collared brown lemur (*Eulemur collaris*) in littoral forest fragments. *Int. J. Primatol.* **2014**, *35*, 957–975. [CrossRef]
6. Marsh, L.K.; Chapman, C.A. *Primates in Fragments: Complexity and Resilience*; Springer: New York, NY, USA, 2013.
7. Ramsay, M.S.; Mercado Malabet, F.; Klass, K.; Ahmed, T.; Muzaffar, S. Consequences of habitat loss and fragmentation for primate behavioral ecology. In *Primates in Anthropogenic Landscapes: Exploring Primate Behavioural Flexibility Across Human Contexts*; Springer International Publishing: Cham, Switzerland, 2023; pp. 9–28.
8. Arroyo-Rodríguez, V.; Mandujano, S. Conceptualization and measurement of habitat fragmentation from the primates' perspective. *Int. J. Primatol.* **2009**, *30*, 497–514. [CrossRef]
9. Tabarelli, M.; Gascon, C. Lessons from fragmentation research: Improving management and policy guidelines for biodiversity conservation. *Conserv. Biol.* **2005**, *19*, 734–739. [CrossRef]
10. Lienert, J. Habitat fragmentation effects on fitness of plant populations—A review. *J. Nat. Conserv.* **2004**, *12*, 53–72. [CrossRef]
11. Umapathy, G.; Kumar, A. Impacts of forest fragmentation on lion-tailed macaque and Nilgiri langur in Western Ghats, south India. In *Primates in Fragments: Ecology and Conservation*; Marsh, L.K., Ed.; Kluwer Academic: Dordrecht, The Netherlands; Plenum Press: New York, NY, USA, 2003; pp. 163–189.
12. Myers, N.; Mittermeier, R.A.; Mittermeier, C.G.; da Fonseca, G.A.B.; Kent, J. Biodiversity hotspots for conservation priorities. *Nature* **2000**, *403*, 853–858. [CrossRef]
13. Perera, M.; Sandun, J. A review of the distribution of Grey Slender Loris (*Loris lydekkerianus*) in Sri Lanka. *Primate Conserv.* **2008**, *23*, 89–96. [CrossRef]
14. Gnanaolivu, S.D.; Kumara, H.N.; Singh, M.; Sudarsanam, D. Ecological determinants of Malabar Slender Loris (*Loris lydekkerianus malabaricus*, Cabrera 1908) occupancy and abundance in Aralam Wildlife Sanctuary, Western Ghats, India. *Int. J. Primatol.* **2020**, *41*, 511–524. [CrossRef]
15. Schulze, H.; Meier, B. The subspecies of *Loris tardigradus* and their conservation status: A review. In *Creatures of the Dark: The Nocturnal Prosimians*; Alterman, L., Doyle, G.A., Izard, M.K., Eds.; Plenum Press: New York, NY, USA, 1995; pp. 193–210.
16. Dittus, W.; Singh, M.; Gamage, S.N.; Kumara, H.N.; Kumar, A.; Nekaris, K.A.I. Loris lydekkerianus (amended version of 2020 assessment). In *The IUCN Red List of Threatened Species*; IUCN Biodiversity Assessment & Knowledge Team: Red List Unit: Cambridge, UK, 2022; p. E.T44722A217741551. [CrossRef]
17. Radhakrishna, S.; Singh, M. Social behaviour of the slender loris (*Loris tardigradus lydekkerianus*). *Folia Primatol.* **2002**, *73*, 181–196. [CrossRef]

18. Nekaris, K.; Bearder, S. The lorisiform primates of Asia and mainland Africa: Diversity shrouded in darkness. In *Primates in Perspective*; Campbell, C.J., Fuentes, A., Mackinnon, K.C., Panger, M., Bearder, S.K., Eds.; Oxford University Press: Oxford, UK, 2007; pp. 24–45.
19. Kumara, H.N.; Singh, M.; Kumar, S. Distribution, habitat correlates, and conservation of *Loris lydekkerianus* in Karnataka, India. *Int. J. Primatol.* **2006**, *27*, 941–969. [CrossRef]
20. Radhakrishna, S.; Kumara, H.N.; Sasi, R. Distribution patterns of slender loris subspecies (*Loris lydekkerianus*) in Kerala, Southern India. *Int. J. Primatol.* **2011**, *32*, 1007–1019. [CrossRef]
21. Gamage, S.N.; Padmalal, U.K.G.K.; Kotagama, S.W. Montane slender loris (*Loris tardigradus nycticeboides*) is a critically endangered primate that needs more conservation attention. *J. Dept. Wildl. Conserv.* **2014**, *2*, 77–83.
22. Gamage, S.; Liyanage, W.; Weerakoon, D.; Gunwardena, A. Habitat quality and availability of the Sri Lanka red slender Loris *Loris tardigradus tardigradus* (Mammalia: Primates: Lorisidae) in the Kottawa Arboretum. *J. Threat. Taxa* **2009**, *1*, 65–71. [CrossRef]
23. Hettiarachchi, C.J.; Gamage, S.N.; Mahanayakage, C.A.; Padmalal, U.K.G.K.; Kotagama, S.W. Habitat suitability modelling for Montane Slender Loris in the Hakgala Strict Nature Reserve: A Geoinformatics approach. *Wildlanka* **2015**, *3*, 144–147.
24. Rasmussen, D.T.; Izard, M.K. Scaling of growth and life history traits relative to body size, brain size, and metabolic rate in lorises and galagos (Lorisidae, Primates). *Am. J. Phys. Anthropol.* **1988**, *75*, 357–367. [CrossRef]
25. Sasi, R.; Kumara, H.N. Distribution and relative abundance of the slender loris *Loris lydekkerianus* in Southern Kerala, India. *Primate Conserv.* **2014**, *28*, 165–170. [CrossRef]
26. Gnanaolivu, S.D.; Erinjery, J.J.; Campera, M.; Singh, M. Distribution and habitat suitability of the Malabar Slender Loris (*Lorislydekkerianus malabaricus*) in the Aralam Wildlife Sanctuary, India. *Land* **2025**, *14*, 872. [CrossRef]
27. Menon, A.R.R. *Vegetation Mapping and Analysis of Aralam Wildlife Sanctuary Using Remote Sensing Techniques*; KFRI Research Report 168; Kerala Forest Research Institute (KFRI): Kerala, India, 1999.
28. Charles-Dominique, P.; Bearder, S.K. Field studies of lorisoid behaviour: Methodological aspects. In *The Study of Prosimian Behaviour*; Doyle, G.A., Martin, R.D., Eds.; Academic Press: New York, NY, USA, 1979; pp. 567–629.
29. Nekaris, K.A. Foraging behaviour of the slender loris (*Loris lydekkerianus lydekkerianus*): Implications for theories of primate origins. *J. Hum. Evol.* **2005**, *49*, 289–300. [CrossRef] [PubMed]
30. Nekaris, K.A.I. Activity budget and positional behavior of the Mysore slender loris (*Loris tardigradus lydekkarianus*): Implications for "slow climbing" locomotion. *Folia Primatol.* **2001**, *72*, 228–241. [CrossRef] [PubMed]
31. Sterling, E.J.; Ramaroson, M.G. Rapid Assessment of the primate fauna of the eastern slopes of the Réserve Naturelle Intégrale d'Andringitra. *Fieldiana Zool.* **1996**, *85*, 293–303.
32. Nekaris, K.A.I. Observations of mating, birthing and parental behaviour in three subspecies of Slender Loris (*Loris tardigradus* and *Loris lydekkerianus*) in India and Sri Lanka. *Int. J. Primatol.* **2003**, *74*, 5–6. [CrossRef]
33. Sutherland, W.J. *Ecological Census Techniques*; Cambridge University Press: Cambridge, UK, 1996.
34. Nekaris, K.A.I.; Liyanage, W.; Gamage, S. Influence of forest structure and composition on population density of the red slender loris *Loris tardigradus tardigradus* in Masmullah proposed forest reserve, Sri Lanka. *Mammalia* **2005**, *69*, 201–210. [CrossRef]
35. Sasidharan, N. *Illustrated Manual on Tree Flora of Kerala Supplemented with Computer-Aided Identification*; KFRI Research Report 282; Kerala Forest Research Institute (KFRI): Kerala, India, 2006.
36. Sasidharan, N. (Ed.) *Forest Trees of Kerala, a Checklist Including Exotics*; KFRI Handbook No. 2; Kerala Forest Research Institute (KFRI): Kerala, India, 2010.
37. Sellers, W. A biomechanical investigation into the absence of leaping in the locomotor repertoire of the slender loris (*Loris tardigradus*). *Folia Primatol.* **1996**, *67*, 1–14. [CrossRef]
38. Braun-Blanquet, J. *Pflanzensociologie: Grundzuge der Vegetationskunde*, 3rd ed.; Springer: Vienna, Austria, 1964.
39. Kent, M.; Coker, P. *Vegetation Description and Analysis: A Practical Approach*; John Wiley and Sons: New York, NY, USA, 1992; pp. 167–169.
40. Svensson, M.S.; Nekaris, K.A.I.; Bearder, S.K.; Bettridge, C.; Butynski, T.; Cheyne, S.M.; Das, N.; de Jong, Y.; Luhrs, A.M.; Luncz, L.; et al. Sleep patterns, daytime predation and the evolution of diurnal sleep site selection in lorisiforms. *Am. J. Phys. Anthropol.* **2018**, *166*, 563–577. [CrossRef]
41. Bearder, S.K.; Nekaris, K.A.I.; Buzzell, C.A. Dangers in the night: Are some nocturnal primates afraid of the dark? In *Eat or Be Eaten: Predator Sensitive Foraging Among Primates*; Miller, L.E., Ed.; Cambridge University Press: Cambridge, UK, 2002; pp. 21–43.
42. Pascal, J.P.; Ramesh, B.R.; Franceschi, D.D. Wet evergreen forest types of the southern Western Ghats, India. *Trop. Ecol.* **2004**, *45*, 281–292.
43. Karuppusamy, S. Vegetation and Forest Types of the Western Ghats. In *Biodiversity Hotspot of the Western Ghats and Sri Lanka*; Apple Academic Press: Palm Bay, FL, USA, 2024; pp. 25–61.
44. Shigwan, B.K.; Kulkarni, A.; Smrithy, V.; Datar, M.N. An overview of tree ecology and forest studies in the Northern Western Ghats of India. *iForest* **2024**, *17*, 213. [CrossRef]

45. Ganesh, T.; Ganesan, R.; Devy, M.S.; Davidar, P.; Bawa, K.S. Assessment of plant biodiversity at a mid elevation evergreen forest of Kalakad–Mundanthurai Tiger Reserve, Western Ghats, India. *Curr. Sci.* **1996**, *71*, 379–392.
46. Murali, K.S.; Shankar, U.; Shaanker, R.U.; Ganeshaiah, K.N.; Bawa, K.S. Extraction of non-timber forest products in the forests of Biligiri Rangan Hills, India. 2. Impact of NTFP extraction on regeneration, population structure, and species composition. *Econ. Bot.* **1996**, *50*, 252–269. [CrossRef]
47. Condit, R.; Ashton, P.S.; Baker, P.; Bunyavejchewin, S.; Gunatilleke, S.; Gunatilleke, N.; Hubbell, S.P.; Foster, R.B.; Itoh, A.; LaFrankie, J.V.; et al. Spatial patterns in the distribution of tropical tree species. *Science* **2000**, *288*, 1414–1418. [CrossRef]
48. Sundarapandian, S.M.; Swamy, P.S. Forest structure in the Western Ghats. *Proc. Indian Natl. Sci. Acad.* **1999**, *109*, 517–529.
49. Parthasarathy, N. Climber diversity in tropical forests. *J. Trop. Ecol.* **1999**, *15*, 315–332.
50. Maguire, D.Y.; Robert, K.; Brochu, K.; Larrivée, M.; Buddle, C.M.; Wheeler, T.A. Vertical stratification of beetles (Coleoptera) and flies (Diptera) in temperate forest canopies. *Environ. Entomol.* **2014**, *43*, 9–17. [CrossRef]
51. Nekaris, K.A.I.; Jayewardene, J. Survey of the slender loris (Primates, Lorisidae) in Sri Lanka. *J. Zool.* **2003**, *259*, 327–334.
52. Gamage, S.; Marikar, F.; Groves, C.; Turner, C.; Padmalal, K.; Kotagama, S. Phylogenetic relationship among slender loris species (Primates, Lorisidae: Loris) in Sri Lanka based on mtDNA CO1 barcoding. *Turk. J. Zool.* **2019**, *43*, 609–616. [CrossRef]
53. Mohandass, D.; Campbell, M.J.; Hughes, A.C.; Mammides, C.; Davidar, P. Edge disturbance drives liana abundance increase and alteration of liana–host tree interactions in tropical forest fragments. *Ecol. Evol.* **2017**, *8*, 4237–4251.
54. Gamage, S.N.; Hettiarachchi, C.J.; Mahanayakage, C.A.; Padmalal, U.K.G.K.; Kotagama, S.W. Factors influencing site occupancy of Montane Slender Loris (*Loris tardigradus nycticeboides*) in Sri Lanka. *Wildlanka* **2015**, *3*, 68–73.
55. Eppley, T.M.; Hoeks, S.; Chapman, C.A.; Ganzhorn, J.U.; Hall, K.; Owen, M.A.; Adams, D.B.; Allgas, N.; Amato, K.R.; Andriamahaihavana, M.; et al. Factors influencing terrestriality in primates of the Americas and Madagascar. *Proc. Natl. Acad. Sci. USA* **2022**, *119*, e2121105119. [CrossRef]
56. Estrada, G.R.; Marshall, A.J. Terrestriality across the primate order: A review and analysis of ground use in primates. *Evol. Anthropol.* **2024**, *33*, e22032. [CrossRef] [PubMed]
57. McGraw, W.S.; Bshary, R. Association of terrestrial mangabeys (*Cercocebus atys*) with arboreal monkeys: Experimental evidence for the effects of reduced ground predator pressure on habitat use. *Int. J. Primatol.* **2002**, *23*, 311–325. [CrossRef]
58. Thiel, S.; Tschapka, M.; Heymann, E.W.; Heer, K. Vertical stratification of seed-dispersing vertebrate communities and their interactions with plants in tropical forests. *Biol. Rev.* **2021**, *96*, 454–469. [CrossRef]
59. Lu, S.; Lin, N.; Huang, A.; Tong, D.; Liang, Y.; Li, Y.; Lu, C. Feeding postures and substrate use of François' langurs (*Trachypithecus francoisi*) in the limestone forest of Southwest China. *Animals* **2024**, *14*, 565. [CrossRef]
60. Li, Y. Terrestriality and tree stratum use in a group of Sichuan snub-nosed monkeys. *Primates* **2007**, *48*, 197–207. [CrossRef]
61. Porter, L.M.; Sterr, S.M.; Garber, P.A. Habitat use and ranging behavior of *Callimico goeldii*. *Int. J. Primatol.* **2007**, *28*, 1035–1058. [CrossRef]
62. Mourthe, I.M.C.; Guedes, D.; Fidelis, J.; Boubli, J.P.; Mendes, S.L.; Strier, K.B. Ground use by northern muriquis. *Am. J. Primatol.* **2007**, *69*, 706–712. [CrossRef]
63. Birot, H.; Campera, M.; Imron, M.A.; Nekaris, K.A.I. Artificial canopy bridges improve connectivity in fragmented landscapes: The case of Javan slow lorises in an agroforest environment. *Am. J. Primatol.* **2020**, *82*, e23076. [CrossRef]
64. Karimloo, L.; Campera, M.; Imron, M.A.; Rakholia, S.; Mehta, A.; Hedger, K.; Nekaris, K.A.I. Habitat Use, Terrestriality and Feeding Behaviour of Javan Slow Lorises in Urban Areas of a Multi-Use Landscape in Indonesia. *Land* **2023**, *12*, 1349. [CrossRef]
65. Hladik, C.M.; Charles-Dominique, P.; Petter, J. Feeding strategies of five nocturnal prosimians in the dry forest of the west coast of Madagascar. In *Nocturnal Malagasy Primates*; Charles-Dominique, P., Cooper, H.M., Hladik, A., Hladik, C.M., Pages, E., Pariente, G.F., Petter-Rousseaux, A., Petter, J.J., Schilling, A., Eds.; Academic Press: New York, NY, USA, 1980; pp. 41–73.
66. Petter, J.J.; Hladik, C.M. Observations on the home range and population density of *Loris tardigradus* in the forests of Ceylon. *Mammalia* **1970**, *34*, 394–409. [CrossRef]
67. Subramoniam, S. Some observations on the habits of the slender loris (*Loris tardigradus*). *J. Bombay Nat. Hist. Soc.* **1957**, *54*, 387–398.
68. Thorpe, S.K.; Holder, R.L.; Crompton, R.H. Origin of human bipedalism as an adaptation for locomotion on flexible branches. *Science* **2007**, *316*, 1328–1331. [CrossRef] [PubMed]
69. Das, N.; Nekaris, K.A.I. Positional behaviour and substrate preference of slow lorises, with a case study of Nycticebus bengalensis in Northeast India. In *Evolution, Ecology and Conservation of Lorises and Pottos*; Nekaris, K., Burrows, A., Eds.; Cambridge University Press: Cambridge, UK, 2020; pp. 210–218.
70. Anderson, J.R. Sleep-related behavioural adaptations in free-ranging anthropoid primates. *Sleep Med. Rev.* **2000**, *4*, 355–373. [CrossRef]
71. Clark, D.B.; Clark, D.A. Distribution and effects on tree growth of lianas and woody hemiepiphytes in a Costa Rican tropical wet forest. *J. Trop. Ecol.* **1990**, *6*, 321–336. [CrossRef]

72. Malizia, A.; Grau, H.R. Liana-host tree associations in a subtropical montane forest of north-western Argentina. *J. Trop. Ecol.* **2006**, *22*, 331–339. [CrossRef]
73. Deepthy, K.B.; Sunil, J.; Manoj, V.S.; Dhanya, M.K.; Maya, T.; Kuriakose, K.P.; Krishnaprasad, K.P. A new report of the myrmecophilous root mealy bug *Xenococcus annandalei* Silvestri (Rhizoecidae: Hemiptera)-a devastating pest. *Entomon* **2017**, *42*, 185–191.
74. Prathapan, K.D.; Chaboo, C.S. Biology of Blepharida-group flea beetles with first notes on natural history of *Podontia congregata* Baly, 1865 an endemic flea beetle from southern India (Coleoptera, Chrysomelidae, Galerucinae, Alticini). *ZooKeys* **2011**, *157*, 95.
75. Wainhouse, D.; Murphy, S.; Greig, B.; Webber, J.; Vielle, M. The role of the bark beetle *Cryphalus trypanus* in the transmission of the vascular wilt pathogen of takamaka (*Calophyllum inophyllum*) in the Seychelles. *For. Ecol. Manag.* **1998**, *108*, 193–199. [CrossRef]
76. Styrsky, J.D.; Eubanks, M.D. Ecological consequences of interactions between ants and honeydew-producing insects. *Proc. R. Soc. B* **2007**, *274*, 151–164. [CrossRef]
77. Sankara, R.K.; Bhat, H.R.; Kulkarni, V.A.; Ramachandra, T.V. Mini Forest—An experiment to evaluate the adaptability of Western Ghats species for afforestation. *Environ. Conserv. J.* **2011**, *121*, 79–83.
78. Ramachandra, T.V.; Setturu, B.; Rajan, K.S.; Subash Chandran, M.D. Modelling the forest transition in Central Western Ghats, India. *Spat. Inf. Res.* **2017**, *25*, 117–130. [CrossRef]
79. Chazdon, R.L. Beyond deforestation: Restoring forests and ecosystem services on degraded lands. *Science* **2008**, *320*, 1458–1460. [CrossRef]

forests

MDPI

Article

Temporal Shifts in Flower-Visiting Butterfly Communities and Their Floral Resources along a Vegetation Type Altered by Anthropogenic Factors

Karla López-Vázquez [1], Carlos Lara [2,*], Pablo Corcuera [3] and Citlalli Castillo-Guevara [2]

[1] Doctorado en Ciencias Biológicas y de la Salud, Universidad Autónoma Metropolitana, Ciudad de México 09310, Mexico; biokmlv@outlook.com
[2] Centro de Investigación en Ciencia Biológicas, Universidad Autónoma de Tlaxcala, Km 10.5 Autopista Tlaxcala-San Martín Texmelucan, San Felipe Ixtacuixtla 90120, Mexico; citlalli.castillo.guevara@gmail.com
[3] Departamento de Biología, Universidad Autónoma Metropolitana, Ciudad de México 09310, Mexico; pcmr@xanum.uam.mx
* Correspondence: carlos.lara.rodriguez@gmail.com

Abstract: Habitat disturbance driven by human activities poses a major threat to biodiversity and can disrupt ecological interactions. Butterfly–plant mutualisms represent an ideal model system to study such anthropogenic impacts, as butterflies exhibit intimate dependencies on larval host plants and adult nectar sources, rendering them highly sensitive to habitat changes affecting the availability of these floral resources. This study examined flower-visiting butterfly communities and their associations with flowering plants in a landscape altered by anthropogenic factors in central Mexico. The study area encompassed a mosaic of vegetation types, including native juniper forests, agricultural lands, and introduced eucalyptus plantations, representing different degrees of human-induced habitat modification. Monthly surveys were conducted over a single year, covering both rainy and dry seasons, to analyze butterfly and plant diversity, community composition, and interactions. Results showed the highest diversity in juniper forests, followed by eucalyptus and agricultural sites. Seasonal turnover was the primary driver of community changes, with habitat-based segregation persisting within seasons. Butterfly diversity strongly correlated with flower abundance, while plant richness played a secondary role. SIMPER and indicator species analyses identified key taxa contributing to compositional dissimilarities among habitats and associated with specific vegetation types and seasons. Our research provides insights into temporal dynamics structuring butterfly–plant interactions across this forest disturbance spectrum, highlighting how habitat changes and seasonality shape these mutualistic communities in changing landscapes.

Keywords: anthropogenic forest disturbance gradient; butterfly communities; habitat disturbance; indicator species; plant–pollinator interactions; temporal dynamics

1. Introduction

Anthropogenic activities such as urbanization, agricultural expansion, and resource extraction have resulted in widespread habitat loss and fragmentation and are recognized as the main drivers of biodiversity decline in ecosystems globally [1]. The disturbance and degradation of natural habitats can profoundly alter ecosystem functioning, community structure, and species interactions, with potential cascading effects across trophic levels [2].

Lepidopterans (butterflies and moths) represent a diverse and ecologically important group of insects in terrestrial ecosystems, acting as pollinators, herbivores, prey for higher trophic levels, and indicators of environmental change [3,4]. Their relatively short life cycles, sensitivity to environmental fluctuations, and well-studied relationships with host plants make them excellent indicators for assessing habitat quality, ecosystem integrity, and impacts of environmental stressors [1,5].

Butterflies exhibit associations with plants when hosting their larvae and when acting as nectar sources for adults, making them particularly vulnerable to habitat disturbances that disrupt the availability and spatial distribution of these essential resources [6,7]. Habitat fragmentation can lead to the isolation of butterfly populations, disrupting gene flow, reducing population sizes, and increasing the risk of local extinctions [8]. However, it can also have positive effects, such as increasing populations of species that prefer open habitats and sustaining some populations that might not survive in larger forests [8,9].

Understanding the responses of butterfly communities and their associated floral resources to varying degrees of habitat disturbance is crucial for developing effective conservation strategies and mitigating biodiversity loss [10]. While several studies have investigated the effects of habitat fragmentation on butterfly diversity, abundance, and community composition, e.g., [9,11–13], few have explicitly examined the concurrent impacts on butterfly community structure and their floral resources along a vegetation type altered by anthropogenic factors [14–17]. Furthermore, the potential influence of seasonality on these relationships remains understudied, despite well-documented seasonal fluctuations in butterfly and plant phenologies, e.g., [18,19].

In the central region of Tlaxcala, Mexico, landscapes encompass a range of habitat types altered by anthropogenic factors, providing an ideal opportunity to investigate the effects of habitat disturbance on butterfly communities and their floral resources. At both local and landscape scales, habitat disturbance can influence butterfly and flowering-plant communities [10]. Disturbance may lead to the loss of specialist species, while favoring generalist species better adapted to disturbed environments, e.g., [13,20]. Changes in environmental conditions, such as microclimate, light availability, and soil properties, can affect the relative abundances and phenologies of flowering plants. Additionally, human activities in the area and pollution can have significant effects on these environmental factors. These combined anthropogenic and environmental changes can potentially disrupt the temporal and spatial availability of resources for butterflies [21,22]. The spatial configuration of vegetation patches can influence butterfly movement and dispersal patterns, affecting gene flow, resource utilization, and metapopulation dynamics [23–26].

Moreover, the responses of butterfly and flowering-plant communities to habitat disturbance may vary across spatial and temporal scales, reflecting the inherent complexities of ecological processes and species-specific traits [10]. For instance, while habitat disturbance may negatively affect specialist species at the local scale during their active seasons, landscape-level factors such as habitat connectivity may be more influential in determining the persistence of wider-ranging generalist species during dispersal periods [27]. Seasonal variations in resource availability, climatic conditions, and life history traits of butterflies and plants may also modulate their responses to disturbance, highlighting the importance of considering temporal dynamics in ecological studies [28,29].

This study aims to elucidate the effects of habitat disturbance on adult butterfly communities and their interactions with flowering-plant resources along a vegetation type altered by anthropogenic factors in central Tlaxcala, Mexico. We addressed the following research questions: (1) How does the degree of habitat disturbance influence the diversity, abundance, and community composition of butterflies and flowering plants across seasons? (2) To what extent do plant richness and flower abundance explain the variation in butterfly community composition across the different vegetation types? (3) Which butterfly species show the highest degree of association with particular vegetation types and seasons, suggesting their potential role as indicator species?

By addressing these questions, our study contributes to a comprehensive understanding of the impacts of habitat disturbance on plant–insect communities in a region experiencing rapid land-use changes, while accounting for seasonal variations. The findings can inform conservation efforts aimed at preserving ecosystem integrity and promoting sustainable management practices.

2. Materials and Methods

2.1. Study Area

Butterfly and plant communities were studied in the municipality of Ixtacuixtla, which is located in the state of Tlaxcala in central Mexico (19°21′ N and 98°22′ W, with an altitude ranging from 2300 to 2350 m above sea level), from February 2023 to January 2024. While this timeframe allowed us to capture seasonal variations within a year, it is important to note that inter-annual variations were not assessed. The study was conducted in an area where the natural habitat of temperate forest (i.e., juniper–oak forest) has been altered by anthropogenic factors over the years, resulting in a mosaic of vegetation types. This landscape, covering an approximate area of 65 hectares, now includes the original juniper–oak forest alongside modified habitats such as agricultural lands and eucalyptus plantations. The study site is located 14.1 km west of the city of Tlaxcala. The original vegetation in the study sites consisted of juniper (*Juniperus deppeana*) and oak (*Quercus* spp.) forests, of which some relicts still remain. However, most of this vegetation was replaced by eucalyptus (*Eucalyptus* sp.) and/or modified to induced cultivation areas with rainfed agricultural areas with scattered trees around them. The climate is considered temperate sub-humid, exhibiting pronounced seasonal variations. The wet season extends from May to October and is characterized by a mean monthly precipitation of 123 mm and an average temperature of 17 °C. Conversely, the period from March to May constitutes the hottest and driest months, with a mean monthly precipitation of 10 mm and an average temperature of 13 °C [30] (Figure 1A).

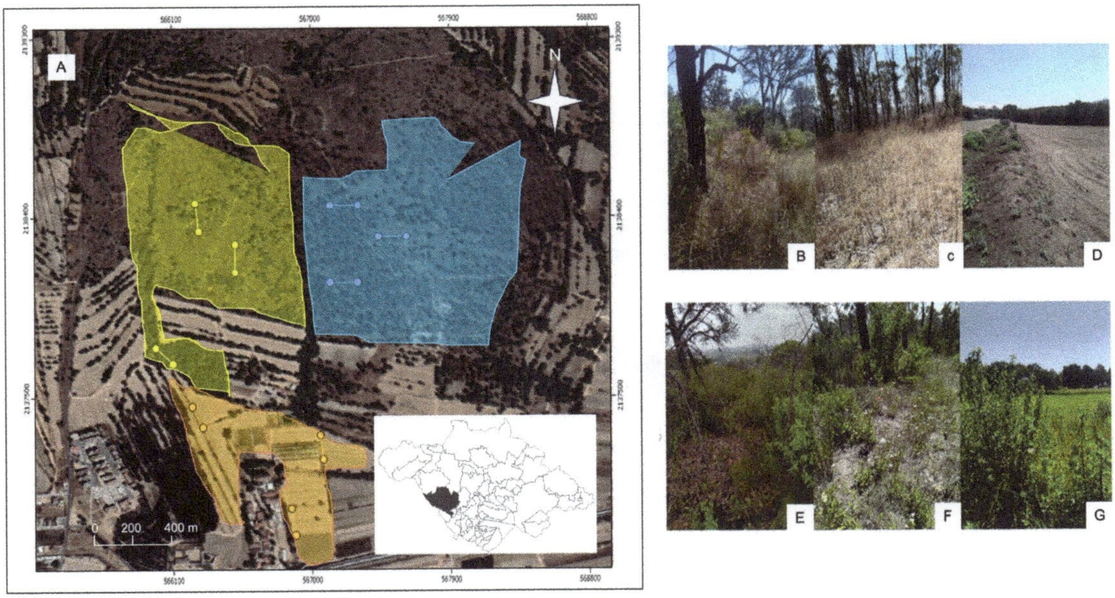

Figure 1. Study site and vegetation characteristics in Tlaxcala, Mexico. (**A**) Satellite image from Google Earth™ with colored overlays depicting juniper forest (yellow), eucalyptus forest (blue), and agricultural land (orange). Colored dots indicate survey transects. (**B–D**) Dry season and (**E–G**) rainy season images of the three vegetation types, demonstrating marked seasonal variation in vegetation cover and lushness.

The classification of vegetation types was established by conducting on-site visits to the locations, using a process known as ground truthing. This method allowed for the verification of land-use patterns through firsthand observations. Subsequently, these findings were further confirmed by examining satellite images of the surveyed areas [31]. In order to cover this landscape, three 100 m transects (with a distance between them of at least 100 m) were established in each vegetation type: Juniper forest (19°20′16.07″ N, 98°22′6.40″ W), where the vegetation is composed of relicts of original vegetation of *Juniperus deppeana* and some individuals of *Quercus* spp. species (Figure 1B,E), within a grassland dominated by native grasses and small shrubs such as *Muhlenbergia implicata*, *Stipa ichu*, and *Aristida schiedeana*, as well as *Rhus standleyi*. Eucalyptus forest (19°20′5.63″ N, 98°21′54.29″ W), an area where the original forest was replaced through induced plantations of *Eucalyptus globulus* trees as the dominant species (Figure 1C,F), and plant species such as *Erythrina coralloides*, *Stevia elatior*, and *Dyssodia papposa* are also present. Lastly, these forested zones are surrounded by agricultural lands (19°19′55.43″ N, 98°21′54.92″ W), used mainly for corn, wheat, and alfalfa crops (Figure 1D,G), and where human settlements are near.

2.2. Surveys of Plant–Butterfly Communities

While we acknowledge the importance of larval and host plant relationships, our research questions focus on the interactions between adult butterflies and flowering plants. Each month, the transects of a vegetation type were sampled (1 day per vegetation type until completing 3 continuous days), covering a schedule from 10:30 to 13:30 h, considering the peak activity times of butterflies. Two trained observers steadily walked the transects performing the censuses to ensure accuracy, data quality, and avoid observer bias, and to accomplish logistical feasibility. Throughout the study, exclusively during clear and sunny conditions, the censuses started from a different vegetation type transect and/or in a different direction to avoid order effects. Any butterfly visiting a flower within 10 m on either side of the transect or flying along it were counted, following the method of Pollard & Yates [32]. The term "flower visit" was used to describe the act of a butterfly species probing for nectar, which was recorded from the instant the proboscis was inserted into the corolla until it was withdrawn. We recorded all butterfly species observed during our surveys, including both sedentary and potentially migratory species. While we did not categorize species based on their mobility or migratory behavior in this study, we acknowledge that these traits may influence the observed patterns of butterfly–plant interactions. Butterfly identification was primarily conducted using Eagle Optics binoculars (Eagle Optics Co., Middleton, WI, USA). In cases where species identification was challenging through this method, individuals were carefully captured using a net and photographed with a Canon EOS Rebel T7 (Canon Inc., Tokyo, Japan). If the species could be identified from the photographs, the butterflies were immediately released. Only specimens that could not be identified in the field were collected and taken to the laboratory for further examination. These collected specimens were subsequently identified using taxonomic keys specific to butterflies, e.g., [33,34]. To ensure accuracy, all records were verified by José de Jesús García Díaz, a taxonomist specializing in butterflies, to corroborate their proper classification. After this process, these specimens were deposited in the Lepidoptera collection of the Centro de Investigación en Ciencias Biológicas at the Universidad Autónoma de Tlaxcala, Mexico. Additionally, samples of the plants that the butterflies interacted with were collected. These plant specimens were identified using specific taxonomic keys [35]. Finally, during each sampling month, all open flowers were quantified within each transect for at least 10 individuals of the plant species visited by butterflies.

2.3. Dynamics on the Richness and Diversity of the Plant–Butterfly Communities

To gain a comprehensive understanding of the ecological dynamics and interactions between butterflies and their flowering-plant communities on these habitat types, seasonal patterns were assessed and data were explored from both the dry and rainy seasons. The number of species found in each habitat type (i.e., combined across surveys over seasons) was used to express the richness of both butterflies and flowering plants. All interacting individuals as well as the flying butterflies recorded during the sampled surveys were included in the calculation of butterfly and flowering-plant richness.

2.4. Data Analysis

To assess the impact of land-use change on butterfly and flowering-plant diversity, we employed the unified framework of Hill numbers [36] to calculate species diversity using the interpolation and extrapolation approach developed by Chao et al. [37]. This method allows for the estimation of diversity at different orders (q), providing a comprehensive understanding of the effective number of species [38] and their relative abundances. We computed the zeroth-order (q = 0), first-order (q = 1), and second-order (q = 2) diversity, which correspond to species richness, the exponential of Shannon's entropy index, and the inverse of Simpson's concentration index, respectively [39]. These diversity measures quantify the number of equally abundant species required to produce the same value of a given diversity index [38] and reflect the degree of evenness in species abundances [40]. To compare the diversity indices across the habitat types, we constructed 95% confidence intervals using the iNEXT package version 3.0.0 [41]. Additionally, we visualized the species abundance distributions using rank-abundance curves to provide further insights into the community structure.

To assess the variability in species composition (beta diversity) among the butterfly and flowering-plant communities sampled across the three vegetation types, we calculated the Bray–Curtis dissimilarity index [42]. This index quantifies the differences in species composition between pairs of communities, considering both the presence and abundance of species. Subsequently, to visualize the patterns of community composition and identify potential clustering based on habitat types and climatic seasons (rainy and dry), we performed a non-metric multidimensional scaling (NMDS) ordination using the Bray–Curtis dissimilarity matrix. NMDS is a robust ordination technique that represents the compositional dissimilarities between communities in a reduced-dimensional space while minimizing the stress of the configuration [43]. The statistical significance of the observed community groupings was then evaluated using the Analysis of Similarities (ANOSIM), which is a non-parametric permutation test that assesses whether the compositional differences between a priori defined groups (in this case, habitat types and seasonality) are greater than expected by chance [44]. The NMDS and ANOSIM analyses were conducted using the vegan version 2.5-6 [45] and MASS version 7.3-61 [46] packages in R [47].

To determine the influence of flower abundance and plant species richness on butterfly community composition, we performed a redundancy analysis (RDA) using the "vegan" package in R [47]. The RDA was conducted on a matrix of butterfly species abundances across the three vegetation types, with flower abundance and plant species richness as explanatory variables. Flower abundance was quantified by counting the number of flowers along the transects at each site, while plant species richness was determined by identifying and counting the number of unique plant species within the same transects. The significance of the RDA axes was assessed using permutation tests with 999 permutations. Pearson's correlation coefficients were calculated to examine the relationships between the RDA axes and the explanatory variables. The RDA ordination plot was generated to visualize the patterns of butterfly community composition in relation to the vegetation types and the explanatory variables.

The compositional dissimilarities between habitats for both butterflies and flowering plants were assessed through the percentage similarity test or SIMPER [48]. SIMPER identifies the species primarily contributing to the observed differences by decomposing the Bray–Curtis dissimilarity index into percentage contributions from each species [49]. In this analysis, a 90% cut-off for low contributions was employed, and the 'simper' function from the vegan package [45] in R was utilized. The test computes the overall sample dissimilarity as well as the dissimilarity contribution of individual species by considering their relative abundances and their dissimilarity across samples [48]. Consequently, SIMPER highlights the key species driving the compositional differences between habitats by accounting for both their abundance patterns and their distinctiveness across samples.

Lastly, to determine the flowering plant and butterfly species that could serve as indicators or representatives of each habitat type, an Indicator Species Analysis (IndVal) was conducted [50]. The IndVal approach combines information on the species' abundance and their occurrence patterns across groups (e.g., habitats) to derive an indicator value for each species in each group [51]. This method identifies species that are both abundant and highly associated with particular groups, making them suitable indicators or representatives of those groups [50]. The analysis was performed using the indval function from the "labdsv" package [52] in the R-project software environment [47].

3. Results

3.1. Richness, Structure, and Seasonality

The composition and abundance of butterfly and flowering-plant communities exhibited pronounced differences across the three habitat types, as predicted. A total of 57 butterfly species (Table 1) and 34 flowering-plant species (Table 2) were recorded throughout the study. The observed number of butterfly species and plant species in the study seemed to reach an asymptote in relation to our sampling effort across the three sampled habitats (a total of 108 hours of evenly distributed observation efforts throughout the study). For butterfly species, we detected 99% sampling completeness for the juniper forest and the eucalyptus forest and 97% for agriculture sites according to the Chao2 estimator, after conducting 12 sampling events for each habitat type throughout the study. Likewise, we achieved 99% sampling completeness for flowering-plant species in the juniper forest, 99% in the eucalyptus forest, and 100% in the agriculture lands, based on the estimated species richness. The effective number of butterfly species varied considerably across both habitat types and seasons (Figure 2A). Butterfly species richness was consistently highest in the juniper forest, intermediate in the eucalyptus forest, and lowest in the agricultural sites. This pattern was maintained across both the rainy and dry seasons. However, the magnitude of the differences among habitats was more pronounced during the dry season compared to the rainy season. Butterfly richness also exhibited marked seasonal variation. Across all three habitat types, the number of butterfly species was highest during the rainy season and lowest during the dry season. The seasonal fluctuations were less pronounced than the differences among habitats. The flowering-plant community showed similar trends to the butterflies in terms of habitat differences (Figure 2B). The juniper forest had the highest effective number of flowering-plant species, followed by the eucalyptus forest, with the agricultural sites having the lowest flowering-plant richness. This pattern was consistent across both the rainy and dry seasons. Flowering-plant species richness was higher during the rainy season compared to the dry season in all habitats, but the magnitude of the seasonal difference varied among vegetation types. The largest seasonal difference in flowering-plant richness was observed in the agricultural sites, while the juniper forest showed a smaller difference between the rainy and dry seasons.

The rank-abundance distributions revealed markedly different butterfly community structures among the habitats (Figure 3A). The agricultural sites exhibited a relatively homogeneous abundance distribution, with few very abundant species and a long tail of rare species. In contrast, the forest habitats displayed a more heterogeneous structure, with abundances more evenly distributed among the species ranks (Table 3). This pattern suggests higher evenness and lower dominance in the forest butterfly communities compared to the disturbed agricultural areas.

Table 1. Butterfly species in each vegetation type recorded throughout the study.

Family	Species	Agriculture Sites	Juniper Forest	Eucalyptus Forest
Papilionidae	*Battus philenor philenor* (Linnaeus, 1771)	X	X	X
	Mimoides thymbraeus (Boisduval, 1836)	X		
	Pterourus garamas garamas (Geyer, Geyer, [1829])		X	
	Pterourus multicaudata multicaudata (Kirby, 1884)	X	X	X
	Heraclides pharnaces (Doubleday, 1846)	X	X	
	Papilio polyxenes asterius (Stoll, 1782)		X	
Pieridae	*Abaeis mexicana mexicana* (Boisduval, 1836)		X	X
	Anteos maerula (Fabricius, 1775)		X	
	Archonias nimbice nimbice (Boisduval, 1836)	X	X	X
	Colias eurytheme (Boisduval, 1852)	X		
	Abaeis salome jamapa (Reakirt, 1866)			X
	Ganyra josephina josepha (Godman & Salvin, 1868)	X	X	X
	Leptophobia aripa elodia (Boisduval, 1836)	X	X	X
	Nathalis iole iole (Boisduval, 1836)	X	X	X
	Phoebis agarithe agarithe (Boisduval, 1836)	X	X	X
	Phoebis argante argante (Fabricius, 1775)		X	X
	Phoebis neocypris virgo (Butler, 1870)	X	X	X
	Phoebis philea philea (Linnaeus, 1763)		X	
	Phoebis sennae marcellina (Cramer, 1777)	X	X	X
	Pieris rapae rapae (Linnaeus, 1758)			X
	Pontia protodice (Boisduval & Le Conte, [1830])	X		X
	Zerene cesonia cesonia (Stoll, 1790)	X	X	X
Nymphalidae	*Adelpha paroeca paroeca* (Bates, 1864)		X	
	Anartia fatima fatima (Fabricius, 1793)	X	X	X
	Anthanassa texana (Edwards, 1863)	X		
	Chlosyne ehrenbergii (Geyer, [1833])	X		
	Chlosyne marina (Geyer, 1837)			X
	Cyllopsis pyracmon pyracmon (Butler,1867)		X	
	Danaus eresimus montezuma Talbot, 1943	X	X	
	Danaus gilippus thersippus (Bates, 1863)	X	X	X
	Danaus plexippus plexippus (Linnaeus, 1758)	X	X	X
	Dione juno huascuma (Reakirt, 1866)	X	X	X
	Dione moneta poeyii Bluter, 1873	X	X	X
	Dione incarnata incarnata (Riley, 1926)	X	X	X
	Euptoieta claudia claudia (Cramer, 1775)	X	X	X
	Euptoieta hegesia meridiania (Stichel, 1938)	X	X	X
	Phyciodes graphica graphica (Felder, 1869)	X		
	Vanessa annabella (Field, 1971)	X	X	X
	Vanessa cardui (Linnaeus, 1758)		X	
Lycaenidae	*Brephidium exilis exilis* (Boisduval, 1852)		X	X
	Echinargus isola (Reakirt, 1867)	X	X	X
	Callophrys spinetorum millerorum (Clench, 1981)		X	
	Leptotes cassius cassidula (Boisduval, 1870)		X	
	Leptotes marina (Reakirt, 1868)		X	X
Hesperiidae	*Amblyscirtes fimbriata fimbriata* (Plötz, 1882)	X	X	
	Atalopedes huron (Edwards, 1863)		X	X
	Burnsius communis albescens (Plötz, 1884)		X	
	Calpodes ethlius (Stoll, 1782)	X		
	Cecropterus cincta (Plötz, 1882)		X	
	Eantis pallida (Felder, 1869)	X	X	X
	Lerema accius (Smith, 1797)		X	
	Oarisma edwardsii (Barnes, 1897)	X	X	
	Telegonus cellus (Boisduval & Le Conte, [1837])	X	X	
Erebidae	*Apantesis proxima* (Guérin-Méneville, 1831)		X	
	Dysschema howardi (Edwards, 1886)			X
Crambidae	*Fissicrambus* sp. (Bleszynski, 1825)		X	
	Pyrausta inornatalis (Fernald, 1885)		X	

Table 2. List of the plant species used by butterflies for nectar feeding and recorded in the three vegetation types throughout the study.

Family	Species	Agriculture Sites	Juniper Forest	Eucalyptus Forest
Phytolaccaceae	*Phytolacca icosandra* L., 1753			X
Papaveraceae	*Argemone platyceras* Link & Otto, 1828	X		
Brassicaceae	*Raphanus sativus* L., 1753	X		
Fabaceae	*Erythrina coralloides* DC., 1825			X
	Mimosa aculeaticarpa Ortega, 1798	X	X	X
	Senna multiglandulosa (Jacq.) H. S. Irwin & Barneby, 1982		X	
Anacardiaceae	*Rhus integrifolia* (Nutt.) Brebner ex W. H. Brewer & S.Watson, 1876		X	X
Onagraceae	*Oenothera gaura* Raf., 1836	X		
Convolvulaceae	*Evolvulus prostratus* Rob., 1962			X
Polemoniaceae	*Loeselia mexicana* (Lam.) Brand, 1907		X	X
Lamiaceae	*Clinopodium multiflorum* (Romo & Delgadillo) B. L. Turner, 1994	X		
Solanaceae	*Physalis philadelphica* Lam., 1793	X		
	Solanum carolinense L., 1753		X	
Rubiaceae	*Bouvardia ternifolia* (Cav.) Schltdl., 1819	X	X	X
Asteraceae	*Amblyopappus pusillus* Hooker & Arnott, 1835		X	
	Baccharis breviseta DC., 1836			X
	Baccharis salicifolia (Ruiz & Pav.) Pers., 1807		X	X
	Barkleyanthus salicifolius (Kunth) H.Rob. & Brettell, 1974	X		
	Bidens odorata Cav., 1794			X
	Brickellia californica (Torr. & A.Gray) A. Gray, 1873	X		
	Dahlia coccinea Cav., 1791	X	X	X
	Dyssodia papposa (Vent.) Hitchc., 1932			X
	Eremosis corymbosa (DC.) Gleason, 1923			X
	Garberia heterophylla (W.M.Wood) Merr. & F. Harper, 1909		X	X
	Heterotheca grandiflora Nutt., 1841	X		
	Pseudognaphalium gaudichaudianum (DC.) Anderb., 2012	X		
	Stevia elatior Kunth, 1818			X
	Stevia serrata Cav., 1794			X
	Tagetes lucida Cav., 1794		X	
	Tithonia tubiformis (Jacq.) Cass., 1825	X		X
	Tridax procumbens L., 1753	X		X
Amaryllidaceae	*Zephyranthes brevipes* (Engelm. ex Buckley) Ingram, 1940	X		
Orchidaceae	*Dichromanthus cinnabarinus* (La Llave & Lex.) Garay, 1982	X		
Phytolaccaceae	*Phytolacca icosandra* L., 1753			X

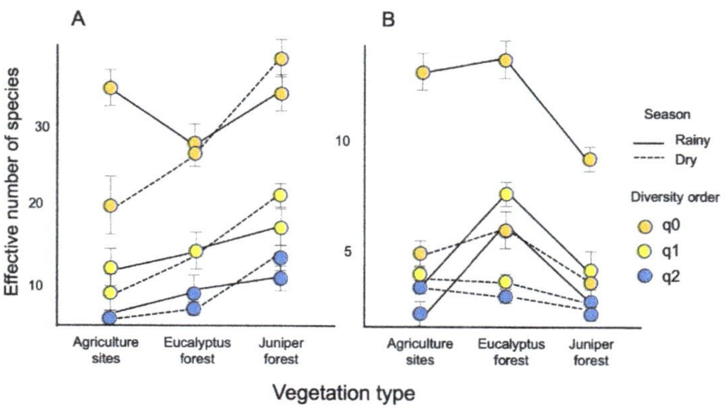

Figure 2. Diversity of butterfly (**A**) and flowering-plant (**B**) species across three vegetation types during the rainy (solid lines) and dry (dashed lines) seasons. Three diversity orders are presented: q = 0 (richness), q = 1 (Shannon entropy), and q = 2 (inverse Simpson), with their 95% confidence intervals. The vegetation types include agricultural sites, juniper forest, and eucalyptus forest. Overlapping intervals suggest no significant differences, while non-overlapping intervals indicate significant differences ($p < 0.05$) in diversity indices.

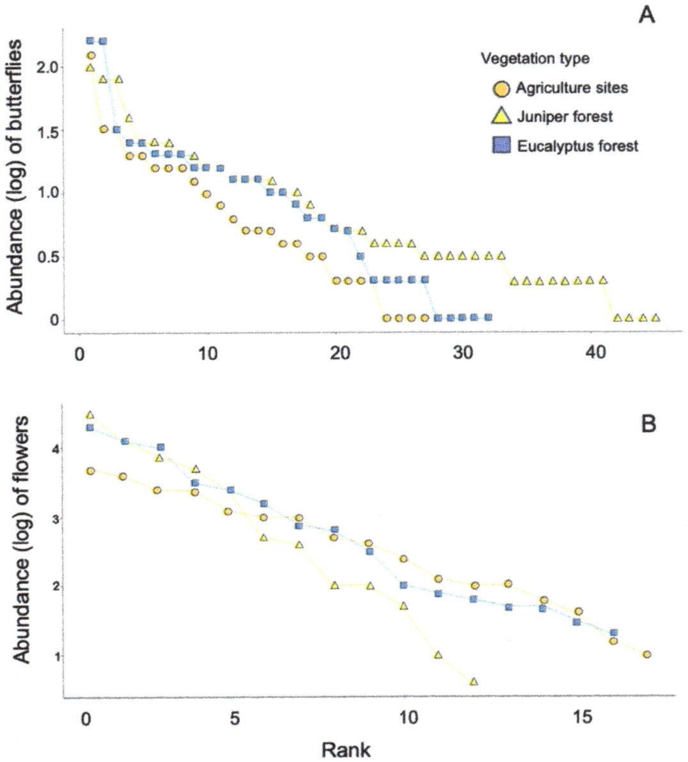

Figure 3. Species rank-abundance plots showing the log-scaled abundance of butterfly species (**A**) and flowering-plant species (**B**) across the three vegetation types monitored. Species are ranked on the x-axis from highest to lowest abundance, with rank 1 (leftmost) representing the most abundant species and increasing rank numbers (moving right) indicating progressively less abundant species. This ranking visualizes both species richness (total number of points) and evenness (slope of the line) for each vegetation type.

Table 3. Butterfly and flowering-plant diversity indicators, including the total number of sampled individuals (number of flowers in the case of plants), number of species, richness (q = 0), common species (q = 1), and dominant species (q = 2) across the three sampled vegetation types. Significant differences ($p < 0.05$), indicated by non-overlapping 95% confidence intervals among the three vegetation types, are marked with asterisks.

Community	Indicator	Agriculture Sites	Juniper Forest	Eucalyptus Forest
Butterfly	Individuals	366	434	590
	Species	32	31	46
	q0	39.20	33.11 *	48.70
	q1	12.49	15.06	20.95 *
	q2	6.34	7.61	12.85 *
Plants	Flowers	9666	25,908	37,279
	Species	18	12	16
	q0	18.1	12.1 *	16.1
	q1	5.20	6.17	3.97 *
	q2	2.91 *	4.43	3.19

The rank-abundance distributions for the flowering-plant communities (Figure 2B) showed similar patterns to those of the butterflies. The agricultural sites had a steeper rank-abundance curve, indicating a plant community with a few highly abundant species and many rare ones. The eucalyptus forest exhibited a slightly more even distribution of flowering-plant species abundances, while the juniper forest had the shallowest rank-abundance curve, suggesting a more balanced plant community with a greater number of species having intermediate abundances (Table 3). The species-accumulation curves for the butterflies (Figure 3A) further corroborated the lower diversity in agricultural sites, with shallower slopes indicating slower accumulation of new species compared to the juniper and eucalyptus forests. Similarly, the plant species-accumulation curves (Figure 3B) showed a slower rate of species accumulation in the agricultural sites compared to the forest habitats, reflecting lower plant diversity in the disturbed areas.

The Bray–Curtis similarity analysis indicated a considerable similarity in butterfly species composition between the juniper forests and eucalyptus forest (62%), as well as between juniper forest and agricultural areas (56%). The eucalyptus forest and agricultural sites also shared high similarity (65%). However, these similarities varied according to the climatic season. In contrast, the similarity in flowering-plant species composition was high between the juniper and eucalyptus forests (63%), but very low between these sites and agricultural areas ($\leq$7%). These similarities also fluctuated according to the climatic season.

3.2. Habitat and Seasonal Drivers of Community Composition

Non-metric multidimensional scaling ordinations based on community data clearly separated samples by season along the primary NMDS1 axis, both for the butterfly (Figure 4A) and flowering-plant (Figure 4B) communities. Their separation along this axis indicates a marked turnover in community composition between these two seasons in all three habitats. However, within each seasonal group, there was also evidence of clustering by habitat type, suggesting that compositional differences among habitats persisted across seasons. This habitat segregation was particularly pronounced for the butterfly communities (Figure 4A), with the three habitat types forming distinct clusters, especially during the dry season. The agricultural sites were the most differentiated, while the juniper and eucalyptus forests showed some overlap but still maintained distinct centroids.

For the flowering-plant communities (Figure 4B), the habitat segregation was less pronounced than for the butterflies but still evident. The agricultural sites were again the most distinct, particularly during the rainy season, indicating these disturbed areas hosted a plant community that differed markedly from the two forest habitats. The juniper and eucalyptus forests exhibited greater overlap in flowering-plant community composition, but their separation indicates persistent differences between these two forest types. Overall, the NMDS ordinations reveal that both butterfly and flowering-plant communities are structured by a combination of seasonal and habitat factors. The seasonal turnover is the primary driver of compositional change, but habitat-specific differences are maintained within each season, indicating that the human-altered vegetation types have a consistent impact on community structure across temporal scales.

The redundancy analysis (RDA) revealed that flower abundance and plant species richness significantly influenced butterfly community composition across the three studied vegetation types (Figure 5). The primary RDA axis (RDA1) accounted for 67.3% of the total variation in butterfly diversity and was strongly correlated with the flower abundance gradient ($r = 0.92$, $p < 0.001$). Along this axis, sites with higher flower abundance harbored greater butterfly diversity, as indicated by the diagonal distribution of data points.

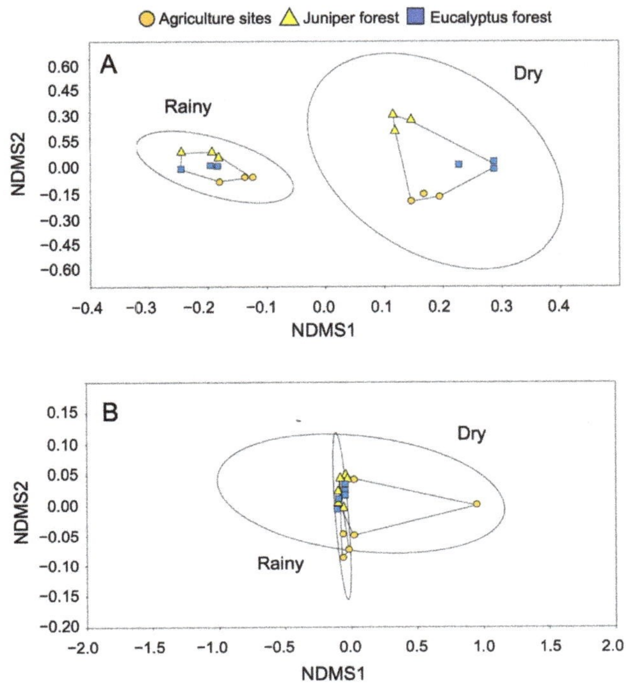

Figure 4. Non-metric multidimensional scaling (NDMS) ordination with solid ellipses encircling vegetation types during rainy and dry seasons based on butterfly (**A**) and flowering-plant (**B**) community composition. One set of ellipses encloses all vegetation types in the rainy season, while another set encloses them in the dry season. Vegetation types are represented by symbols: agriculture sites (yellow circles), juniper forest (orange triangles), and eucalyptus forest (blue squares). The closer the symbols within each seasonal ellipse, the more similar the community composition across vegetation types for that season within each panel. The statistical significance of the differences observed between groups was assessed using ANOSIM (Analysis of Similarities). For butterflies, the global R statistic was $R = 0.85$ ($p = 0.0001$), indicating a strong separation between community compositions across the different vegetation types and seasons. For plants, the global R statistic was $R = 0.32$ ($p = 0.0002$), showing a moderate but significant separation in community composition.

The secondary RDA axis (RDA2), explaining an additional 18.5% of the variation, was associated with plant species richness ($r = 0.78$, $p < 0.01$). The eucalyptus forest site was positively correlated with this axis, reflecting its higher plant richness compared to the other vegetation types. The RDA ordination clearly separated the three vegetation types, suggesting distinct butterfly communities among habitats. Agricultural sites clustered at the lower end of the flower abundance gradient, indicating lower butterfly diversity due to reduced floral resources in these disturbed areas. Juniper forest sites exhibited intermediate levels of butterfly diversity, despite relatively low flower abundance, suggesting the influence of additional factors not included in this analysis.

In contrast, the eucalyptus forest site was associated with both high plant richness and moderately high flower abundance, potentially contributing to its relatively high butterfly diversity. Overall, the RDA results suggest that the drivers of adult butterfly community composition differ across the studied vegetation types, with flower abundance and plant richness playing varying roles in structuring these communities.

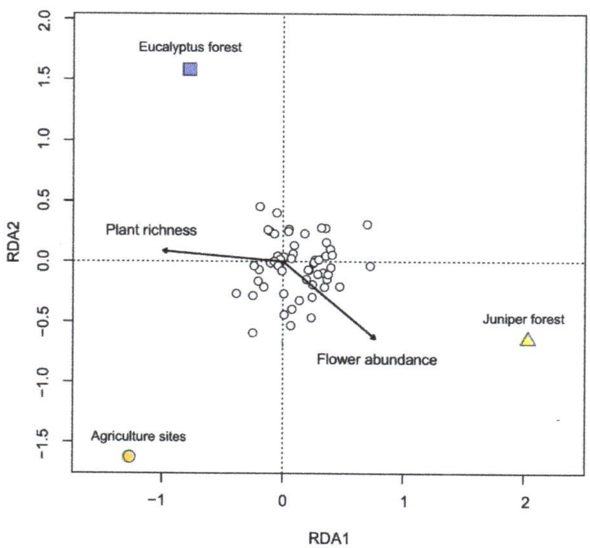

Figure 5. Redundancy analysis (RDA) ordination biplot with RDA1 and RDA2 axes relating butterfly community data to plant community variables (plant richness and flower abundance). Open circles represent individual butterfly species, and color-coded symbols represent the vegetation type. The arrows indicate the direction of increasing values for plant richness and flower abundance. Butterfly species in eucalyptus forests are more closely associated with higher flower abundance, while those in juniper forests are more closely associated with higher plant richness. Butterfly species in agriculture sites show a weaker association with both explanatory variables. The RDA model was statistically significant with a permutation test ($F = 4.56$, $p = 0.001$).

3.3. Key Indicator Species

The SIMPER analysis identified the butterfly and plant species contributing most to dissimilarity in community composition among the three vegetation types (Table 4) and between the rainy and dry seasons (Table 5). For butterflies (Table 4A), the species driving the greatest dissimilarity between vegetation types were *Zerene cesonia cesonia* (15.32% contribution), *Battus philenor philenor* (13.63%), and *Dione incarnata incarnata* (0.19%). *Z. cesonia cesonia* was overwhelmingly more abundant in the eucalyptus forest compared to the other habitats, while *B. philenor philenor* peaked in agricultural sites but was common across all vegetation types. *D. incarnata incarnata* exhibited highest abundance in the juniper forest.

Among flowering plants (Table 4B), the species contributing most to vegetation dissimilarity were *Rhus integrifolia* (37.86%), *Loeselia mexicana* (24.61%), and *Mimosa aculeaticarpa* (6.38%). *R. integrifolia* and *L. mexicana* showed very high flower abundance in the eucalyptus forest but were nearly absent from agricultural areas. *M. aculeaticarpa* occurred across all habitats but attained greatest abundance in the eucalyptus forest.

For seasonal dissimilarity in butterflies (Table 5A), *B. philenor philenor* (20.58%), *Z. cesonia cesonia* (10.87%), and *Nathalis iole iole* (6.68%) were the top contributors, being far more abundant during the rainy season compared to the dry season across vegetation types. The flowering-plant species accounting for most of the seasonal dissimilarity (Table 5B) were *R. integrifolia* (23.25%), *L. mexicana* (13.6%), and *M. aculeaticarpa* (13.47%). These species exhibited very low flower abundance or were absent during the rainy season but became highly abundant sources of floral resources in the dry season.

Table 4. Results of the SIMPER analysis showing (**A**) butterfly and (**B**) flowering-plant species contributing most (in order of decreasing percentage) to dissimilarity between vegetation types. For plant species, mean abundance measures refer specifically to flowers. All dissimilarity values shown were statistically significant ($p < 0.05$).

(A) Butterfly Species	Agriculture Sites Mean Abundance	Juniper Forest Mean Abundance	Eucalyptus Forest Mean Abundance	% Contribution to Dissimilarity
Zerene cesonia cesonia	11.7	26.3	50.7	15.32
Battus philenorphilenor	43.3	31.3	47.3	13.63
Agraulisincarnata incarnata	4	29.3	2.33	10.19
Echinargus isola	0.667	4.33	12.3	4.576
Leptophobia aripaelodia	11	1.67	1.67	4.165
Nathalis ioleiole	3.67	13.3	8.33	4.081
Dione junohuascuma	6	9	10.3	3.902
Leptotes marina	0	5.67	6.67	3.162
Fissicrambus sp.	0.667	8	3.67	3.059
Euptoieta claudiaclaudia	1	7.67	4.67	2.894
Ganyra josephinajosepha	0.333	1.33	7	2.488
Pterourus multicaudata multicaudata	7	7.33	6.33	2.331
Colias eurytheme	5.33	0	0	2.303
(B) Plant Species				
Rhus integrifolia	0	1000	4660	37.86
Loeselia mexicana	0	4320	5950	24.61
Mimosa aculeaticarpa	1790	2640	3350	6.384
Raphanus sativus	1350	0	0	5.706
Tithonia tubiformis	919	0	6.33	3.87
Eremosis corymbosa	0	0	1030	3.811
Salvia polystachya	0	909	0	3.215
Bouvardia ternifolia	884	1610	823	2.84

Table 5. Results of the SIMPER analysis showing (**A**) butterfly and (**B**) flowering-plant species contributing most (in order of decreasing percentage) to dissimilarity between the rainy and dry seasons. For plant species, mean abundance measures refer specifically to flowers. All dissimilarity values shown were statistically significant ($p < 0.05$).

(A) Butterfly Species	Rainy Mean Abundance	Dry Mean Abundance	% Contribution to Dissimilarity
Battus philenorphilenor	30.9	9.78	20.58
Zerenecesonia cesonia	14.4	2.78	10.87
Nathalis ioleiole	7.89	0.556	6.678
Dione incarnata incarnata	7.67	3.44	6.148
Dione junohuascuma	7.11	1	5.872
Pterorurus multicaudata multicaudata	5.67	1.22	4.315
Dione monetapoeyyi	4	0.333	3.309
Phoebis sennae marcellina	3.33	0.111	3.154
Echinargus isola	1.67	3.67	2.808
Danaus gilippusthersippus	2.89	0	2.761
Leptophobia aripaelodia	2.78	1.78	2.718
Ganyra josephinajosepha	2.78	0.111	2.485
Leptotes marina	2.56	1.56	2.197
(B) Plant species			
Rhus integrifolia	0	3030	23.25
Loeselia mexicana	142	1330	13.6
Mimosa aculeaticarpa	0	1400	13.47
Raphanus sativus	449	153	10.43
Bouvardia ternifolia	234	12.4	7.651
Salvia polystachya	251	0	5.187
Rhus integrifolia	0	3030	23.25
Loeselia mexicana	142	1330	13.6

Based on the indicator species analysis (IndVal), several butterfly and flowering-plant species exhibited strong associations with particular vegetation types or seasons (Figures 6 and 7). For butterflies (Figure 6A), *Leptophobia aripa elodia* emerged as a strong indicator of agricultural sites. *Zerene cesonia cesonia* and *Battus philenor philenor* were among the top indicators for the eucalyptus forest. Meanwhile, *Dione incarnata incarnata* was the strongest indicator species for the juniper forest vegetation type. Regarding seasonal indicators, butterflies like *Battus philenor philenor*, *Dione juno*, *Phoebis sennae marcellina*, and *Danaus gilippus thersippus* exhibited high indicator values for the rainy season across vegetation types. No butterfly species emerged as a particularly strong indicator of the dry season. Turning to flowering plants (Figure 6B), *Raphanus sativus* was a top indicator of agricultural areas during the rainy season. *Rhus integrifolia* and *Loeselia mexicana* were among the strongest indicators of the dry season across habitats. *Mimosa aculeaticarpa* also showed a high indicator value for this dry period. For associations with specific vegetation types, *Rhus integrifolia* and *Loeselia mexicana* again had the highest indicator values, pointing to their strong ties as indicator species for the eucalyptus forest habitat. *Salvia polystachya* emerged as an indicator species most associated with the juniper forest vegetation type.

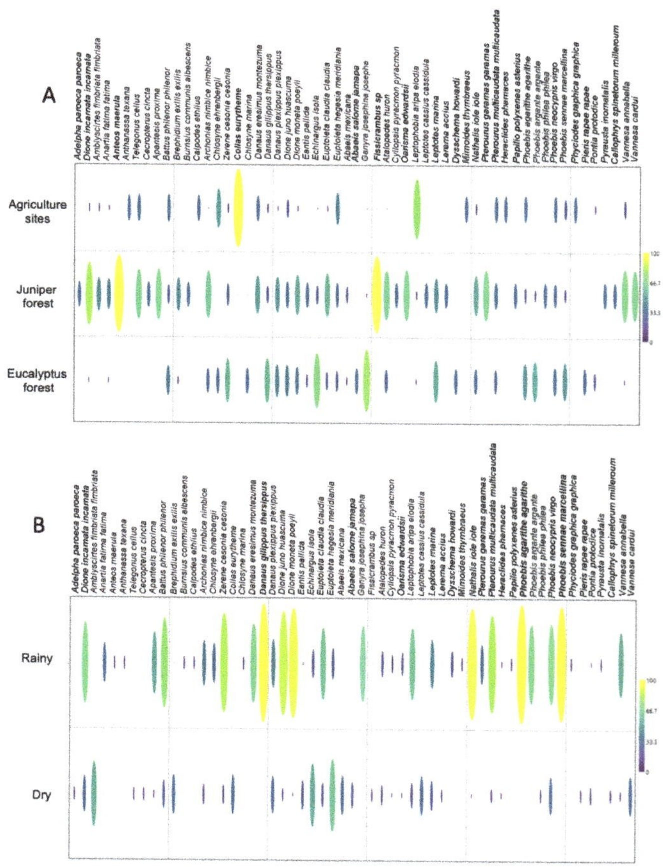

Figure 6. Indicator values for butterfly species in each (**A**) vegetation type and (**B**) season. Taller bars indicate stronger association between a species and that vegetation type or season. The color gradient from yellow to purple represents increasing indicator values, with purple being the strongest indicators.

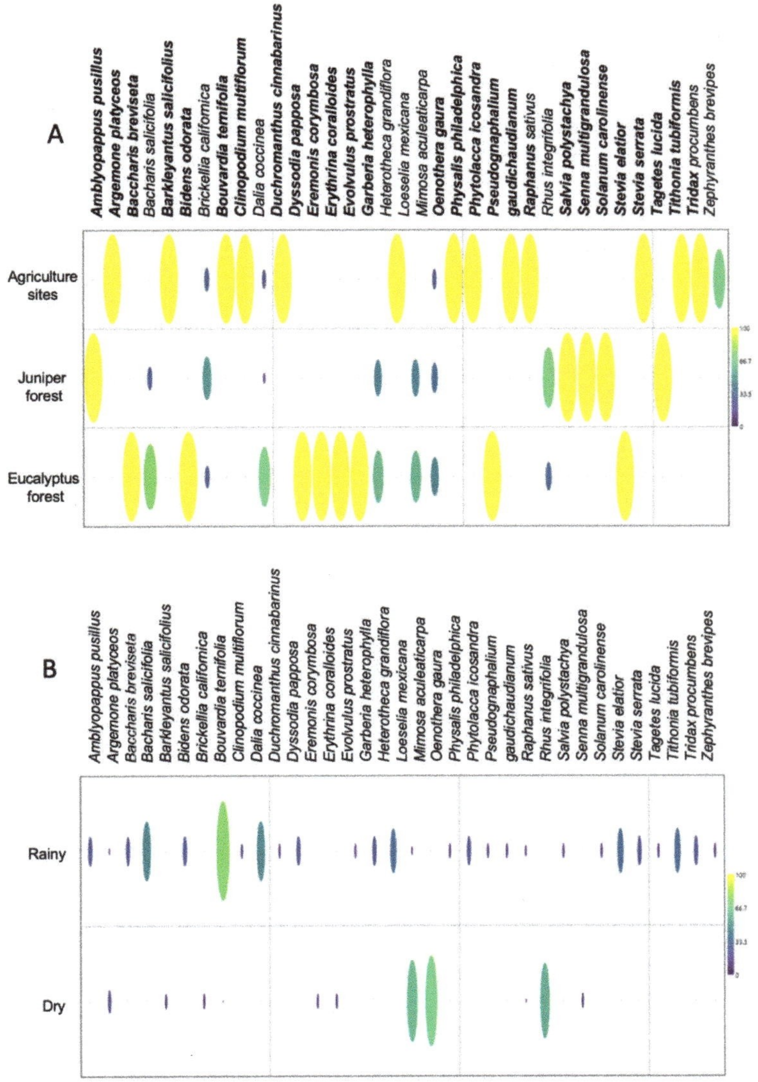

Figure 7. Indicator values displayed for plant species across (**A**) vegetation types and (**B**) seasons. Bar height corresponds to the strength of association between a given plant species and the specified vegetation type or season. Taller bars signify stronger indicator values. The color ramp transitions from yellow to purple, representing increasing indicator values.

4. Discussion

This study elucidates how adult butterfly visitation patterns and flowering-plant community composition, abundance, and structure respond to anthropogenic habitat changes and seasonal fluctuations in the studied landscape. The findings reinforce the extensive literature documenting biodiversity losses and biotic homogenization in human-modified environments [1,53,54]. However, it also reveals nuanced responses across different intensities and types of disturbance, as well as to seasonal resource dynamics, underscoring the complexities of biodiversity's responses to multiple interacting stressors.

The juniper forest sites, representing relatively undisturbed reference conditions, consistently harbored the highest plant and butterfly diversity levels. This accords with a wealth of studies globally reporting greater species richness and abundance in more intact, heterogeneous natural habitats [55,56]. In contrast, the less diverse communities observed in intensive agricultural areas reflect changes in the composition and abundance of flowering-plant species, as well as alterations in seasonal patterns. These changes in floral resources and phenology can significantly impact butterfly diversity and abundance. While our study primarily focused on these factors, it is worth noting that such changes in plant communities and seasonality are often consequences of broader landscape modifications, including habitat conversion and simplification [57–59]. These alterations can affect resource availability and quality for butterflies, potentially influencing their population dynamics and community structure [60–62].

The eucalyptus plantations occupied an intermediate position, corroborating studies across regions showing diminished biodiversity compared to intact native ecosystems, even for plantations of native tree species [63–65]. Beyond simplified vegetation structure and altered microclimate, factors like reduced food-plant diversity, intensive management practices, and temporal resource fluctuations appear to limit their value in sustaining diverse communities [66,67]. While representing an improvement over agricultural matrices, production forestry does not provide a substitute for preserving primary forests.

Our study revealed distinct associations between adult butterfly communities and floral resources across the different vegetation types, specifically in the context of flower visitation. In the eucalyptus plantations, butterflies were more closely associated with higher flower abundance, while in the juniper forests, they were more closely linked to higher plant richness. These findings suggest that the relative importance of resource quantity versus diversity in shaping butterfly communities may vary depending on the habitat context. Previous studies have also documented such context-dependent responses. For example, Valtonen et al. [68] found that butterfly species richness was more strongly correlated with nectar plant abundance in managed grasslands, while nectar plant diversity was a better predictor in semi-natural grasslands. Similarly, Pöyry et al. [69] reported that butterfly abundance was more sensitive to floral abundance in agricultural landscapes, whereas species richness was more responsive to floral diversity in forested landscapes. The weaker associations observed in the agricultural sites in our study likely reflect the overall resource scarcity and degraded conditions limiting both butterfly and plant communities, as seen in other intensively cultivated systems [70,71]. The species-level responses further underscore the importance of resource heterogeneity in supporting diverse butterfly assemblages, with different species linked to varying combinations of floral abundance and plant diversity. This aligns with the concept of niche complementarity, whereby a greater array of resources enables the coexistence of more species with different resource requirements [72,73]. Collectively, these nuanced patterns highlight the need to consider both the quantity and diversity of floral resources, as well as the habitat context, when assessing drivers of butterfly community structure across vegetation types altered by anthropogenic factors.

The tight positive correlation between plant richness and floral abundance matches well-established ecological theory. Floral resources are key drivers of pollinator diversity, abundance, and community composition patterns [74–76]. The scarcity of flowering plants in the most disturbed agricultural habitats likely contributed directly to the impoverished butterfly communities documented, mirroring patterns across degraded environments globally [77,78]. When resources are limited, competitive displacement by disturbance-adapted generalists overshadows rare specialists, driving homogenization of abundance distributions and diversity losses [13,27,79].

The pronounced seasonal turnover fits expectations of phenological resource tracking by multivoltine species in highly seasonal tropical/subtropical regions [80–82]. However, the distinct compositional clustering by habitat within each season reveals that local environmental filters remain paramount in structuring communities, as supported by other studies [16,67,83]. The agricultural matrices displayed the strongest habitat signature during the resource-flush rainy season, suggesting these areas become even more restrictive as ephemeral resources peak. Many forest species may be unable to effectively exploit these resources across the degraded agricultural landscape [17,84].

Our indicator species analysis revealed that certain butterfly species are strongly associated with particular vegetation types, serving as potential indicator species. In the agriculture sites, *Colias eurytheme* and *Leptophobia aripa elodia* were identified as the main indicator species. Previous studies have found that species in the genus *Colias* are often associated with open, disturbed habitats such as agricultural fields and pastures [70,85]. Similarly, species in the genus *Leptophobia* are known to occur in open, sunny areas and are often found in agricultural landscapes [86,87].

In the juniper forest, *Anteos maerula* and *Fissicrambus* sp. emerged as the primary indicator species. Butterflies in the genus *Anteos*, which belong to the family Pieridae, have been associated with woodland and forest habitats in previous studies [88,89]. Species in the genus *Fissicrambus*, which are geometrid moths, have been used as indicators of forest health and have been found to respond to changes in forest structure and composition [90, 91].

In the eucalyptus forest, *Ganyra josephina josepha* was identified as the main indicator species. This species belongs to the family Pieridae, which has been shown to have strong associations with forest habitats in various studies [92,93]. Butterflies in the genus *Ganyra* have been identified as potential indicators of forest disturbance and have been found to respond to changes in forest management practices [94,95].

Furthermore, our analysis revealed that *Danaus gilippus thersippus* and *Phoebis agarithe* were the primary indicator species during the rainy season. *Danaus gilippus* belongs to the family Nymphalidae, and species in this genus have been widely used as indicators of environmental change and have been found to respond to seasonal variations in resource availability [93,96]. *Phoebis agarithe*, a member of the family Pieridae, has been associated with seasonal changes in tropical and subtropical regions [64,97]. These findings underscore the importance of considering both spatial and temporal factors when using butterfly species as ecological indicators. The presence and abundance of these indicator species can provide valuable insights into the ecological integrity, conservation value, and seasonal dynamics of different habitats [89,98].

A key strength was the integrated examination of both plant and pollinator communities, rare for local studies, providing a holistic biodiversity perspective. Furthermore, extending beyond simple diversity metrics to analyze composition, structure, and explicit incorporation of seasonality offers novel insights into multidimensional impacts of habitat modification, aligning with recent ecological frameworks [99–101]. Ongoing analyses identifying robust indicator taxa for each habitat disturbance–seasonality scenario will prove invaluable for monitoring, modeling biodiversity change, and informing conservation strategies and restoration efforts [102–104].

Lepidoptera behavioral diversity is crucial in interpreting our results. Species-specific behaviors, especially migratory patterns, may influence observed interactions. For example, the presence of migratory species like *Vanessa cardui* might reflect temporary resource use rather than habitat preference. Species mobility also affects responses to habitat changes and resource distribution. Future studies should categorize species based on behavioral traits to better understand how different Lepidoptera groups respond to habitat changes and interact with plant communities. This approach could help distinguish local habitat effects from broader landscape-level or seasonal factors influencing butterfly–plant interactions.

Our focus on adult butterfly–flower interactions provides valuable insights into one aspect of butterfly ecology, but it is important to consider these findings within the broader context of plant–lepidopteran relationships. The strong associations we observed between adult butterflies and floral resources may not necessarily reflect the overall habitat suitability for butterfly populations, as larval host-plant availability is also crucial. Future studies integrating both adult and larval resource requirements would provide a more comprehensive understanding of how habitat changes affect butterfly communities across their entire life cycle. Additionally, to gain a more complete picture of these ecological dynamics, multi-year studies are essential. Long-term research would allow us to account for inter-annual variations in climate, vegetation, and butterfly populations, providing a more robust context for understanding these complex interactions and their responses to anthropogenic changes over time.

5. Conclusions

Our findings reinforce the paramount importance of preserving intact forest ecosystems to sustain diverse ecological communities in the long-term. While plantations afford some habitat value versus intensive agriculture, they represent a markedly degraded state. Seasonal resource pulses can temporarily boost diversity but cannot compensate fully for chronic habitat degradation. Maintaining heterogeneous landscapes, promoting connectivity, and mitigating anthropogenic pressures like habitat conversion will be crucial for facilitating dispersal, enhancing community resilience, and conserving biodiversity amid accelerating global change [54,105,106]. This integrated investigation advances our understanding of biodiversity responses to different human disturbance regimes, while underscoring needs for additional targeted research developing robust mitigation strategies to sustain at-risk ecological communities in human-dominated environments.

Author Contributions: Conceptualization, K.L.-V. and C.L.; methodology, K.L.-V., C.L. and P.C.; software, K.L.-V., C.L. and P.C.; validation, K.L.-V., C.L. and P.C.; formal analysis, K.L.-V., C.L., P.C. and C.C.-G.; investigation, K.L.-V. and C.L.; resources, K.L.-V. and C.L.; data curation, K.L.-V. and C.L.; writing—original draft preparation, K.L.-V. and C.L.; writing—review and editing, K.L.-V., C.L., P.C. and C.C.-G. All authors have read and agreed to the published version of the manuscript.

Funding: This research was funded by CONAHCyT, Mexico through a doctoral scholarship to K.L.-V. (1003512).

Data Availability Statement: The raw data supporting the conclusions of this article will be made available by the authors on request.

Acknowledgments: We gratefully acknowledge Mauro Piedras, Deysi Muñoz and Mariana Cuautle for field assistance. This work constitutes partial fulfillment of Karla María López-Vázquez's doctorate degree requirements at UAM. The authors thank José de Jesús García Díaz for their support with the butterfly species identification. We thank the Academic Editor and three anonymous reviewers for their invaluable feedback regarding the manuscript.

Conflicts of Interest: The authors declare no conflicts of interest.

References

1. Sánchez-Bayo, F.; Wyckhuys, K.A.G. Worldwide decline of the entomofauna: A review of its drivers. *Biol. Conserv.* **2019**, *232*, 8–27. [CrossRef]
2. Istvánovics, V. Ecosystem health and integrity concepts revisited in the light of global environmental changes. *Ecol. Indic.* **2022**, *135*, 108564.
3. Kawahara, A.Y.; Plotkin, D.; Espeland, M.; Meusemann, K.; Toussaint, E.F.; Donath, A.; Gimnich, F.; Frandsen, P.B.; Zwick, A.; dos Reis, M.; et al. Phylogenomics reveals the evolutionary timing and pattern of butterflies and moths. *Proc. Natl. Acad. Sci. USA* **2019**, *116*, 22657–22663. [CrossRef] [PubMed]
4. Wagner, D.L. Insect declines in the Anthropocene. *Annu. Rev. Entomol.* **2020**, *65*, 457–480. [CrossRef] [PubMed]
5. Rákosy, L.; Schmitt, T. Are butterflies and moths suitable ecological indicator systems for restoration measures of semi-natural calcareous grassland habitats? *Ecol. Indic.* **2011**, *11*, 1040–1045. [CrossRef]
6. Thompson, J.N.; Pellmyr, O. Evolution of oviposition behavior and host preference in Lepidoptera. *Annu. Rev. Entomol.* **1991**, *36*, 65–89. [CrossRef]
7. Dennis, R.L. Butterfly habitats, broad-scale biotope affiliations, and structural exploitation of vegetation at finer scales: The matrix revisited. *Ecol. Entomol.* **2004**, *29*, 744–752. [CrossRef]
8. Scriven, S.A.; Beale, C.M.; Benedick, S.; Hill, J.K. Barriers to dispersal of rain forest butterflies in tropical agricultural landscapes. *Biotropica* **2017**, *49*, 206–216. [CrossRef]
9. Brückmann, S.V.; Krauss, J.; Steffan-Dewenter, I. Butterfly and plant specialists suffer from reduced connectivity in fragmented landscapes. *J. Appl. Ecol.* **2010**, *47*, 799–809. [CrossRef]

10. Aguirre-Gutiérrez, J.; WallisDeVries, M.F.; Marshall, L.; van't Zelfde, M.; Villalobos-Arámbula, A.R.; Boekelo, B.; Bartholomeus, H.; Franzén, M.; Biesmeijer, J.C. Butterflies show different functional and species diversity in relationship to vegetation structure and land use. *Glob. Ecol. Biogeogr.* **2017**, *26*, 1126–1137. [CrossRef]
11. Steffan-Dewenter, I.; Tscharntke, T. Butterfly community structure in fragmented habitats. *Ecol. Lett.* **2000**, *3*, 449–456. [CrossRef]
12. Öckinger, E.; Smith, H.G. Landscape composition and habitat area affects butterfly species richness in semi-natural grasslands. *Oecologia* **2006**, *149*, 526–534. [CrossRef] [PubMed]
13. Börschig, C.; Klein, A.M.; von Wehrden, H.; Krauss, J. Traits of butterfly communities change from specialist to generalist characteristics with increasing land-use intensity. *Basic Appl. Ecol.* **2013**, *14*, 547–554. [CrossRef]
14. Quintero, C.; Morales, C.L.; Aizen, M.A. Effects of anthropogenic habitat disturbance on local pollinator diversity and species turnover across a precipitation gradient. *Biodivers. Conserv.* **2010**, *19*, 257–274. [CrossRef]
15. Nyafwono, M.; Valtonen, A.; Nyeko, P.; Roininen, H. Butterfly community composition across a successional gradient in a human-disturbed afro-tropical rain forest. *Biotropica* **2014**, *46*, 210–218. [CrossRef]
16. Perović, D.; Gámez-Virués, S.; Börschig, C.; Klein, A.M.; Krauss, J.; Steckel, J.; Rothenwöhrer, C.; Erasmi, S.; Tscharntke, T.; Westphal, C. Configurational landscape heterogeneity shapes functional community composition of grassland butterflies. *J. Appl. Ecol.* **2015**, *52*, 505–513. [CrossRef]
17. Sambhu, H.; Northfield, T.; Nankishore, A.; Ansari, A.; Turton, S. Tropical rainforest and human-modified landscapes support unique butterfly communities that differ in abundance and diversity. *Environ. Entomol.* **2017**, *46*, 1225–1234. [CrossRef]
18. Stefanescu, C.; Peñuelas, J.; Filella, I. Effects of climatic change on the phenology of butterflies in the northwest Mediterranean Basin. *Glob. Chang. Biol.* **2003**, *9*, 1494–1506. [CrossRef]
19. Posledovich, D.; Toftegaard, T.; Wiklund, C.; Ehrlén, J.; Gotthard, K. The developmental race between maturing host plants and their butterfly herbivore--the influence of phenological matching and temperature. *J. Anim. Ecol.* **2015**, *84*, 1690–1699. [CrossRef]
20. Kitahara, M.; Sei, K.; Fujii, K. Patterns in the structure of grassland butterfly communities along a gradient of human disturbance: Further analysis based on the generalist/specialist concept. *Popul. Ecol.* **2000**, *42*, 135–144. [CrossRef]
21. Cleland, E.E.; Chuine, I.; Menzel, A.; Mooney, H.A.; Schwartz, M.D. Shifting plant phenology in response to global change. *Trends Ecol. Evol.* **2007**, *22*, 357–365. [CrossRef] [PubMed]
22. Schenk, M.; Krauss, J.; Holzschuh, A. Desynchronizations in bee-plant interactions cause severe fitness losses in solitary bees. *J. Anim. Ecol.* **2018**, *87*, 139–149. [CrossRef] [PubMed]
23. Dover, J.W.; Settele, J. The influences of landscape structure on butterfly distribution and movement: A review. *J. Insect Conserv.* **2009**, *13*, 3–27. [CrossRef]
24. Baguette, M.; Blanchet, S.; Legrand, D.; Stevens, V.M.; Turlure, C. Individual dispersal, landscape connectivity and ecological networks. *Biol. Rev.* **2013**, *88*, 310–326. [CrossRef] [PubMed]
25. Nowicki, P.; Vrabec, V.; Binzenhöfer, B.; Feil, J.; Zakšek, B.; Hovestadt, T.; Settele, J. Butterfly dispersal in inhospitable matrix: Rare, risky, but long-distance. *Landsc. Ecol.* **2014**, *29*, 401–412. [CrossRef]
26. Villemey, A.; van Halder, I.; Ouin, A.; Barbaro, L.; Chenot, J.; Tessier, P.; Calatayud, F.; Martin, H.; Roche, P.; Archaux, F. Mosaic of grasslands and woodlands is more effective than habitat connectivity to conserve butterflies in French farmland. *Biol. Conserv.* **2015**, *191*, 206–215. [CrossRef]
27. Gámez-Virués, S.; Perović, D.J.; Gossner, M.M.; Börschig, C.; Blüthgen, N.; De Jong, H.; Simons, N.K.; Klein, A.-M.; Krauss, J.; Maier, G.; et al. Landscape simplification filters species traits and drives biotic homogenization. *Nat. Commun.* **2015**, *6*, 8568. [CrossRef]
28. Carnicer, J.; Stefanescu, C.; Vila, R.; Dincă, V.; Font, X.; Peñuelas, J. A unified framework for diversity gradients: The adaptive trait continuum. *Glob. Ecol. Biogeogr.* **2013**, *22*, 6–18. [CrossRef]
29. Pla-Narbona, C.; Stefanescu, C.; Pino, J.; Cabrero-Sañudo, F.J.; García-Barros, E.; Munguira, M.L.; Melero, Y. Butterfly biodiversity in the city is driven by the interaction of the urban landscape and species traits: A call for contextualised management. *Landsc. Ecol.* **2022**, *7*, 81–92. [CrossRef]
30. Hudson, R.; Rodríguez-Martínez, L.; Distel, H.; Cordero, C.; Altbacker, V.; Martínez-Gómez, M. A comparison between vegetation and diets records from wet and dry season in the cottontail rabbit Sylvilagus floridanus at Ixtacuixtla, central Mexico. *Acta Theriol.* **2005**, *50*, 377–389. [CrossRef]
31. Olofsson, P.; Foody, G.M.; Herold, M.; Stehman, S.V.; Woodcock, C.E.; Wulder, M.A. Good practices for estimating area and assessing accuracy of land change. *Remote Sens. Environ.* **2014**, *148*, 42–57. [CrossRef]
32. Pollard, E.; Yates, T.J. *Monitoring Butterflies for Ecology and Conservation: The British Butterfly Monitoring Scheme*; Chapman & Hall: London, UK, 1993.
33. Opler, P.A. *Peterson First Guide to Butterflies and Moths of North America*; Houghton Mifflin Harcourt: Boston, MA, USA, 1994.
34. Glassberg, J. *A Swift Guide of the Butterflies of Mexico and Central America*; Sunstreak Books: Morristown, NJ, USA, 2007.
35. Calderón de Rzedowski, G.; Rzedowski, J. *Flora Fanerogámica del Valle de México*, 2nd ed.; Instituto de Ecología, A.C. y Comisión Nacional para el Conocimiento y Uso de la Biodiversidad: Pátzcuaro, México, 2005.
36. Hill, M.O. Diversity and evenness: A unifying notation and its consequences. *Ecology* **1973**, *54*, 427–432. [CrossRef]
37. Chao, A.; Gotelli, N.J.; Hsieh, T.C.; Sander, E.L.; Ma, K.H.; Colwell, R.K.; Ellison, A.M. Rarefaction and extrapolation with Hill numbers: A framework for sampling and estimation in species diversity studies. *Ecol. Monogr.* **2014**, *84*, 45–67. [CrossRef]
38. Jost, L. Entropy and diversity. *Oikos* **2006**, *113*, 363–375. [CrossRef]

39. Hsieh, T.C.; Ma, K.H.; Chao, A. iNEXT: An R package for rarefaction and extrapolation of species diversity (Hill numbers). *Methods Ecol. Evol.* **2016**, *7*, 1451–1456. [CrossRef]
40. Magurran, A.E. *Measuring Biological Diversity*; Blackwell Science Ltd.: Oxford, UK, 2004.
41. Hsieh, T.C.; Ma, K.H.; Chao, A. iNEXT: Interpolation and Extrapolation for Species Diversity. R Package Version 2.0.20. Available online: http://chao.stat.nthu.edu.tw/wordpress/software_download/ (accessed on 27 May 2024).
42. Krebs, C.J. *Ecology: The Experimental Analysis of Distribution and Abundance*, 6th ed.; Benjamin Cummings: San Francisco, CA, USA, 2009.
43. Minchin, P.R. An evaluation of the relative robustness of techniques for ecological ordination. *Vegetatio* **1987**, *69*, 89–107. [CrossRef]
44. Anderson, M.J.; Walsh, D.C.I. Permanova, anosim, and the Mantel test in the face of heterogeneous dispersions: What null hypothesis are you testing? *Ecol. Monogr.* **2013**, *83*, 557–574. [CrossRef]
45. Oksanen, J.; Blanchet, F.G.; Friendly, M.; Kindt, R.; Legendre, P.; McGlinn, D.; Minchin, P.R.; O'Hara, R.B.; Simpson, G.L.; Solymos, P.; et al. Vegan: Community Ecology Package. R Package Version 2.5-6. Available online: https://CRAN.R-project.org/package=vegan (accessed on 27 May 2024).
46. Ripley, B.; Venables, B.; Bates, D.M.; Hornik, K.; Gebhardt, A.; Firth, D.; Ripley, M.B. Package 'Mass'. R Package Version 7.3-52. Available online: https://CRAN.R-project.org/package=mass (accessed on 27 May 2024).
47. R Core Team. *R: A Language and Environment for Statistical Computing*; R Foundation for Statistical Computing: Vienna, Austria, 2014.
48. Clarke, K.R. Non-parametric multivariate analyses of changes in community structure. *Aust. J. Ecol.* **1993**, *18*, 117–143. [CrossRef]
49. Clarke, K.R.; Warwick, R.M. *Change in Marine Communities: An Approach to Statistical Analysis and Interpretation*, 2nd ed.; PRIMER-E: Plymouth, UK, 2001.
50. Dufrêne, M.; Legendre, P. Species assemblages and indicator species: The need for a flexible asymmetrical approach. *Ecol. Monogr.* **1997**, *67*, 345–366. [CrossRef]
51. De Cáceres, M.; Legendre, P. Associations between species and groups of sites: Indices and statistical inference. *Ecology* **2009**, *90*, 3566–3574. [CrossRef]
52. Roberts, D.W. labdsv: Ordination and Multivariate Analysis for Ecology. R Package Version 1.8-0. Available online: https://cran.r-project.org/web/packages/labdsv/labdsv.pdf (accessed on 27 May 2024).
53. Haddad, N.M.; Brudvig, L.A.; Clobert, J.; Davies, K.F.; Gonzalez, A.; Holt, R.D.; Lovejoy, T.E.; Sexton, J.O.; Austin, M.P.; Collins, C.D.; et al. Habitat fragmentation and its lasting impact on Earth's ecosystems. *Sci. Adv.* **2015**, *1*, e1500052. [CrossRef] [PubMed]
54. Newbold, T.; Hudson, L.N.; Contu, S.; Hill, S.L.; Beck, J.; Liu, Y.; Meyer, C.; Phillips, H.R.P.; Scharlemann, J.P.; Purvis, A. Widespread winners and narrow-ranged losers: Land use homogenizes biodiversity in local assemblages worldwide. *PLoS Biol.* **2018**, *16*, e2006841. [CrossRef] [PubMed]
55. Gibson, L.; Lee, T.M.; Koh, L.P.; Brook, B.W.; Gardner, T.A.; Barlow, J.; Peres, C.A.; Bradshaw, C.J.A.; Laurance, W.F.; Lovejoy, T.E.; et al. Primary forests are irreplaceable for sustaining tropical biodiversity. *Nature* **2011**, *478*, 378–381. [CrossRef] [PubMed]
56. Arroyo-Rodríguez, V.; Melo, F.P.L.; Martínez-Ramos, M.; Bongers, F.; Chazdon, R.L.; Meave, J.A.; Norden, N.; Santos, B.A.; Leal, I.R.; Tabarelli, M. Multiple successional pathways in human-modified tropical landscapes: New insights from forest succession, forest fragmentation and landscape ecology research. *Biol. Rev.* **2017**, *92*, 326–340. [CrossRef]
57. Benton, T.G.; Bryant, D.M.; Cole, L.; Crick, H. Linking agricultural practice to insect and bird populations: A historical study over three decades. *J. Appl. Ecol.* **2002**, *39*, 673–687. [CrossRef]
58. Fischer, J.; Lindenmayer, D.B. Landscape modification and habitat fragmentation: A synthesis. *Glob. Ecol. Biogeogr.* **2007**, *16*, 265–280. [CrossRef]
59. Seibold, S.; Gossner, M.M.; Simons, N.K.; Blüthgen, N.; Müller, J.; Ambarli, D.; Ammer, C.; Bauhus, J.; Fischer, M.; Habel, J.C.; et al. Arthropod decline in grasslands and forests is associated with landscape-level drivers. *Nature* **2019**, *574*, 671–674. [CrossRef]
60. Öckinger, E.; Schweiger, O.; Crist, T.O.; Debinski, D.M.; Krauss, J.; Kuussaari, M.; Petersen, J.D.; Pöyry, J.; Settele, J.; Summerville, K.S.; et al. Life-history traits predict species responses to habitat area and isolation: A cross-continental synthesis. *Ecol. Lett.* **2010**, *13*, 969–979. [CrossRef]
61. Hallmann, C.A.; Sorg, M.; Jongejans, E.; Siepel, H.; Hofland, N.; Schwan, H.; Stenmans, W.; Müller, A.; Sumser, H.; Hörren, T.; et al. More than 75 percent decline over 27 years in total flying insect biomass in protected areas. *PLoS ONE* **2017**, *12*, e0185809. [CrossRef]
62. Zioga, E.; Kelly, R.; White, B.; Stout, J.C. Plant protection product residues in plant pollen and nectar: A review of current knowledge. *Environ. Res.* **2020**, *189*, 109873. [CrossRef]
63. Marsden, S.J.; Whiffin, M.; Galetti, M. Bird diversity and abundance in forest fragments and Eucalyptus plantations around an Atlantic forest reserve, Brazil. *Biodivers. Conserv.* **2001**, *10*, 737–751. [CrossRef]
64. Barlow, J.; Overal, W.L.; Araujo, I.S.; Gardner, T.A.; Peres, C.A. The value of primary, secondary and plantation forests for fruit-feeding butterflies in the Brazilian Amazon. *J. Appl. Ecol.* **2007**, *44*, 1001–1012. [CrossRef]
65. Francesconi, W.; Nair, P.K.R.; Levey, D.J.; Daniels, J.; Cullen, L. Butterfly distribution in fragmented landscapes containing agroforestry practices in Southeastern Brazil. *Agrofor. Syst.* **2013**, *87*, 1321–1338. [CrossRef]
66. Carnus, J.M.; Parrotta, J.; Brockerhoff, E.; Arbez, M.; Jactel, H.; Kremer, A.; Lamb, D.; O'Hara, K.O.; Walters, B. Planted forests and biodiversity. *J. For.* **2006**, *104*, 65–77. [CrossRef]

67. Brockerhoff, E.G.; Jactel, H.; Parrotta, J.A.; Quine, C.P.; Sayer, J. Plantation forests and biodiversity: Oxymoron or opportunity? Biodivers. *Conserv.* **2008**, *17*, 925–951.
68. Valtonen, A.; Hirka, A.; Szőcs, L.; Ayres, M.P.; Roininen, H.; Csóka, G. Long-term species loss and homogenization of moth communities in Central Europe. *J. Anim. Ecol.* **2017**, *86*, 730–738. [CrossRef]
69. Pöyry, J.; Paukkunen, J.; Heliölä, J.; Kuussaari, M. Relative contributions of local and regional factors to species richness and total density of butterflies and moths in semi-natural grasslands. *Oecologia* **2009**, *160*, 577–587. [CrossRef]
70. Ekroos, J.; Heliölä, J.; Kuussaari, M. Homogenization of lepidopteran communities in intensively cultivated agricultural landscapes. *J. Appl. Ecol.* **2010**, *47*, 459–467. [CrossRef]
71. Curtis, R.J.; Brereton, T.M.; Dennis, R.L.; Carbone, C.; Isaac, N.J. Butterfly abundance is determined by food availability and is mediated by species traits. *J. Appl. Ecol.* **2015**, *52*, 1676–1684. [CrossRef]
72. Tylianakis, J.M.; Laliberté, E.; Nielsen, A.; Bascompte, J. Conservation of species interaction networks. *Biol. Conserv.* **2010**, *143*, 2270–2279. [CrossRef]
73. Bartual, A.M.; Sutter, L.; Bocci, G.; Moonen, A.C.; Cresswell, J.; Entling, M.; Giffard, B.; Jacot, K.; Jeanneret, P.; Holland, J.; et al. The potential of different semi-natural habitats to sustain pollinators and natural enemies in European agricultural landscapes. *Agric. Ecosyst. Environ.* **2019**, *279*, 43–52. [CrossRef]
74. Ebeling, A.; Klein, A.M.; Schumacher, J.; Weisser, W.W.; Tscharntke, T. How does plant richness affect pollinator richness and temporal stability of flower visits? *Oikos* **2008**, *117*, 1808–1815. [CrossRef]
75. Varo, J.P.; Mudri-Stojnić, S.; Riedinger, V.; Rundlöf, M.; Scheper, J.; Wickens, J.B.; Wickens, V.J.; Bommarco, R.; Kleijn, D.; Potts, S.G.; et al. Mass-flowering crops dilute pollinator abundance in agricultural landscapes across Europe. *Ecol. Lett.* **2016**, *19*, 1228–1236.
76. Grab, H.; Blitzer, E.J.; Danforth, B.; Loeb, G.; Poveda, K. Temporally dependent pollinator competition and facilitation with mass flowering crops affects yield in co-blooming crops. *Sci. Rep.* **2017**, *7*, 45296. [CrossRef] [PubMed]
77. Öckinger, E.; Eriksson, A.K.; Smith, H.G. Effects of grassland abandonment, restoration and management on butterfly communities. *Biol. Conserv.* **2006**, *130*, 365–375.
78. Sykes, L.; Santini, L.; Etard, A.; Newbold, T. Effects of rarity form on species' responses to land use. *Conserv. Biol.* **2020**, *34*, 688–696. [CrossRef]
79. Wagner, D.L.; Grames, E.M.; Forister, M.L.; Berenbaum, M.R.; Stopak, D. Insect decline in the Anthropocene: Death by a thous and cuts. *Proc. Natl. Acad. Sci. USA* **2021**, *118*, e2023989118. [CrossRef]
80. Ribeiro, D.B.; Freitas, A.V. The effect of reduced-impact logging on fruit-feeding butterflies in Central Amazon, Brazil. *J. Insect Conserv.* **2012**, *16*, 733–744. [CrossRef]
81. Abrahamczyk, S.; Kluge, J.; Gareca, Y.; Reichle, S.; Kessler, M. The influence of climatic seasonality on the diversity of different tropical pollinator groups. *PLoS ONE* **2011**, *6*, e27115. [CrossRef]
82. Aleixo, K.P.; Menezes, C.; Imperatriz Fonseca, V.L.; da Silva, C.I. Seasonal availability of floral resources and ambient temperature shape stingless bee foraging behavior (*Scaptotrigona* aff. depilis). *Apidologie* **2017**, *48*, 117–127. [CrossRef]
83. Aduse-Poku, K.; Molleman, F.; Oduro, W.; Oppong, S.K.; Lohman, D.J.; Etienne, R.S. Relative contribution of neutral and deterministic processes in shaping fruit-feeding butterfly assemblages in Afrotropical forests. *Ecol. Evol.* **2018**, *8*, 296–308. [CrossRef] [PubMed]
84. Mahood, S.P.; Lees, A.C.; Peres, C.A. Amazonian countryside habitats provide limited avian conservation value. *Biodivers. Conserv.* **2012**, *21*, 385–405. [CrossRef]
85. Kuussaari, M.; Heliölä, J.; Pöyry, J.; Saarinen, K. Contrasting trends of butterfly species preferring semi-natural grasslands, field margins and forest edges in northern Europe. *J. Insect Conserv.* **2007**, *11*, 351–366. [CrossRef]
86. Koh, L.P.; Sodhi, N.S. Importance of reserves, fragments, and parks for butterfly conservation in a tropical urban landscape. *Ecol. Appl.* **2004**, *14*, 1695–1708. [CrossRef]
87. Horner-Devine, M.C.; Daily, G.C.; Ehrlich, P.R.; Boggs, C.L. Countryside biogeography of tropical butterflies. *Conserv. Biol.* **2003**, *17*, 168–177. [CrossRef]
88. Sparrow, H.R.; Sisk, T.D.; Ehrlich, P.R.; Murphy, D.D. Techniques and guidelines for monitoring neotropical butterflies. *Conserv. Biol.* **1994**, *8*, 800–809. [CrossRef]
89. Fleishman, E.; Thomson, J.R.; Mac Nally, R.; Murphy, D.D.; Fay, J.P. Using indicator species to predict species richness of multiple taxonomic groups. *Conserv. Biol.* **2005**, *19*, 1125–1137. [CrossRef]
90. Schmidt, B.C.; Roland, J. Moth diversity in a fragmented habitat: Importance of functional groups and landscape scale in the boreal forest. *Ann. Entomol. Soc. Am.* **2006**, *99*, 1110–1120. [CrossRef]
91. Summerville, K.S.; Ritter, L.M.; Crist, T.O. Forest moth taxa as indicators of lepidopteran richness and habitat disturbance: A preliminary assessment. *Biol. Conserv.* **2004**, *116*, 9–18. [CrossRef]
92. Hill, J.K.; Hamer, K.C.; Tangah, J.; Dawood, M.M. Ecology of tropical butterflies in rainforest gaps. *Oecologia* **2003**, *135*, 294–302. [CrossRef]
93. Bonebrake, T.C.; Ponisio, L.C.; Boggs, C.L.; Ehrlich, P.R. More than just indicators: A review of tropical butterfly ecology and conservation. *Biol. Conserv.* **2010**, *143*, 1831–1841. [CrossRef]

94. Uehara-Prado, M.; Brown, K.S.; Freitas, A.V.L. Species richness, composition and abundance of fruit-feeding butterflies in the Brazilian Atlantic Forest: Comparison between a fragmented and a continuous landscape. *Glob. Ecol. Biogeogr.* **2007**, *16*, 43–54. [CrossRef]
95. Filgueiras, B.K.; Melo, D.H.; Leal, I.R.; Tabarelli, M.; Freitas, A.V.; Iannuzzi, L. Fruit-feeding butterflies in edge-dominated habitats: Community structure, species persistence and cascade effect. *J. Insect Conserv.* **2011**, *15*, 485–495. [CrossRef]
96. Pozo, C.; Luis-Martínez, A.; Llorente-Bousquets, J.; Salas-Suárez, N.; Maya-Martínez, A.; Vargas-Fernández, I.; Warren, A.D. Seasonality and phenology of the butterflies (Lepidoptera: Papilionoidea and Hesperioidea) of Mexico's Calakmul region. *Fla. Entomol.* **2008**, *91*, 407–422. [CrossRef]
97. Grotan, V.; Lande, R.; Chacon, I.A.; DeVries, P.J. Seasonal cycles of diversity and similarity in a Central American rainforest butterfly community. *Ecography* **2014**, *37*, 509–516. [CrossRef]
98. Munyuli, T. Drivers of species richness and abundance of butterflies in coffee-banana agroforests in Uganda. *Int. J. Biodivers. Sci. Ecosyst. Serv. Manag.* **2013**, *9*, 298–310. [CrossRef]
99. Ferreira, P.A.; Boscolo, D.; Carvalheiro, L.G.; Biesmeijer, J.C.; Rocha, P.L.; Viana, B.F. Responses of bees to habitat loss in fragmented landscapes of Brazilian Atlantic Rainforest. *Landsc. Ecol.* **2015**, *30*, 2067–2078. [CrossRef]
100. Soga, M.; Kawahara, T.; Fukuyama, K.; Sayama, K.; Kato, T.; Shimomura, M.; Itoh, T.; Yoshida, T.; Koike, S. Landscape versus local factors shaping butterfly communities in fragmented landscapes: Does host plant diversity matter? *J. Insect Conserv.* **2015**, *19*, 781–790. [CrossRef]
101. Grass, I.; Jauker, B.; Steffan-Dewenter, I.; Tscharntke, T.; Jauker, F. Past and potential future effects of habitat fragmentation on structure and stability of plant-pollinator and host-parasitoid networks. *Nat. Ecol. Evol.* **2018**, *2*, 1408–1417. [CrossRef]
102. Mulwa, R.K.; Böhning-Gaese, K.; Schleuning, M. High bird species diversity in structurally heterogeneous farmland in western Kenya. *Biotropica* **2012**, *44*, 801–809. [CrossRef]
103. Brito, M.M.; Ribeiro, D.B.; Raniero, M.; Hasui, É.; Ramos, F.N.; Arab, A. Functional composition and phenology of fruit-feeding butterflies in a fragmented landscape: Variation of seasonality between habitat specialists. *J. Insect Conserv.* **2014**, *18*, 547–560. [CrossRef]
104. Burivalova, Z.; Towsey, M.; Boucher, T.; Truskinger, A.; Apelis, C.; Roe, P.; Game, E.T. Using soundscapes to detect variable degrees of human influence on tropical forests in Papua New Guinea. *Conserv. Biol.* **2018**, *32*, 205–215. [CrossRef] [PubMed]
105. Öckinger, E.; Bergman, K.O.; Franzen, M.; Kadlec, T.; Krauss, J.; Kuussaari, M.; Pöyry, J.; Smith, H.G.; Steffan-Dewenter, I.; Bommarco, R. The landscape matrix modifies the effect of habitat fragmentation in grassland butterflies. *Landsc. Ecol.* **2012**, *27*, 121–131. [CrossRef]
106. Brennan, A.; Naidoo, R.; Greenstreet, L.; Mehrabi, Z.; Ramankutty, N.; Kremen, C. Functional connectivity of the world's protected areas. *Science* **2022**, *376*, 1101–1104. [CrossRef]

forests

MDPI

Article

How to Count the Uncountable? An Attempt at Wild Boar *Sus scrofa*, Linnaeus, 1758 Monitoring in an Urbanized Area

Jakub Gryz [1], Karolina D. Jasińska [2], Michał Brach [3], Mateusz Jackowiak [4] and Dagny Krauze-Gryz [2,*]

1 Department of Forest Ecology, Forest Research Institute, Braci Leśnej 3, Sękocin Stary, 05-090 Raszyn, Poland; j.gryz@ibles.waw.pl
2 Department of Forest Zoology and Wildlife Management, Institute of Forest Sciences, Warsaw University of Life Sciences WULS—SGGW, Nowoursynowska 159, 02-776 Warsaw, Poland; karolina_jasinska@sggw.edu.pl
3 Department of Geomatics and Land Management, Institute of Forest Sciences, Warsaw University of Life Sciences WULS—SGGW, Nowoursynowska 159, 02-776 Warsaw, Poland; michal_brach@sggw.edu.pl
4 Institute of Environmental Protection—National Research Institute, Słowicza 32, 02-170 Warsaw, Poland; mateusz.jackowiak@ios.gov.pl
* Correspondence: dagny_krauze_gryz@sggw.edu.pl

Abstract: In order to minimize human–wildlife conflicts, long-term and data-based management plans need to be developed. We aimed at assessing wild boar (*Sus scrofa*, Linnaeus, 1758) density within selected areas of Warsaw, the capital city of Poland. Wild boar abundance was estimated based on snow tracking conducted within 19 well-defined sectors, varying in habitat structure and location within the city. Moreover, in two forest reserves, wild boar population density was assessed with the use of camera traps and a random encounter model. The wild boar density index (n tracks/100 m/24 h) in Warsaw varied from 0 to 3.58 depending on the sector (0 to 4.0 ind./100 ha). In turn, it was 1.8 ind./100 ha for all the forested areas, which did not seem high compared to the other wild boar populations in Poland. The REM-derived estimates were higher than the snow-tracking-based estimates. Wild boars were mostly present in forests, especially those with oak stands, while the species was absent from central districts and mostly urbanized areas. We argue that the snow tracking method reflected the general distribution of wild boars across the city and in different habitats.

Keywords: random encounter model; snow tracking; urban area; density assessment; habitat use

1. Introduction

As cities expand, their impacts on natural ecosystems and the wildlife residing within them increase [1]. In general, increased urbanization leads to a loss in mammalian diversity [2]. Nevertheless, the use of urban areas by wildlife and urban species population numbers have increased over the years [3,4]. In general, an expansion of ungulates to new areas has been observed in Europe [5]. Indeed, wild boar (*Sus scrofa* Linnaeus, 1758) populations have grown steadily across Europe in the past few decades, resulting in more frequent human–wild boar conflicts [6–8]. This species occurs throughout a wide spectrum of habitats, including urban and suburban areas [6,9–13]. The impact of wild boars on economic interests may be diverse, including the risk of vehicle collisions and damage to crops and amenities (see [6]). Wild boars may also act as reservoirs for many important infectious diseases in domestic animals and also people [14,15]. Some animals that persist in urban environments demonstrate behaviors distinct from their non-urban counterparts [1]. For example, the flight distance of urban wild boars is shorter [16]; thus, wild boar–human encounters are more frequent than in non-urban areas. This may cause anxiety and fear of wild boar attacks or aggressiveness towards people or domestic animals [11,13,17,18].

In order to minimize human–wildlife conflicts, effective long-term management plans need to be developed [6,11]. Nevertheless, the management of wild boar populations remains a challenge for scientists and city governors [7,11,17,19–23]. Also, no data-driven strategy for urban wild boars has been implemented in Warsaw.

To control ASF spread, the Polish Ministry of Agriculture and Rural Development issued a regulation to hunters to reduce the population density of wild boars across the country to the level of 0.1–0.5 individuals per 100 ha [24]. In the following years, this resulted in very intensive wild boar population control, with an increase from 100 to 200 thousand to as many as 350 thousand individuals being shot yearly in Poland [25]. According to the official hunting data, the wild boar population across Poland decreased from around 200–250 thousand individuals in 2013–2018 to 50–100 thousand individuals in the subsequent seasons, with no population rise being recorded [25]. Wild boar density estimation is difficult [7], and in Poland, the number of wild boars in hunting districts is mainly determined by hunters arbitrarily and is often guessed rather than scientifically estimated [26]. Thus, the real population abundance in hunting grounds in Poland and in urban areas, is unknown.

At the same time, media coverage of wild boars' presence in urban areas and their increasing abundance in Warsaw has been high for several years [27,28]. Despite the lack of monitoring, wild boar population control commenced in 2008. At that time, eight animals were trapped and removed (i.e., translocated from the city) using four traps. In turn, in 2015, 27 traps located in various parts of the city were used, and 594 animals were trapped (this amounted to 1.15 ind. trapped per 100 ha of the city). In 2014, wild boars started to be shot in Warsaw [29]. Additionally, in 2017, the first case of African Swine Fever (ASF) was recorded in Warsaw (the first ASF case in Poland was recorded a few years earlier in 2014 [30]). Due to the virus itself [31] and the intensive culling of the species, the population was reduced significantly all over the country [25] and in Warsaw, too. After a few years' break, wild boar population control in Warsaw was again undertaken in 2023. According to a decision issued by the President of Warsaw [32], there are between 1500 and 2500 wild boars in the whole city (which gives a density of 3 to 5 ind./100 ha). This assumption supported the decision to remove 500 individuals (through trapping with further euthanasia) over 19 months (June 2024–December 2025). However, the official data from hunting districts covering Warsaw (and neighboring areas) shows a very low population density, around 0.02 to 0.5 ind./100 ha [33].

The reason for our study was to provide science-based data, which may help to develop a data-driven strategy for urban wild boar management in Warsaw. This research was conducted in selected Warsaw areas but also focused on Wawer—the biggest, least urbanized, and most forested district of Warsaw and assumingly among the areas with the most frequent wild boar–human encounters.

We, therefore, aimed to (i) assess the current (2023/2024) density of wild boar in the selected areas of Warsaw, varying in terms of habitat features and location within the city, and compare current data with data for the winter seasons 2020–2022, to show which areas and habitats are used by wild boars and whether the population density changes; (ii) compare winter density assessments obtained by two different methods (snow tracking and camera traps) in two major forest reserves in Warsaw to check whether these density estimates are consistent; and (iii) to conduct complex wild boar density assessment in the most forested district in Warsaw (Wawer district) to deliver an estimation of the absolute wild boar abundance in this part of the city.

2. Methods

2.1. Study Area

The research was conducted in Warsaw (52°13′47″ N 21°00′42″ E), the capital city of Poland, the second largest (517,100 ha) and the most populous (1,862,000 inhabitants, 3599 inhabitants/100 ha) city of the country [34]. The city is situated at an altitude of around 113 m above sea level, with an annual rainfall of about 500 mm and an average annual temperature of 7.7 °C. The Vistula River flows through Warsaw, dividing the city into two parts. The left bank of the river is characterized by a high degree of anthropogenic transformation, while along the right bank, natural riparian forests, included in the Natura 2000 network and constituting an essential ecological corridor, have been preserved (Figure 1). Forests account for about 15% of the city area and are located mainly on the outskirts. A high proportion of various green areas makes Warsaw easily penetrable for many mammal and bird species. Its green areas are also important refuges for numerous animal species, e.g., [35,36].

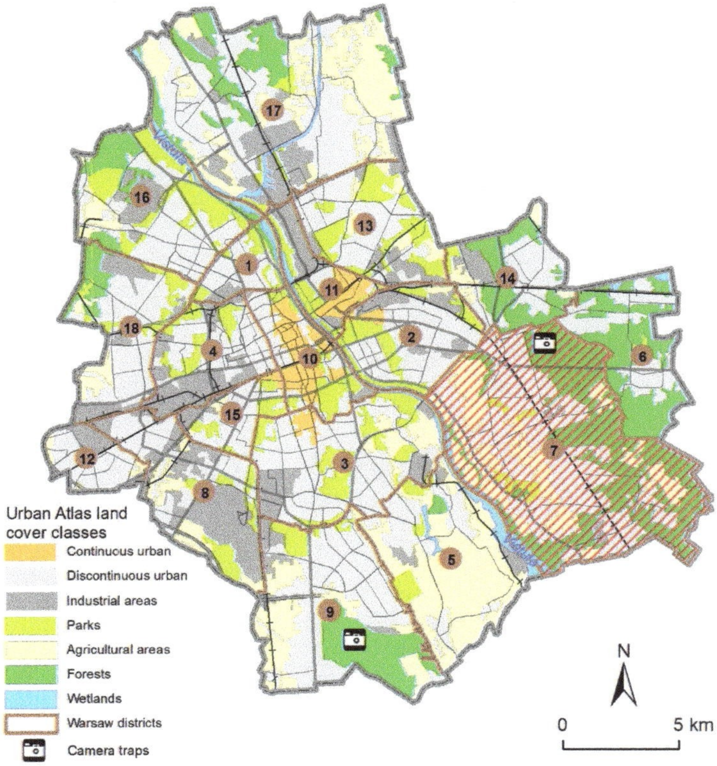

Figure 1. Distribution of green areas and other habitats in Warsaw. The numbers refer to city districts: 1—Żoliborz, 2—Praga-Południe, 3—Mokotów, 4—Wola, 5—Wilanów, 6—Wesoła, 7—Wawer, 8—Włochy, 9—Ursynów, 10—Śródmieście, 11—Praga-Północ, 12—Ursus, 13—Targówek, 14—Rembertów, 15—Ochota, 16—Bielany, 17—Białołęka, 18—Bemowo. The research was conducted in city districts no. 2, 3, 7, 9, 10, 14, 15, 16. The dashed area shows Wawer, the city district where complex wild boar density estimation was conducted. Study plots where camera traps for REM density estimation (see Section 2.4) were located are shown.

Warsaw is divided into 18 districts, among which Wawer (Figure 1) is the biggest (7971 ha, 16% of the area of the city) and the most forested. Wawer is the peripheral district, which was included in the capital city in 1951. Officially, it has over 87,000 inhabitants (1095 inhabitants/100 ha). Built-up areas cover almost 2800 ha, arable lands (partially fallow) over 2000 ha, and forests nearly 3000 ha [34,37]. Wawer's forest cover is the largest in Warsaw, surpassing 35%. Forests are located primarily in the eastern part of the district and are dominated by pine stands planted after World War 2. Around 150–200-year-old oaks dominate just a tiny fraction of the forests. The biodiversity hot spots are riparian forests at the Vistula River (also the western part of the city district) and alder swamp forests of the Zakole Wawerskie. A considerable part of the area is under some form of nature protection.

2.2. Study Sectors

Fieldwork was performed in 8 out of 18 city districts, 4 inner (Ochota, Śródmieście, Mokotów, and Praga Południe) and 4 outer districts (Wawer, Rembertów, Ursynów, and Bielany), and within 19 sectors, located in different districts (Figure 1). Seven sectors were located in Wawer. Five were located in Bielany, two in Ursynów, and the rest in the other investigated districts (Table 1). The borders of the sectors were major streets, the Vistula River, or compact built-up areas, which made them easily identifiable during the fieldwork. The sectors differed in terms of their area and habitat structure. Ten sectors were forest-dominated; the rest were covered by other habitats (parks, riparian areas, or built-up areas). The forest-dominated sectors differed regarding stands, age, composition, and the area's status. A detailed description of the sectors can be found in Appendix A.

Table 1. Description of the sectors in which the fieldwork was conducted. The total transect length on which snow tracking (see Section 2.3) was carried out is given. nd—no snow tracking data.

Sector No.	City District	Area (ha)	Habitat	Transect Length (m)	
				2020–2022	2023–2024
1	Wawer	527.1	forest	34,330	12,480
2		388.2	forest	9820	10,400
3		830	forest	4240	15,040
4		1113	forest	nd	13,600
5		1322	other	nd	11,200
6		340.6	other	12,960	7040
7		540.4	forest	5600	13,760
8	Rembertów	155.5	forest	14,720	4320
9	Praga Południe	184.9	other	nd	10,000
10	Ursynów	1108.5	other	12,560	12,310
11		89.4	other	5770	2320
12	Mokotów	96.4	other	1400	3360
13	Śródmieście	456.6	other	10,650	10,000
14	Ochota	86.9	other	7600	1000
15	Bielany	291.8	forest	2310	4760
16		129.5	forest	1820	5320
17		56.7	forest	2240	2240
18		233.5	forest	4270	3600
19		57.0	other	1400	900
Total		5404.0		131,690	143,650

As one of the aims was to estimate wild boar abundance in a sample city district, i.e., Wawer, most of the district (65%) was covered by sectors (Appendix B). In the remaining part of the district, conducting an inventory was impossible. The area of the sectors was assumed to be at least as big as an estimated home range size of urban wild boar (100 ha, [9]).

We used two methods to assess the density of wild boars in Warsaw during winter: (1) snow tracking on transect routes, and (2) camera traps and random encounter model REM.

2.3. Snow Tracking on Transect Routes

Wild boar abundance was estimated based on snow tracking in the 2023/2024 winter season. In each sector, snow tracking was conducted in all habitats. In 16 out of the 19 sectors in Warsaw, snow tracking was also conducted in the past (2020–2022), which allowed us to compare wild boar density indices between the two periods (Table 1).

Snow tracking was conducted on sparsely distributed, linear transects of different lengths divided into 100 m sections. Tracking was carried out one to four days after snowfall. The number of wild boars' trails crossing each 100 m of the transect was registered, and then the relative density index was calculated (N tracks/100 m/24 h) [38]. The number of trails registered was divided by the number of days from the last snowfall to adjust the number to 24 h after the snowfall.

The total area of all the districts where snow tracking was performed in the 2023/2024 winter season was 5404 ha, accounting for over 10% of the city. The total transect length was 143.65 km. The total transect length in 2020–2022 was 131.69 km (Table 1).

For all the sectors in Warsaw, the relative density index was calculated and used to compare those two periods. For the Wawer district, to estimate the total abundance of wild boar in a sector, the absolute density was calculated in each sector by using the FMP (Formozov–Malyshev–Pereleshin) formula [39]:

$$D\ (\text{N ind.}/100\ \text{ha}) = 1.57 \times (\text{N tracks/km}/24\ \text{h})/\text{DMD},$$

where D stands for density, 1.57 is a constant value, and DMD indicates the daily movement distance. No data on the DMD of wild boar in Warsaw was available, so we used the values (DMD in winter = 14 km) obtained for another big city in Poland, Cracow [9]. The mean value from seven sectors was adopted for the part of Wawer, which was not covered by snow tracking. Based on this, the absolute abundance of wild boar was calculated for the whole district.

Searching for the relation between the habitat structure of the sectors and the wild boar density index, we calculated the percentage share of forests and agricultural land (according to Urban Atlas) in the seven sectors (sectors no. 1 to 7) in Wawer (Appendix B). For five forest-dominated sectors (no. 1, 8, 10, 16, and 18), detailed data on stand composition and forest habitat was available [33]. Thus, we could check if there was a relation between the percentage share of oaks *Quercus* spp. older than 50 years (and already producing seeds) and share of deciduous forest habitats within a sector and wild boar density index.

2.4. Camera Traps and Random Encounter Model (REM)

In the two forest reserves (Sobieski Forest, sector no 1, and Kabaty Forest, sector no 10, Figure 1), wild boar population density was assessed with the use of camera traps (Reconyx™ Hyperfire™ PC800, PC900 or PC850, Reconyx Inc., Holmen, WI, USA, Browning® Spec Ops Advantage, Dark Ops Apex, Prometheus Group, Birmingham, AL, USA). Eight camera traps were used in sector no. 1, and nine in sector no. 10 (17 camera traps in total). In the Sobieski Forest, they were placed in the reserve, in its part being surrounded by non-protected forest and as far as possible from streets. In the Kabaty Forest, the traps were located in its central part. The distance between the camera traps was 200–400 m. Data were collected from November 2023 to March 2024 to reflect the density of wild boar in winter. The number of trap days was 788 and 622 in Sobieski and Kabaty Forest, respectively. The camera traps were placed in trees at approximately 50 cm height. We recorded each wild boar appearing in the images without distinguishing between individuals. A new observation was considered if a minimum of 15 min elapsed between subsequent photos or a series of photos showing an animal/animals. This rule was abandoned only when an animal in the photo was different in age, sex, body size, or fur pattern (i.e., spotted), indicating clearly it was a different individual than the one previously registered [36]. We used the random encounter model (REM) to calculate wild boar density.

$$D = y/t \quad \pi/(vr(2 + \theta))$$

We used the number of independent photo captures (y) and camera effort (t = trap days). We used wild boar daily range (v = 14 km/day as assessed for winter) after [9]. The detection distance (r = 5.4 m) was adopted after [40]. The detection angle (θ, 0.69, and 0.96, depending on the camera trap type, in radians) was taken from the camera trap manual.

2.5. Statistical Analyses

Differences in the wild boar density index between the sectors were studied using the Kruskal–Wallis test (the data did not follow a normal distribution, Shapiro–Wilk test, $p < 0.05$). A two-sample paired Wilcoxon test was used to compare density indices between the two study periods. Again, the data did not follow normal distribution (Shapiro–Wilk test, $p < 0.05$).

The correlation between the wild boar density and habitat variables was checked using the Pearson correlation analysis.

3. Results

3.1. Wild Boar Density in Warsaw

The wild boar density index (n tracks/100 m/24 h) in Warsaw varied from 0 to 3.58 (0 to 4.0 ind./100 ha). A higher wild boar density was recorded in the city's outer rather than the inner districts (Kruskal–Wallis test, H = 4.84, $p = 0.03$) and mainly in Warsaw's east–south and north parts. The wild boar density was higher in the natural areas (i.e., forests, 1.6 tracks/100 m/24 h; 1.8 ind./100 ha) than in the other habitats (H = 8.17, $p = 0.005$, Figure 2).

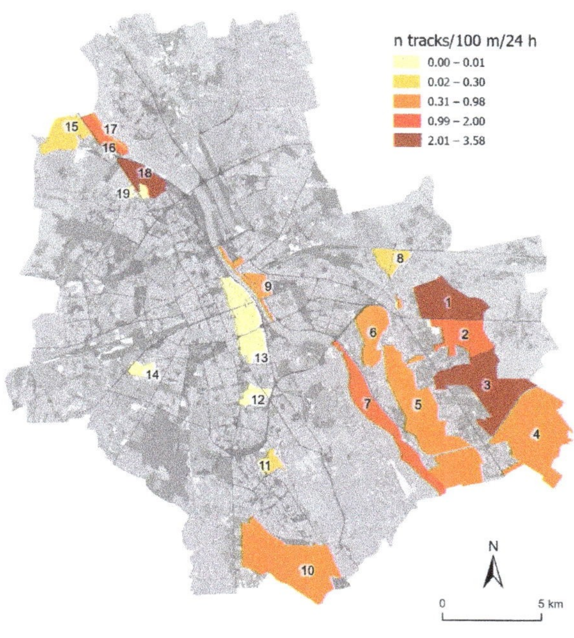

Figure 2. Wild boar density index in different areas of Warsaw (i.e., sectors) as assessed by snow tracking on transect routes in winter season 2023/2024. Sector numbers as in Table 1.

The mean wild boar density index in the current period (2023/2024) was higher than in the earlier period (0.47 vs. 1.07 tracks/100 m/24 h) (two-sample paired Wilcoxon test, W = 5, $p = 0.005$). Nevertheless, this increase was evident only in part of the sectors, while in the northern part of Warsaw (sectors no. 16–19), the wild boar abundance remained at a similar level (Figure 3).

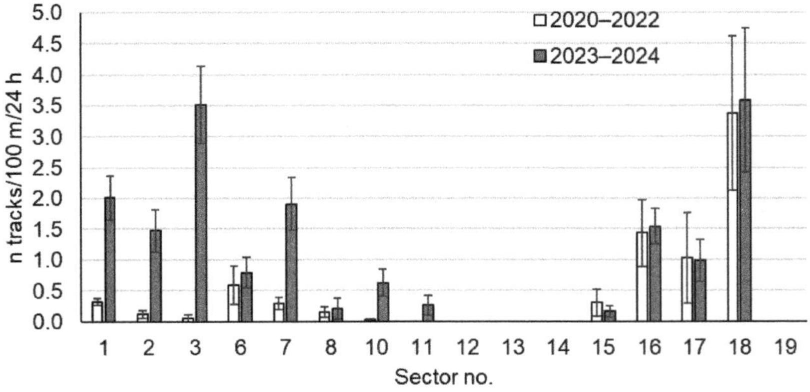

Figure 3. Changes in the mean (±SE) wild boar density index in different areas of Warsaw (sector numbers as in Figure 2, Table 1) as assessed by snow tracking on transect routes. Only the sectors where snow tracking was performed in the two study periods are included.

Wild boars seemed more abundant in the forests with a high proportion of oak stands; no relation was seen for the deciduous forest habitats. However, no statistical significance was found (Pearson correlation; oaks r = 0.69, $p = 0.194$; deciduous forest habitats, r = −0.27, $p = 0.657$, Figure 4a,b).

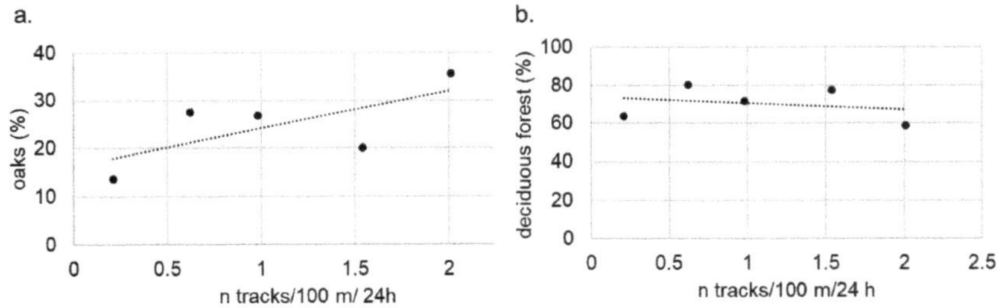

Figure 4. Relation between wild boar density index assessed based on snow tracking (winter 2023/2024) and (**a**) percentage share of oaks older than 50 years; (**b**) deciduous forest habitats in a given forest area. Forest-dominated sectors were considered (1, 8, 10, 16, and 18, see Figure 2, Table 1).

3.2. *Wild Boar Density Estimate with Random Encounter Model (REM)*

In the two forest reserves (Sobieski Forest, sector no. 1, and Kabaty Forest, sector no. 10), we calculated the wild boar density based on a random encounter model (REM). The results obtained by this method were consistent with the snow tracking method. Yet, the values were much higher (Figure 5) than those obtained based on snow tracking (2.3 and 0.7 ind./100 ha for Sobieski and Kabaty Forests, respectively).

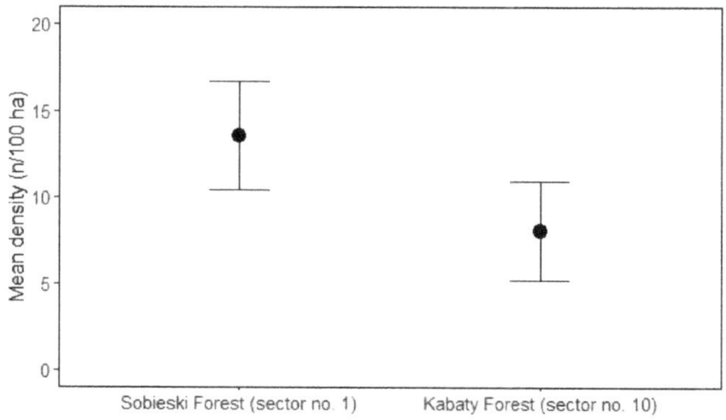

Figure 5. Mean (±SE) wild boar density estimated by random encounter model (REM) in two forest reserves in Warsaw (Sobieski Forest, sector no. 1 and Kabaty Forest, sector no. 10) in winter 2023/2024.

3.3. Wild Boar Abundance Assessment in One of the City Districts (Wawer)

In Wawer, the city district in which complex wild boar density assessment was performed, the density index (n tracks/100 m/24 h) ranged from 0.79 in sector no. 6 to 3.44 in sector no. 3. The estimated wild boar density ranged from 0.9 to 3.9 ind./100 ha. The mean estimates for the whole district equaled 1.6 tracks/100 m/24 h and 1.8 ind./100 ha. The total abundance of wild boar in the entire city district was assessed at 103 individuals (Table 2).

Table 2. Abundance assessment of wild boar in a city district of Warsaw (Wawer), divided into seven distinct sectors, in which snow tracking on transect routes was performed (2023/2024). To show the total abundance of wild boar in the whole district, the mean density index from sectors 1 to 7 was adopted as the value for the remaining part of the district not covered by snow tracking.

Sector No.	Area (ha)	Density Index (±SE) (n Tracks/100 m/24 h)	Density (n ind./100 ha)	Abundance (n ind.)
1	527.1	2.01 (0.32)	2.3	12
2	388.2	1.47 (0.31)	1.6	6
3	830.0	3.44 (0.56)	3.9	32
4	1113.3	0.98 (1.03)	1.1	12
5	1321.7	0.9 (0.25)	1.0	13
6	340.6	0.79 (0.22)	0.9	3
7	540.4	1.87 (0.39)	2.1	11
remaining area	2909.7	1.60	1.8	13
Total	7971.0			103

There was a significant difference between the values obtained for the different sectors (Kruskal–Wallis test, $H = 22.97$, $p = 0.000$). The wild boar density tended to be higher in the sectors with large forest cover and lower when arable lands dominated; however, this was not statistically significant (Pearson correlation, forest cover, $r = 0.52$, $p = 0.235$ and arable land, $r = -0.45$, $p = 0.306$, Figure 6a,b).

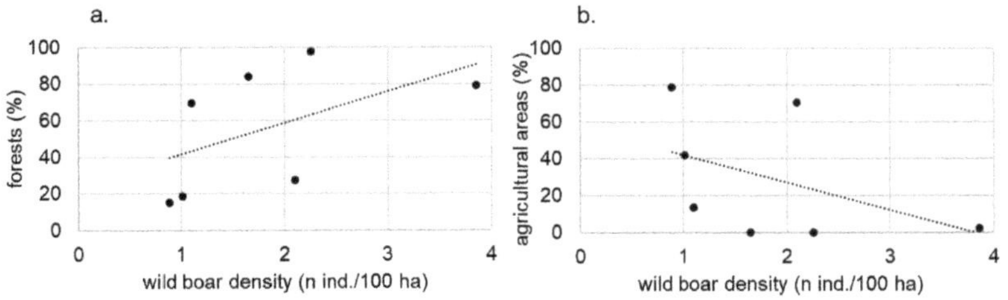

Figure 6. Relation between the wild boar density and (**a**) forest cover, and (**b**) share of farmland in a given sector (sectors no. 1 to 7, located in one city district, Wawer) as assessed based on snow tracking on transect routes in winter 2023/2024.

4. Discussion

In this study, we showed wild boar density in selected areas of Warsaw. This ungulate species was mainly present in forests. This is in line with the strong preference of wild boar for natural landscapes and areas with high primary productivity [16], as the availability of food resources and protective cover are among the factors that determine wild boar abundance [41]. In our study, the highest densities were obtained for oak-dominated forests, offering plenty of natural food (in the form of acorns) and located in the outer city districts. Also, in Berlin, wild boars mainly inhabited forests [11]. In turn, in another Polish city, Cracow, grasslands and natural open areas were mostly preferred [10]. The wild boars in Warsaw seemed to avoid built-up areas; the same pattern was found in Cracow [10]. They were also less abundant in the areas with arable land domination. Indeed, arable lands were not intensively used for grubbing in Cracow [23]. The two wild boar hot spots (i.e., Sobieski Forest in the south and Bielany Forest in the north of Warsaw) were connected to forests outside the city (including Kampinos National Park) directly or via the Vistula River. Indeed, rivers and their associated riparian forests act as migration corridors for wildlife [12,42,43]. According to this study and other published data, wild boars were absent from the central districts and mostly urbanized areas; this included the biggest Warsaw urban parks, i.e., Mokotów Field [44], Skaryszew Park [45], Szczęśliwice Park, or Royal Łazienki Park (this study). Yet, some accidental presence of wild boars in such green spaces was recorded, with Krasiński Garden, a small, historical urban park, being an example [46]. In general, such centrally located green areas are less appropriate habitats for wild boars, providing fewer hiding places and fewer natural food sources for woody species [8]. Comparing the current data with that collected two years earlier, we see that the wild boar densities increased. Nevertheless, the species was and is still mostly present in areas of high naturalness.

It is hard to claim whether wild boars colonize urban areas or if the city spreads over wild boar habitats [11]. The urbanization process is ongoing, with urban green areas being built up. This makes wildlife search for new, even suboptimal, habitats, leading to more frequent human–wildlife conflicts. Urban forests seem to be crucial as wild boar hot spots. Indeed, wild boar inhabiting urban forests in Berlin formed genetically distinct and geographically coherent clusters. Nevertheless, if these populations exceed carrying capacity, dispersers from urban populations would invade built-up areas [11]. Thus, the regular monitoring of the urban population is necessary.

Wild boar density assessment is difficult [7], and wild boar densities in Poland have recently become very variable due to ASF and increased hunting mortality [31]. For example, in the Roztocze National Park, drive counts showed densities of 16 ind./100 ha, while in the subsequent seasons, this value dropped to 2.2 ind./100 ha and again increased to 12 ind./100 ha [47]. In turn, in the Białowieża Primeval Forest, the wild boar density was estimated at around 7 ind./100 ha based on drive counts. After ASF was recorded, these values dropped to less than 2 ind./100 ha [31]. When another method was applied (i.e., pellet group counts) the assessed density was even much lower (Gryz J., unpbl.). Wild boar inventory using camera traps in northeast Poland (Iława County) showed a density of 2.5 ind./100 ha of the forest [48]. Inventory in western Poland (Lubusz Voivodeship, areas managed by State Forests) conducted using the same method showed a density of 1.6 ind./100 ha [49]. A complex inventory of wild boars conducted in state forests, using drive counts (autumn 2016), showed a density of around 2 ind./100 ha [50]. However, counting wild boars on a large regional scale is considered unfeasible, and estimations should rather be performed locally and in specific habitats [51]. Besides all this uncertainty over real population abundance, our results on wild boar density in Warsaw do not seem to be high. The mean wild boar density assessed for different areas of the city was from 0 to 4.0 ind./100 ha. In turn, for all the forested areas and for the Wawer district, it was 1.8 ind./100 ha.

Owing to the wild boar population increase and an ongoing urban sprawl, this ungulate has become a new urban species, finding optimal habitats in urban forests that offer abundant food sources. This food supply, combined with higher temperatures in the city (urban heat island phenomenon UHI, [52]), may lead to population increase thanks to reduced juvenile mortality and boosted reproductive success [53]. Additionally, as an omnivorous ungulate, wild boars are often attracted by easily accessible anthropogenic food, including communal waste [7], review in: [8], especially when access to natural resources is limited [54]. They may even enter urban areas looking for pet (i.e., cat) food [12], making them more probable to be recorded by city inhabitants and generating human–wild boar conflicts.

Our results on the wild boar densities need to be taken with caution. Snow tracking in urban areas can be challenging, especially in sectors with a high share of built-up areas or inaccessible plots. The more built-up areas in a sector, the longer stretch at which snow tracking is not possible, e.g., in sector no. 5, snow tracking was possible only along 11 km out of 17 km of our walking distance. In the case of big wild boar packs, it may be difficult to quantify the number of trails crossing the transect. Some sectors were close or adjacent so that the same individuals could travel between the sectors. They were partially separated by streets and built-up areas, but those did not constitute a definite barrier so that the same individuals could be present in different sectors. Moreover, animals from the forests located outside this city district could have entered our study sectors. Besides these shortcomings, our estimations reflect the general distribution of wild boar over the city and in different habitats. We believe the snow tracking method has many pros. It is non-invasive, cost-effective, and can be used over long-term periods and in various habitats, even the most urbanized, where camera trap placement may be difficult. The method may estimate the absolute abundance or deliver the density index, allowing for monitoring population trends and changes in habitat use [38]. On the other hand, it is only applicable with sufficient snow cover; thus, the future use of the method remains questionable. Urban habitats are patchy, with dispersed food or resting sites. This makes wild boars move more to meet their daily requirements [9]. Therefore, the DMD value may reflect local habitat structure, and variation in this value will affect the obtained density value. DMD value should be ideally obtained for a given area or adopted from a study conducted in similar conditions. Thus, the method should be supplemented with other techniques to monitor long-term population trends, such as camera trapping or pellet group counts. The first allows for density assessment (either using the relative index or the absolute density using the REM) but also gives quite precise data on group size, productivity, or activity patterns, parameters that in urban animals may shift as an adjustment to urban conditions [2,36]. In our case, REM gave much higher results than the snow tracking estimations, yet consistent with a general difference between the two forests (i.e., Sobieski and Kabaty Forests). A similar pattern was found in the Białowieża Primeval Forest, where the wild boar density was estimated based on drive counts, and the REM method was compared [31]. REM is suggested to be the most promising methodology for wild boar density assessment [55], yet in urban areas, it may be prone to overestimation [56]. One of the key assumptions of REM is that the pattern of activity of individuals should be random as related to camera trap placement [57]. This may be violated in heavily penetrated urban forests, where the avoidance of humans may influence wild boar foraging decisions [23]. Moreover, the day range is a crucial REM parameter to obtain unbiased density [57]. This value may differ greatly from that obtained in rural areas and depends on the habitat structure in urban areas [9]. Drive counts are impossible to conduct in heavily penetrated small urban forests. Moreover, they give very variable results, probably related to the highly gregarious behavior of wild boar [55]. A molecular analysis aimed at population assessment [56] and genetic differentiation may help better understand migration patterns between urban and suburban populations [11].

The attitudes of city inhabitants towards wild boar may be very variable [58], with some claiming wild boar eradication/relocation from any urban areas [13] and others being very affectionate about this ungulate species, with attempts to feed wild boars and/or approach not being rare [17,27,28]. Even those who believe wild boar should return to their natural habitats are against wild boar culling [58]. Many people are strictly against wild boar killing, which makes managing wild boar populations in urban areas difficult. In Warsaw, unofficial groups that sabotage the attempts of wild boar population control (e.g., letting wild boar escape from the traps, disrupting wild boar hunts, etc.) are organized. There were also attempts to destroy or steal wild boar traps [29]. Our results did not point to a high wild boar density in Warsaw. Nevertheless, between June 2023 and the time of our study (winter 2023/2024), over 20 wild boars had been culled just in the Wawer district. Unleashed dogs may magnify the conflict between wild boar and humans in urban areas. This domestic carnivore is perceived as a persisting threat by wild boars [16]. Thus, when a dog attacks a wild boar, this ungulate responds, resulting in the dog owner feeling threatened. Indeed, during our study, unleashed dogs were recorded frequently (own observations).

5. Conclusions

Wild boar management should start with society's education on wild boar behavior and ecology so that the consequences of certain human actions (like feeding, approaching animals, etc.) are clearly understood. It is clear that a long-term general strategy should be implemented. This should be based on the regular monitoring of wild boar population in the sampled locations, aimed at assessing wild boar population trend, distribution within the city, and population productivity. This should also include hunting districts neighboring the city. At the same time, the protection of forested areas, especially old oak stands, which serve as refuges for wildlife, needs to be safeguarded. Also, regulations (on limited access to people and dogs) must be respected. Wild boar-proof garbage bins are necessary to keep wild boar from built-up areas. Any population regulations should be based on the data delivered by population monitoring, with wild boar trapping and further euthanasia probably being a better solution than wild boar shooting in urban areas.

Author Contributions: Conceptualization, J.G. and D.K.-G.; Methodology, J.G., D.K.-G., K.D.J. and M.J.; Formal Analysis, D.K.-G. and K.D.J.; Investigation, J.G., K.D.J., M.J. and D.K.-G.; Data Curation, J.G., D.K.-G. and K.D.J.; Writing—original draft preparation, J.G. and D.K.-G.; Writing—review and editing, K.D.J., M.J. and M.B.; Visualization, D.K.-G., M.B. and K.D.J. All authors have read and agreed to the published version of the manuscript.

Funding: This research received no external funding.

Data Availability Statement: Data is available from the corresponding author upon request.

Conflicts of Interest: The authors declare no conflicts of interest.

Appendix A. Description of Study Sectors

Sector no. 1 (527.1 ha) was dominated by forest (98%), part of which (116 ha) was legally protected as a nature reserve (Sobieski Forest). Stands were diverse but dominated by 70–80-year-old pine and oak stands. Besides, there were natural 150–200-year-old oak stands in the reserve. Most of the forest is managed by the Municipal Forests of Warsaw.

Sector no. 2 (388.2 ha) was also forest-dominated (84%, Barciucha Forest) with a share of built-up (primarily detached or semi-detached houses) areas (13%). Private forest plots are now being built up intensively, including old oak stands, with new roads and other infrastructure being established. The Wawerski Canal runs through the sector.

Sector no. 3 (830 ha) was also mostly covered by forest (79%, Wilanowski Forest), with built-up areas accounting for 14%. Warsaw's southern bypass runs at its southern border. Almost 50% of the forests are or will soon be built up.

Sector no. 4 (1113 ha) was mostly covered by forest (70%), and built-up areas accounted for 13% (primarily detached and semi-detached houses). Its pine stands are young, mostly planted after World War II. Warsaw's southern bypass was its northern border.

Sector no. 5 (1322 ha) was the most urbanized (built-up areas accounted for 27% of the area). Agricultural lands (mostly fallow lands) accounted for 42% of the area; forests accounted for 19% of the area. The whole area is undergoing fast urbanization, i.e., plots are fenced, and new buildings are erected. Along its western border (Miedzeszyński Embankment), there are some wet areas, reed, and willow *Salix* spp. covered.

Sector no. 6 (340.6 ha) encompassed former Vistula floodplains. Forests (15%) were primarily located in its northern part, were alder-dominated, and were protected as a Nature and Landscape Complex (Zakole Wawerskie). The rest of the area is covered by reeds, meadows, and fallow land in various stages of succession (79%). Built-up areas are scarce and account for 6% of the whole area.

Sector no. 7 (540.4 ha) encompassed the Vistula floodplains and was bordered by the river from the west and Miedzeszyński Embankment from the east. The landscape was a mixture of riparian forests (27%) and open areas (meadows and fallow lands, 71%). There are old river beds and swamps in some parts. Built-up areas accounted for less than 2% of the area and were primarily located in the southern and northern parts of the sector. Some recreational areas also existed (like golf courses, boating clubs, and beaches).

Sector no. 8 (155.5 ha, located in Rembertów district) encompassed a woodland cut into two separated areas by a heavy traffic road. Its stands primarily comprise Scots pine (*Pinus sylvestris* L.), oaks, and birch (*Betula pendula* Roth), though most of this site was afforested in the 1950s. Half of the forest (70 ha) is protected as a nature reserve (Kawęczyn reserve) with limited access.

Sector no. 9 (184.9 ha) was located in an inner district (Praga Południe). It encompassed about a 5 km long riparian area (with remnants of willow and poplar forest and some municipal beaches along) restricted by flood bank and the Vistula River. Also, a nearby urban park (Skaryszewski), the National Stadium surrounded by its open areas, and an old river port (now undergoing housing development) were included in this study site. The whole area is intensively used for recreational activities.

Sector no. 10 (1108.5 ha, Ursynów) encompassed the largest compact forest complex in Warsaw (more than 900 ha) and has been protected as a reserve (Kabaty Forest) since 1980. The multi-species tree stands are dominated by Scots pines and oaks, which grow in fertile habitats. Some stands are more than 120 years old. This reserve is contiguous with other Warsaw green areas via the Vistula Scarp. It is surrounded by developed areas, except in the east, where the reserve is adjacent to arable land, an urban park, and a botanical garden. In the protected area, visitors can move only along tourist trails, where walking and biking are allowed. Forest management is highly limited. In the reserve and an adjected urban park, special places are dedicated to leisure activity, barbecue areas, fireplaces, etc. Thus, there are plenty of anthropogenic food remnants that attract wild boars.

Sector no. 11 (89.4 ha, Ursynów) was located on the Warsaw University of Life Sciences campus. Three different habitats can be distinguished: a modern campus with new buildings, an old campus being a historical park, and a nature reserve (Ursynów Scarp). The reserve protects the scarp rising above what was once the proglacial valley that carried meltwaters from the ice sheets. This valley is now considered to be the Vistula

Valley. Protection was provided to a 23 ha area here in 1996, and the reserve includes a multi-species stand of broadleaved woodland. However, the woodland and scrub habitat only accounts for 30%–40% of the protected area. The remainder of the reserve is composed of meadowland and ruderal plant associations. The scarp runs south of the reserve and is connected to woods at the southern edge of Warsaw agglomeration. To the southeast, the reserve borders have more extensive open areas, although these areas are destined for more urban development.

Sector no. 12 (96.4 ha, Mokotów) encompassed two parks (Królikarnia, a historical park adjacent to a historic classicist palace, and Arkadia, now being protected as the Arcadia Nature and Landscape Complex), allotment gardens, sports areas, and XIX fort, all located among highly urbanized parts of the city.

Sector no. 13 (456.6 ha, Śródmieście) covered green areas in a strict city center. These were the historical Royal Łazienki Park, Botanical Garden, Agrykola Park, the surroundings of Sejm and Senate Complex, Marshal Edward Rydz-Śmigły Park, Kazimierzowski Park, and smaller parks and squares located within the built-up areas of the city center and surrounded by busy streets. Some parts of this site are built up by historical and modern buildings. The sector was limited by the Vistula River from the east. Yet, this part of the river bank is highly anthropogenically transformed and deprived of riparian forests and other natural habitats.

Sector no. 14 (86.9 ha, Ochota) included the neighboring Szczęśliwice Park, remnants of Szczęśliwice Fort (part of the 19th century Warsaw Fortress) and adjacent allotment gardens. This municipal park was established in the 1960s in the area, which used to serve as a debris and rubbish dump and clay excavation site (now turned into ponds). The soil is highly degraded, so pioneer species (like poplars) have a significant share among the plants. There is also Szczęśliwice Hill, created initially as a dump for rubble from the destruction of Warsaw during the war, then covered with earth and turned into a slope with a ski lift. The area is a popular recreational site for inhabitants.

Sector no. 15 (291.8 ha, Bielany) was a forest (Młociny Forest) adjacent to sector no. 16 (Młociny Park). This site is directly linked with Kampinos National Park. In the south, it is restricted by one of the biggest Warsaw cemeteries (Northern Communal Cemetery). Tree stands are dominated by Scots pine and oaks with an admixture of birch, black alder (*Alnus glutinosa* (L.) Gaertn), and other deciduous species.

Sector no. 16 (129.5 ha, Bielany) was a forest complex (Młociny Park) with a meadow in the southern part (so-called forest park). Stands are dominated by Scots pine and oaks. The central part of the forest is swampy and dominated by black alder and poplars *Populus* spp. Some pine and oak stands are over 150 years old.

Sector no. 17 (56.7 ha, Bielany) was located along the Vistula River on one side and Młociny Park on the other side. It encompassed a ca. three-kilometer-long riparian habitat mostly covered by willow and poplar forests or bushes. Meadows and oxbow lake complemented the landscape.

Sector no. 18 (233.5 ha, Bielany) was an urban forest (Bielany Forest) surrounding the university campus and historical abbey complex. In its north part, a sports club is located. Most of the forest has been protected as a nature reserve since 1978. Its forest habitats include oak–lime–hornbeam forest, riparian forest, and alder carr. The oldest oak stands are more than 250 years old. Two small watercourses cross the reserve area. On the eastern side, the reserve is partly separated from the Vistula River by a 3-lane expressway (yet it partly runs on a flyover). At the same time, the northernmost limits are connected to the Młociny Park. In contrast, dense urban development is present to the west and south.

Sector no. 19 (57.0 ha, Bielany) encompassed the University of Physical Education in Warsaw campus and a small forest complex (Linde Forest). Stands aged 60–80 are dominated by Scots pine, black locust, birch, and oaks.

Appendix B. Habitat Composition of Study Sectors (no. 1–7) Located in Wawer City District

Table A1. Percentage share of habitats (Urban Atlas Categories) within study sectors (no. 1–7) located in Wawer district.

Habitat Category	**Sector no (% of Area)**						
	1	**2**	**3**	**4**	**5**	**6**	**7**
Continuous urban fabric	0.1	4.6	5.0	1.9	20.5	1.1	0.1
Discontinuous urban fabric	0.0	7.3	7.2	9.7	5.7	3.6	1.0
Discontinuous medium-density urban fabric		1.4	1.8	1.4	0.5	0.8	0.5
Industrial and commercial units	0.3	0.3	0.4	1.2	4.4	0.0	0.3
Roads	0.3	2.5	3.1	2.0	3.3	0.4	0.2
Railways			0.3	0.0			
Construction sites				0.2	0.7	0.0	0.1
Green urban areas			0.8	0.0	3.9		
Agricultural and semi-natural areas			2.2	13.6	42.1	78.8	70.5
Forests	97.8	83.9	79.1	69.5	18.6	15.3	27.4
Water				0.4			0.0
Cemetery	1.6				0.2		
Total area (ha)	527.0	388.2	830.0	1113.3	1321.7	340.6	540.4

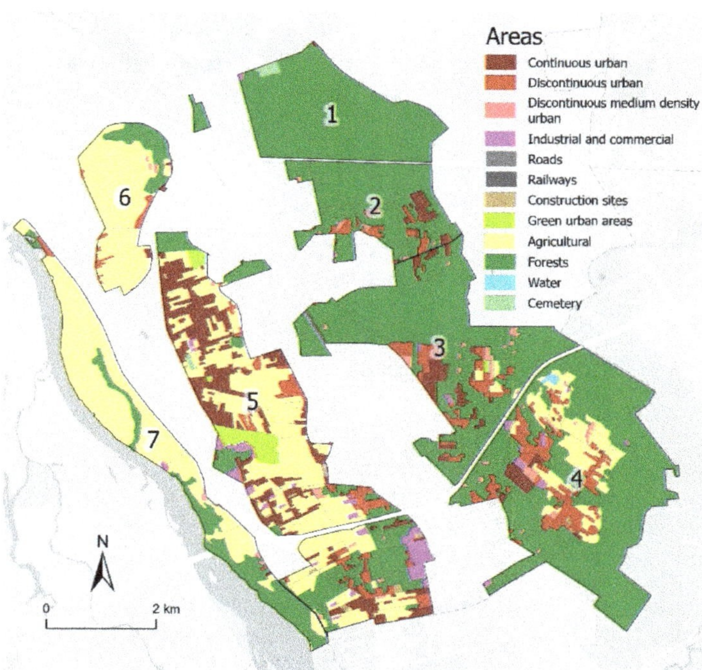

Figure A1. Habitat structure of seven sectors (sectors no. 1–7) in which Wawer (city district of Warsaw) was divided to estimate wild boar density.

References

1. Ritzel, K.; Gallo, T. Behavior Change in Urban Mammals: A Systematic Review. *Front. Ecol. Evol.* **2020**, *8*, 576665. [CrossRef]
2. McCleery, R. Urban Mammals. In *Urban Ecosystem Ecology*; Aitkenhead-Peterson, J., Volder, A., Eds.; American Society of Agronomy: Madison, WI, USA, 2010. [CrossRef]
3. Vuorisalo, T.; Andersson, H.; Hugg, T.; Lahtinen, R.; Laaksonen, H.; Lehikoinen, E. Urban development from an avian perspective: Causes of hooded crow (*Corvus corone cornix*) urbanization in two Finnish cities. *Landsc. Urban. Plan.* **2003**, *62*, 69–87. [CrossRef]
4. Baker, P.J.; Harris, S. Urban mammals: What does the future hold? An analysis of the factors affecting patterns of use of residential gardens in Great Britain. *Mam. Rev.* **2007**, *37*, 297–315. [CrossRef]
5. Acevedo, P.; Farfán, M.Á.; Márquez, A.L.; Delibes-Mateos, M.; Real, R.; Vargas, J.M. Past, present and future of wild ungulates in relation to changes in land use. *Landsc. Ecol.* **2011**, *26*, 19–31. [CrossRef]
6. Massei, G.; Kindberg, J.; Licoppe, A.; Gačić, D.; Šprem, N.; Kamler, J.; Baubet, E.; Hohmann, U.; Monaco, A.; Ozoliņš, J.; et al. Wild boar populations up, numbers of hunters down? A review of trends and implications for Europe. *Pest. Manag. Sci.* **2015**, *71*, 492–500. [CrossRef]
7. Tack, J. *Wild Boar (Sus scrofa) Populations in Europe: A Scientific Review of Population Trends and Implications for Management*; European Landowners' Organization: Brussels, Belgium, 2018; p. 56.
8. Sütő, D.; Heltai, M.; Katona, K. Quality and use of habitat patches by wild boar (*Sus. scrofa*) along an urban gradient. *Biol. Futura* **2020**, *71*, 69–80. [CrossRef] [PubMed]
9. Podgórski, T.; Baś, G.; Jędrzejewska, B.; Sönnichsen, L.; Śnieżko, S.; Jędrzejewski, W.; Okarma, H. Spatiotemporal behavioral plasticity of wild boar (*Sus. scrofa*) under contrasting conditions of human pressure: Primeval forest and metropolitan area. *J. Mam.* **2013**, *94*, 109–119. [CrossRef]
10. Baś, G.; Bojarska, K.; Śnieżko, S.; Król, W.; Kurek, K.; Okarma, H. Habitat use by wild boars *Sus. scrofa* in the city of Kraków. *Chron. Przyr. Ojcz.* **2017**, *73*, 354–362.
11. Stillfried, M.; Fickel, J.; Börner, K.; Wittstatt, U.; Heddergott, M.; Ortmann, S.; Kramer-Schadt, S.; Frantz, A.C. Do cities represent sources, sinks or isolated islands for urban wild boar population structure? *J. Appl. Ecol.* **2017**, *54*, 272–281. [CrossRef]
12. Castillo-Contreras, R.; Carvalho, J.; Serrano, E.; Mentaberre, G.; Fernández-Aguilar, X.; Colom, A.; González-Crespo, C.; Lavín, S.; López-Olvera, J.R. Urban wild boars prefer fragmented areas with food resources near natural corridors. *Sci. Total Environ.* **2018**, *615*, 282–288. [CrossRef]
13. Basak, S.M.; Hossain, M.S.; O'Mahony, D.T.; Okarma, H.; Widera, E.; Wierzbowska, I.A. Public perceptions and attitudes toward urban wildlife encounters—A decade of change. *Sci. Total Environ.* **2022**, *834*, 155603. [CrossRef] [PubMed]
14. Meng, X.J.; Lindsay, D.S.; Sriranganathan, N. Wild boars as sources for infectious diseases in livestock and humans. *Philos. Trans. R. Soc. Lond. B Biol. Sci.* **2009**, *364*, 2697–2707. [CrossRef] [PubMed]
15. Barrios-Garcia, M.N.; Ballari, S.A. Impact of wild boar (*Sus. scrofa*) in its introduced and native range: A review. *Biol. Inv.* **2012**, *14*, 2283–2300. [CrossRef]
16. Stillfried, M.; Gras, P.; Börner, K.; Göritz, F.; Painer, J.; Röllig, K.; Wenzler, M.; Hofer, H.; Ortmann, S.; Kramer-Schadt, S. Secrets of Success in a Landscape of Fear: Urban Wild Boar Adjust Risk Perception and Tolerate Disturbance. *Front. Ecol. Evol.* **2017**, *5*, 157. [CrossRef]
17. Kotulski, Y.; König, A. Conflicts, crises and challenges: Wild boar in the Berlin City–A social empirical and statistical survey. *Nat. Croat.* **2008**, *17*, 233–246.
18. Goulding, M.J.; Roper, T.J. Press responses to the presence of free-living Wild Boar (*Sus. scrofa*) in southern England. *Mam. Rev.* **2002**, *32*, 272–282. [CrossRef]
19. Hafeez, S.; Abbas, M.; Khan, Z.H.; Rehman, E.U. Preliminary analysis of the diet of wild boar (*Sus. scrofa* L., 1758) in Islamabad, Pakistan. *Turk. J. Zool.* **2011**, *35*, 115–118. [CrossRef]
20. Cahill, S.; Llimona, F.; Cabañeros, L.; Calomardo, F. Characteristics of wild boar (*Sus. scrofa*) habituation to urban areas in the Collserola Natural Park (Barcelona) and comparison with other locations. *Anim. Biodivers. Conserv.* **2012**, *35*, 221–233. [CrossRef]
21. Dudzińska, M.; Dawidowicz, A. Severity of Socio-Spatial Conflicts Involving Wild Boars in the City Using Social Media Data. *Sensors* **2021**, *21*, 8215. [CrossRef]
22. Zakład Lasów Poznańskich. *Program Działań Profilaktycznych Miasta Poznania Wobec Zagrożeń Bezpieczeństwa Ludzi Spowodowanych Migracją Dzików na Tereny Zurbanizowane*; Załącznik do uchwały Nr LX/1128/VIII/2022 Rady Miasta Poznania z dnia 8 marca 2022 r; Zakład Lasów Poznańskich: Poznań, Poland, 2021; p. 54.
23. Ciach, M.; Tetkowski, P.; Fedyń, I. Local-scale habitat configuration makes a niche for wildlife encroaching into an urban landscape: Grubbing sites of wild boar *Sus. scrofa* in a city matrix. *Urban Ecosyst.* **2023**, *26*, 629–639. [CrossRef]
24. Jurgiel, K. Rozporządzenie Ministra Rolnictwa I Rozwoju Wsi z dnia 19 lutego 2016 w sprawie zarządzenia odstrzału sanitarnego dzików. *J. Laws* **2016**, *229*.
25. Panek, M.; Budny, M. *Sytuacja Zwierząt Łownych w Polsce—Wyniki Monitoringu*; Stacja Badawcza PZŁ: Czempiń, Poland, 2022.
26. Bobek, B.; Furtek, J.; Wojciuch-Płoskonka, M. Using data from collective hunts to estimate the wild boar (*Sus. scrofa*) population density in north-eastern Poland. *Mamm. Biol.* **2021**, *101*, 933–940. [CrossRef]
27. Gryz, J. Czy należy bać się dzika? *Gaz. Wawerska* **2022**, *55*, 20–21.
28. Gryz, J. Sąd nad dzikami. *Gaz. Wawerska* **2023**, *62*, 6.

29. Kowal, P.; Jasińska, K.; Werka, J.; Ajdysiński, J.; Mierzwiński, J. Odłów dzików jako metoda ograniczenia ich liczebności na terenie Warszawy. *Badania I Rozw. Młodych Nauk. W Polsce. Nauk. Przyr. Część III* **2016**, *5*, 78–85.
30. Migdał, W.; Zając, M.; Rutkowska-Mazur, A.; Migdał, Ł. Dzik—Zagrożenie ASF i zagospodarowanie mięsa. *Przegl. Hod.* **2018**, *4*, 1–5.
31. Morelle, K.; Bubnicki, J.; Churski, M.; Gryz, J.; Podgórski, T.; Kuijper, D. Disease-Induced Mortality Outweighs Hunting in Causing Wild Boar Population Crash After African Swine Fever Outbreak. *Front. Vet. Sci.* **2020**, *7*, 378. [CrossRef]
32. President of Warsaw. *Decision no. 203/OŚ/2024*; President of Warsaw: Warsaw, Poland, 2024.
33. Forest Data Bank. 2024. Available online: https://www.bdl.lasy.gov.pl/portal/en (accessed on 12 July 2024).
34. Zegar, T. (Ed.) *Rocznik Statystyczny Warszawy*; Statistical Office in Warsaw: Warsaw, Poland, 2023.
35. Jackowiak, M.; Gryz, J.; Jasińska, K.; Brach, M.; Bolibok, L.; Kowal, P.; Krauze-Gryz, D. Colonization of Warsaw by the red fox *Vulpes vulpes* in the years 1976–2019. *Sci. Rep.* **2021**, *11*, 13931. [CrossRef]
36. Jasińska, K.D.; Krauze-Gryz, D.; Jackowiak, M.; Gryz, J. Changes in roe deer (*Capreolus capreolus*) daily activity patterns in Warsaw during the COVID-19 pandemic. *Eur. Zool. J.* **2022**, *89*, 870–876. [CrossRef]
37. Zegar, T. (Ed.) *Panorama Dzielnic Warszawy*; Statistical Office in Warsaw: Warsaw, Poland, 2021.
38. Krauze-Gryz, D.; Jackowiak, M.; Klich, D.; Gryz, J.; Jasińska, K.D. Following urban predators—Long-term snow-tracking data reveals changes in their abundance and habitat use. *J. Zool.* **2024**, *323*, 213–224. [CrossRef]
39. Stephens, P.A.; Zaumyslova, O.Y.; Miquelle, D.G.; Myslenkov, A.I.; Hayward, G.D. Estimating population density from indirect sign: Track counts and the Formozov–Malyshev–Pereleshin formula. *Anim. Conserv.* **2006**, *9*, 339–348. [CrossRef]
40. Hofmeester, T.R.; Rowcliffe, J.M.; Jansen, P.A. A simple method for estimating the effective detection distance of camera traps. *Remote Sens. Ecol. Conserv.* **2017**, *3*, 81–89. [CrossRef]
41. Merli, E.; Meriggi, A. Using harvest data to predict habitat-population relationship of the wild boar *Sus. scrofa* in Northern Italy. *Acta Theriol.* **2006**, *51*, 383–394. [CrossRef]
42. Romanowski, J. Vistula river valley as the ecological corridor for mammals. *Pol. J. Ecol.* **2007**, *55*, 805–819.
43. Ciach, M.; Fröhlich, A. Ungulates in the city: Light pollution and open habitats predict the probability of roe deer occurring in an urban environment. *Urban. Ecosyst.* **2019**, *22*, 513–523. [CrossRef]
44. Krauze-Gryz, D.; Gryz, J.; Wrotek, N.; Zosicz, M. Occurrence of selected mammal species in the Pole Mokotowskie Park in Warsaw. *Stud. Geogr.* **2020**, *61*, 97–103.
45. Krauze-Gryz, D.; Lesiński, G.; Pieniążek, A.; Romanowski, J.; Owsianka, M.; Prus, M. Ssaki Parku Skaryszewskiego w Warszawie. In *Park Skaryszewski w Warszawie—Przyroda i Użytkowanie*; Romanowski, J., Ed.; OKSW: Warsaw, Poland, 2016; pp. 215–226.
46. Redakcja Warszawa. Dziki w Ogrodzie Krasińskich w Centrum Warszawy, *Nasze Miasto*. 2023. Available online: https://warszawa.naszemiasto.pl/dziki-w-ogrodzie-krasinskich-w-centrum-warszawy-miasto/ar/c1-9300841 (accessed on 12 July 2024).
47. Mysłajek, R.W.; Stachyra, P.; Figura, M.; Nowak, S. Drive counts fail to accurately estimate the population sizes of wild ungulates. *J. Vert. Biol.* **2024**, *73*, 1–7. [CrossRef]
48. Taxus. *Inwentaryzacja Zwierzyny Grubej na Terenie Powiatu Iławskiego z Wykorzystaniem Foto-Pułapek—Raport z Wykonania Usługi*; Taxus SI: Warszawa, Poland, 2022; pp. 1–27. Available online: https://www.olsztyn.lasy.gov.pl/documents/688368/42993017/Raport+inwentaryzacja/1e0c240d-96f9-e4b4-8dcd-57a9ac4d8d75 (accessed on 12 July 2024).
49. Gniewek-Juszczak, E. Lubuskie Prekursorem w Kraju. To u Nas Pilotażową Metodą Policzono Dziki. *Gazeta Lubuska*. 2023. Available online: https://gazetalubuska.pl/lubuskie-prekursorem-w-kraju-to-u-nas-pilotazowa-metoda-policzono-dziki-inwentaryzacja-jest-konieczna-aby-skutecznie-walczyc-z/ar/c8-17387233 (accessed on 25 December 2023).
50. Ministry of Environment. Inwentaryzacja Dzików Zakończona. 2016. Available online: https://www.lasy.gov.pl/pl/informacje/aktualnosci/inwentaryzacja-dzikow-zakonczona (accessed on 12 July 2024).
51. Enetwild Consortium; Keuling, O.; Sange, M.; Acevedo, P.; Podgorski, T.; Smith, G.; Scandura, M.; Apollonio, M.; Ferroglio, E.; Body, G.; et al. Guidance on estimation of wild boar population abundance and density: Methods, challenges, possibilities. *EFSA Support. Publ.* **2018**, *15*, EN-1449. [CrossRef]
52. Błażejczyk, K.; Kuchcik, M.; Milewski, P.; Dudek, W.; Kręcisz, B.; Błażejczyk, A.; Szmyd, J.; Degórska, B.; Pałczyński, C.M. *Miejska Wyspa Ciepła w Warszawie: Uwarunkowania Klimatyczne i Urbanistyczne*; Wydawnictwo Akademickie SEDNO Spółka z oo.: Warsaw, Poland, 2014.
53. Geisser, H.; Reyer, H.-U. The influence of food and temperature on population density of wild boar *Sus scrofa* in the Thurgau (Switzerland). *J. Zool.* **2005**, *267*, 89–96. [CrossRef]
54. Stillfried, M.; Gras, P.; Busch, M.; Börner, K.; Kramer-Schadt, S.; Ortmann, S. Wild inside: Urban wild boar select natural, not anthropogenic food resources. *PLoS ONE* **2017**, *12*, e0175127. [CrossRef]
55. Guerrasio, T.; Brogi, R.; Marcon, A.; Apollonio, M. Assessing the precision of wild boar density estimations. *Wildl. Soc. Bull.* **2022**, *46*, e1335. [CrossRef]
56. Lee, S.M.; Song, E.; Choi, T.Y. Comparison of three indirect methods for density estimation of the wild boar (*Sus. scrofa*) in different habitats in South Korea. *Eur. Zool. J.* **2024**, *91*, 674–686. [CrossRef]

57. ENETwild Consortium. Harmonization of the use of hunting statistics for wild boar density estimation in different study areas. *EFSA Support. Publ.* **2019**, *16*, EN-1706. [CrossRef]
58. Conejero, C.; Castillo-Contreras, R.; González-Crespo, C.; Serrano, E.; Mentaberre, G.; Lavín, S.; López-Olvera, J.R. Past experiences drive citizen perception of wild boar in urban areas. *Mamm. Biol.* **2019**, *96*, 68–72. [CrossRef]

forests

MDPI

Article

The Impacts of Sunlight on the Lichen Scots Pine Forest Community

Patrycja Fałowska [1,2], Patrycja Dziurowicz [1,2], Karolina Waszkiewicz [1,2], Paulina Wietrzyk-Pełka [1] and Michał Hubert Węgrzyn [1,*]

[1] Laboratory of Polar Research, Institute of Botany, Faculty of Biology, Jagiellonian University, Gronostajowa 3, 30-387 Kraków, Poland; patrycja.falowska@uj.edu.pl (P.F.); patrycja.dziurowicz@uj.edu.pl (P.D.); karolinaa.waszkiewicz@uj.edu.pl (K.W.); paulina.wietrzyk@gmail.com (P.W.-P.)

[2] Doctoral School of Exact and Natural Sciences, Jagiellonian University, Prof. S. Łojasiewicza 11, 30-348 Kraków, Poland

* Correspondence: michal.wegrzyn@uj.edu.pl

Abstract: The habitat of the lichen Scots pine forest is currently one of the communities that requires attention and active protection due to its rapid disappearance. In our study, we identified factors that can be modified to preserve this habitat. The primary focus of our research was on the significance of tree height and sunlight in fostering the development of a lichen-rich layer within the lichen Scots pine forest. Additionally, we investigated the environmental factors influencing the presence of specific species of lichens and bryophytes, including those that form communities and those that displace species characteristic of the lichen Scots pine forest community. Our study was conducted in Bory Tucholskie National Park (N-W Poland) using 20 experimental sites and 248 point-frame plots from spring 2021 to winter 2022. Fieldwork involved species surveys, measurements of photosynthetic activity in lichens and bryophytes, and collection of environmental data. Through parametric testing, modeling, and mapping, our main findings confirmed that light availability, influenced by tree height, is a critical factor in maintaining a well-preserved lichen-rich layer and facilitating habitat reintroduction.

Keywords: *Cladonio-Pinetum* complex; tree height; environmental factors; Bory Tucholskie National Park; photosynthetic activity; habitat reintroduction; lichens; bryophyte

1. Introduction

Photosynthesis is regulated by numerous environmental factors that not only limit steady-state photosynthesis but also constrain the speed of response to environmental fluctuations [1–3]. Under natural conditions, photosynthesis occurs in fluctuating irradiance. We define this process as dynamic photosynthesis, regulated by the kinetics of underlying processes and interspecific variation in response to fluctuating irradiance [2,4].

The effect of light on photoautotrophic species can be assessed in direct terms (i.e., how light directly affects the photosynthetic system of the plant) and indirectly (i.e., how this solar energy also in the form of thermal energy shapes the ecosystem conditions, including water conditions). The amount of energy that reaches the ecological niche in a community is largely shaped by the structure of the plant community itself. In the case of the community of the lichen Scots pine forest, it is a specifically developed strongly deformed Scots pine stand on a poor, sandy base with a lichen-rich undergrowth layer. The lichens that occur there are light-requiring organisms, adapted to high values of sunlight intensity, as evidenced by the presence of the same species of lichens in open habitats, for example, alpine or tundra communities [5,6]. Previous studies have shown that an important environmental factor limiting the occurrence of undergrowth is the thickness of organic matter on the forest substrate [7]. However, an equally important factor is the height of the pine stand, which changes a lichen-rich undergrowth layer into a bryophyte-rich undergrowth layer. This relationship was believed to be related to the limiting impact of the pine stand on the availability of solar energy in the forest undergrowth layer [7]; however, this has not been empirically proven in research to date.

Since 2015, studies have been carried out in Bory Tucholskie National Park on 13 hectares of lichen Scots pine forest habitat belonging to the *Cladonio-Pinetum* Juraszek 1927 community [7–9]. In 2020, a pilot study was conducted focusing on the determination of the degree of viability of selected species of lichens and bryophytes with respect to environmental conditions, using the method of assessing photosynthetic activity [9].

The region where the research took place is marked by severely impoverished and arid conditions, linked to a particular substrate created during the last ice age (consisting of sandy soils in the valleys of major rivers and post-glacial outwash sands). Over time, this environment has been influenced by various human activities, including cattle grazing, the collection of litter (such as pine needles, pinecones, and small branches), deforestation, and clear-cutting [7,8].

On the one hand, previous studies [9] have observed a certain repeatability of photosynthetic activity of the studied species in relation to the seasons and time of day; on the other hand, clearly the conservation status of the lichen undergrowth was correlated with the height of the trees. So, in the current study, which has been conducted through the use of automatic measuring devices (loggers) distributed in dozens of sites, an attempt was made to clarify the mechanisms of the effect of the sun per year in relation to the height of the trees on the photosynthetic activity of lichens and bryophytes in the undergrowth of the brushwood forest.

The aim of the presented research was to study how sunlight identified as a source of PAR (Photosynthetic Active Radiation) and thermal energy shape the lichen-rich undergrowth layer in the lichen Scots pine forest community. Due to the structure of this community with pine stands of different age and height, the amount of light reaching the undergrowth differs. Therefore, we assume that both the height of the stand and the amount of sunlight reaching the forest floor influence the development of the lichen-rich undergrowth layer. Apart from environmental conditions, lichen photosynthesis is closely related to the photosynthetic abilities of the photobiont, i.e., an autotrophic prokaryotic or eukaryotic alga [10–13]. On the other hand, in bryophytes, it occurs in the gametophyte, which is the dominant generation, additionally nourishing the sporophyte [14]. Furthermore, we studied the influence of environmental variables on certain species of lichens and bryophytes present in the study area. We set the following hypotheses: (1) an increase in the percentage cover of the lichen-rich undergrowth layer is directly proportional to the increased amount of light reaching the substrate, which is dependent on the tree height; (2) an increase in the amount of light decreases the percentage cover of the bryophyte-rich undergrowth layer.

2. Material and Methods

2.1. Study Area

Fieldwork was carried out for one year from spring 2021 to winter 2022 in the 'Bory Tucholskie' National Park (Tuchola Forest region, Nort-West Poland) (Figure 1). In 2017 in the north part of the National Park, a study area (13 ha) of the lichen Scots pine forest was designated that included 20 experimental localities designated in 2017 and an additional 248 point-frame plots analyzed in 2021 (Figure 1).

2.2. Lichen Undergrowth Layer and Trees Height Mapping

In 2018, an aerial LIDAR scan of the stand was performed, which was used to make a tree crown model [7,15]. Based on the model, tree heights (THs) were calculated, which were grouped into three height ranges (1–9 m, 10–13 m, and 14–18 m) and presented on the map (Figure 2A). Data for crown areas (CAs) and tree cover (TC-2017) were obtained from the same tree crown model. These data were used to perform later statistical analysis.

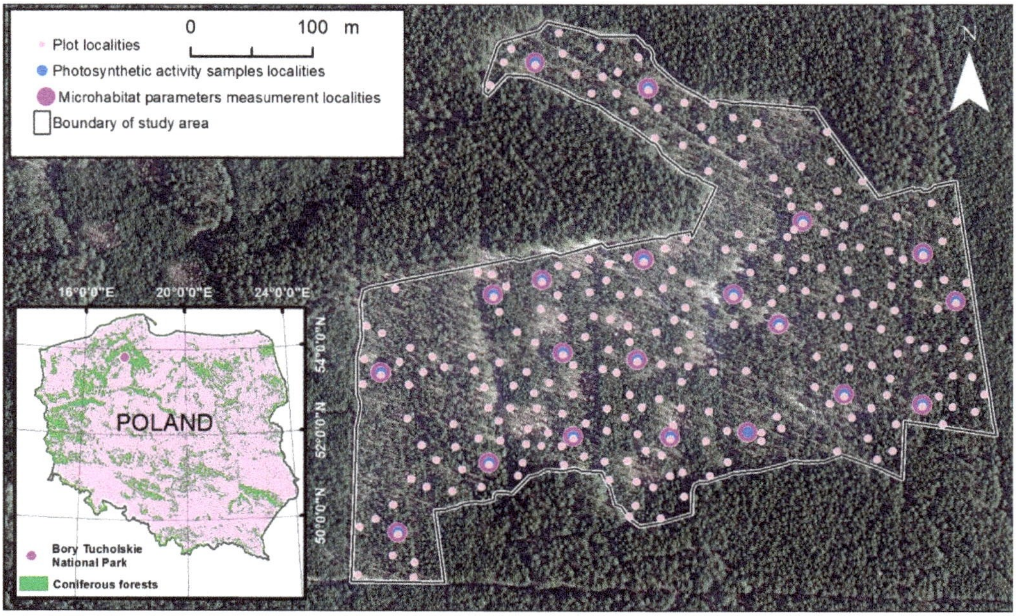

Figure 1. Study area in the north part of Bory Tucholskie National Park in Poland (white line) of the lichen Scots pine forest community with 20 experimental localities (violet dots) having automatic substrate temperature and light intensity loggers; among them, 20 localities carried out the measurement of the chosen cryptogams species' photosynthetic activity [Fv/Fm] (the maximal quantum yields of photosystem II) (blue dots) and 248 new localities were point-framing plots (pink dots). Orthophotomap layer by © The Head Office of Geodesy and Cartography (www.geoportal.gov.pl 2022).

Utilizing the random points tool within the ArcGIS software 10.8.2 (ESRI, Redlands, CA, USA, 2020) across the study area, with the exclusion of zones where all undergrowth cover was cleared in 2018 as part of a prior project, a total of 248 sites were delineated, forming the shapefile layer of the point-frame (Figure 1). In the field, randomly determined positions were located with a precision of 1 m using an iPad device equipped with the MapPlus 2.8.20 application (Miocool Inc., 2021). For this purpose, an aluminum frame was used, divided by lines into 100 squares with dimensions of 5 by 5 cm. The frame was placed at the designated site and then each species that was at the intersection of the lines was noted as one percent. The records for each species were summed, and thus the percentage cover of each species in each plot was calculated. In the case that there were no species at the intersection, the zero value was observed. Then, for each plot, the percentage cover for all lichen and bryophyte species was summed. Based on the obtained results, a geodatabase of the point-frame shapefile layer was created. This layer was used to create the percentage cover of lichens in the Scots pine forest undergrowth layer model (Figure 2B). Both models are presented as maps (Figure 2) for visual comparison.

2.3. Measurement Microhabitat Variables

Each of the 20 experimental localities had automatic loggers (HOBO MX2202 Underwater Temp/Light, Version Number: 140.59) measuring the temperature of the substrate [°C] and the intensity of sunlight [lux] at 6:00 a.m., 9:00 a.m., 12:00 a.m., 3:00 p.m., and 6:00 p.m. from 1 June 2021 to 31 February 2022. For presenting the results of statistical analyses, measurements taken on the days of measuring the photosynthetic activity of selected species of lichens and bryophytes were used. For selected microhabitat variables that had a statistically significant effect, we carried out an interpolation modeling process using the Kriging Tool in the ArcGIS ArcMap application.

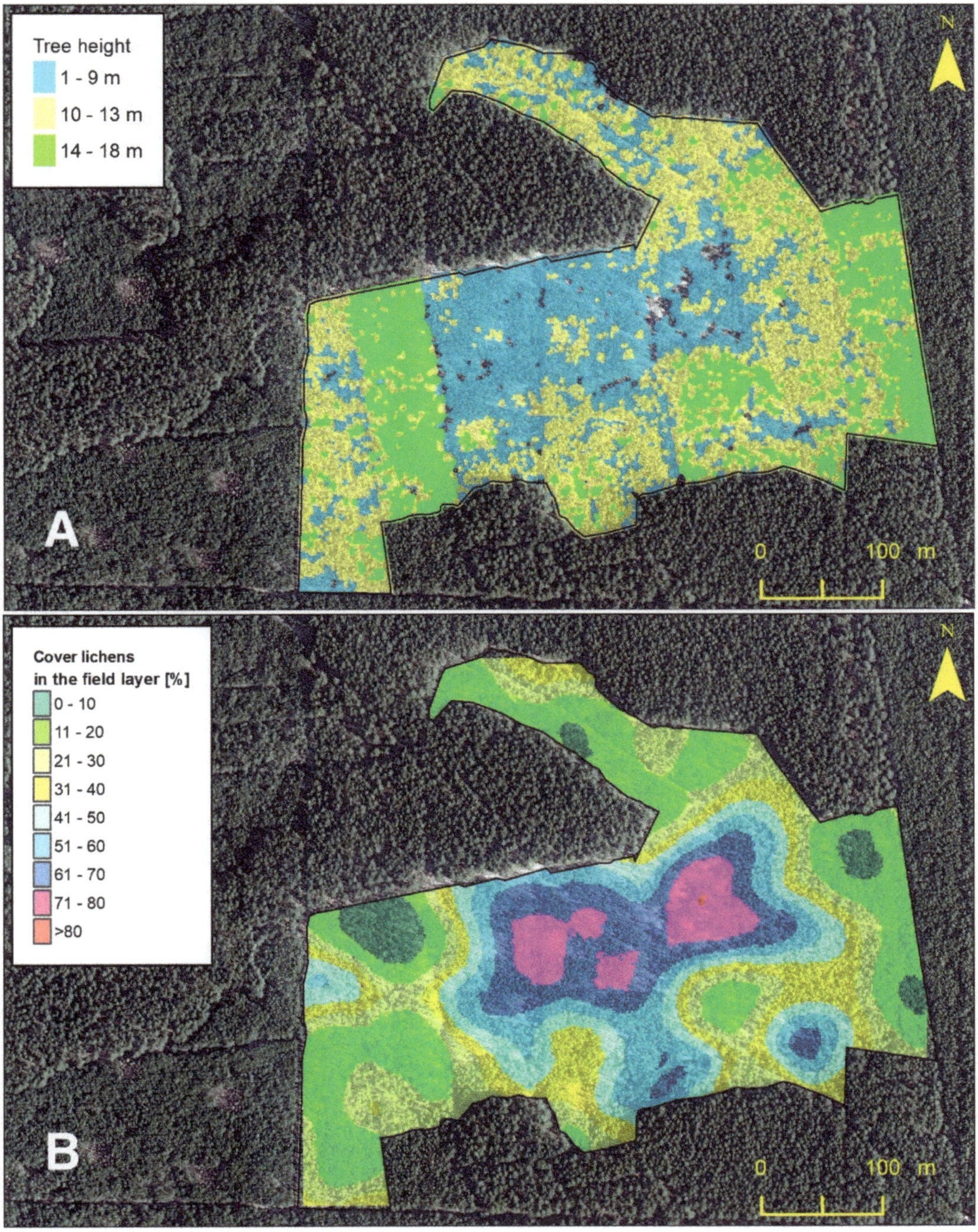

Figure 2. Map of the tree height model (**A**) and map of the percentage cover lichens in the Scots pine forest undergrowth layer model (**B**).

2.4. Measurement of Photosynthetic Activity

All 20 experimental localities contained 4 selected species of lichens (*Cladonia mitis* Sandst; *Cladonia uncialis* (L.) Weber ex F.H. Wigg.) and bryophytes (*Pleurozium schreberi* (Willd. ex Brid.) Mitt.; *Dicranum scoparium* (L.) Hedw.) (Figure 1). We chose 3 days each in the months representing the four seasons found in this part of Europe. The choice of days was based on weather conditions, so that the days of conducting field work were preceded by stabilized weather characteristic of the season. In these places in the outlined seasons (spring (23–25 April 2021); summer (25–27 June 2021); autumn (1–3 October 2021); and winter (4–6 February 2022), we randomly selected individuals from species within the plot and carried out measurements of photosynthetic activity [Fv/Fm] (the maximal quantum yields of photosystem II) at designated times of the day (6:00 a.m.; 9:00 a.m.; 12:00 a.m.; 3:00 p.m.; and 6:00 p.m.). Twenty samples of each species were simultaneously collected on a designated day and at a specified time. Each sample, approximately 2 g in weight, was carefully deposited into conical tubes containing 5 milliliters of rainwater. Subsequently, the samples were subjected to a hydration period lasting 2 h. The objective of this thallus watering procedure was to ensure a uniform hydration level across all samples, thus standardizing the experimental conditions.

Measurements were conducted utilizing a Handy PEA+ fluorometer manufactured by Hansatech Instrument Ltd., located in King's Lynn, Norfolk, UK. The prompt fluorescence (PF) method [16] was employed for data acquisition. Prior to measurements, each sample underwent a dark adaptation phase lasting approximately 15 min [17], facilitating the quenching of the light phase of photosynthesis reactions. Upon completion of the adaptation phase, samples were subjected to continuous light exposure with wavelengths shorter than 670 nm. Chlorophyll fluorescence emitted within the range of 680 to 760 nm was then recorded by the photodetector. Post-measurement, the Fv/Fm ratio was determined, representing the maximum photochemical efficiency of photosystem II. This ratio is widely acknowledged as a robust indicator of the photochemical activity within the photosynthetic apparatus [18].

2.5. Statistical Analysis

The following variables were selected to determine the relationship between tree height and lichens: (a) percentage cover of lichen ($N = 248$); (b) percentage cover of *Cladonia mitis* ($N = 248$); (c) percentage cover of *C. gracilis* (L.) Willd. ($N = 248$) (both lichen species characteristic of the *Cladonio-Pinetum* complex); (d) percentage cover of *P. schreberi* ($N = 248$) (bryophyte species were additionally taken into account due to their expansive nature causing it to occupy areas after retreating lichens). Pearson's r correlation analysis was performed for an equal sample size ($p < 0.05$). Scatterplots were presented to show the relationship between the various variables (Figure 3).

To investigate the relationship between lichen cover and microhabitat variables ($N = 248$), i.e., substrate temperature (Table 1a) and light intensity (Table 1b) in a seasonal approach (summer, autumn, and winter), multiple regression analysis was performed for an equal sample size ($p < 0.05$). The results obtained are presented in Table 1.

To investigate the relationship between the photosynthetic activity of selected species (*C. mitis*, *C. uncialis*, *D. scoparium*, and *P. schreberi*) in a seasonal approach (summer, autumn, and winter) at different times (9:00 a.m., 12:00 a.m., and 15:00 p.m.); the temperature in a seasonal approach (summer, autumn, and winter) at different times (9:00 a.m., 12:00 a.m., and 15:00 p.m.); light intensity in a seasonal approach (summer, autumn, and winter) at different times (9:00 a.m., 12:00 a.m., and 15:00 p.m.); and the height of trees, multiple regression analysis was performed for an equal sample size ($p < 0.05$). Modeling was not performed for the spring period due to the lack of measurements of microhabitat parameters. A total of 36 models were used (for the photosynthetic activity value of four species and three variables) where one data series counted $N = 248$ (total $N = 11,904$). The results obtained are presented in Supplementary Table S1.

Table 1. Results of stepwise multiple regression analysis for the effect of two microhabitat parameters: (a) substrate temperature (r = 0.61039, r^2 = 0.37258371, F = 48.2988, *p < 0.001*) and (b) light intensity (r = 0.599672, r^2 = 0.359607, F = 68.79, *p < 0.001*) in a seasonal approach (summer, autumn, and winter) on lichen cover. The constant tree height parameter was also included in both analyses. Variables with significant effect (*p < 0.05*) are provided in bold. Variables are listed according to *p*-value.

	Variables	B	*t*	*p*
(a)	Tree height	**−7.74**	**−10.147**	***p* < 0.001**
	ASTA	**65.31**	**3.0355**	***p* = 0.003**
	ASTW	**−95.19**	**−2.419**	***p* = 0.016**
(b)	Tree height	**−8.15**	**−11.123**	***p* < 0.001**
	ALIA	**0.009**	**2.025**	***p* = 0.044**

As a result of photosynthetic activity measurements, the collected results (N = 2880) were statistically analyzed by a two-way analysis of variance (ANOVA) test followed by Tukey's HSD (Honestly Significant Difference) test for an equal sample size ($p < 0.05$). The test was performed to reveal Fv/Fm differences between the following pairs of variables: (a) species (*C. mitis, C. uncialis, D. scoparium,* and *P. schreberi*) and season (spring, summer, autumn, and winter), and (b) species and time of day depending on the season (9:00, 12:00, and 15:00) (Table 2). Before the analysis, the normality of the distribution was verified using the Kolmogorov–Smirnov test ($p > 0.05$) and the Levene test ($p > 0.05$) to assess equality of variance. Box-and-whisker plot charts were presented to illustrate the differences (Figure 4).

Table 2. Results of the two-way ANOVA assessing the effects of the following variables on cryptogams' Fv/Fm: (a) species (*Cladonia mitis* Sandst; *Cladonia uncialis* (L.) Weber ex F.H. Wigg., *Dicranum scoparium* (L.) Hedw., *Pleurozium schreberi* (Willd. ex Brid.) Mitt.) and seasons (spring, summer, autumn, and winter); (b) species and daytime (8:00 a.m., 12:00 p.m., and 3:00 p.m.) during spring, (c) summer, (d) autumn, and (e) winter. Significant effects ($p < 0.05$) are denoted in bold.

	Variables	F	*p*
(a)	Species	**150.6**	***p* < 0.001**
	Season	**460.2**	***p* < 0.001**
	Species x season	**16.4**	***p* < 0.001**
(b)	Species	**96.2**	***p* < 0.001**
	Daytime in spring	**12.15**	***p* < 0.001**
	Species x daytime in spring	**7.61**	***p* < 0.001**
(c)	Species	**81.18**	***p* < 0.001**
	Daytime in summer	0.76	*p* = 0.47
	Species x daytime in summer	**3.33**	***p* = 0.003**
(d)	Species	**30.6**	***p* < 0.001**
	Daytime in autumn	2.7	*p* = 0.069
	Species x daytime in autumn	**11.0**	***p* < 0.001**
(e)	Species	**13.93**	***p* < 0.001**
	Daytime in winter	**197.01**	***p* < 0.001**
	Species x daytime in winter	0.41	*p* = 0.875

To investigate the relationship between species of lichens and bryophytes, forest variables, and microhabitat variables, a canonical correspondence analysis (CCA), preceded by random Monte Carlo permutation tests, was performed using Canoco for Windows Version 4.51 (C) 1997–2003 Biometric—Plant Research International Wageningen, The Netherlands, and CANOCO CanocoDraw for Windows 4.1 (C) 1999–2003 Petr Smilauer. Statistical tests were performed using the StatSoft STATISTICA program (data analysis software system), version 13.3. The abbreviations of variables are explained in Table 3.

3. Results

3.1. Influence of Microhabitat Parameters on Photosynthetic Activity and Cover of Lichens and Bryophytes

By comparing the graphic interpretation of the lichen/bryophyte undergrowth layer model based on the quantitative data of the species of lichens and bryophytes and the tree height model, some relationships were found visually showing the presence of a large percentage cover of lichens in the undergrowth layer in low (1–9 m) and medium (10–13 m) places of the pine stand (Figure 2). On the other hand, the bryophyte-rich undergrowth layer was dominant in places where the pine stand reached the highest values (14–18 m) (Figure 2).

The linear relationship between tree height value, lichen cover value, and selected species of lichen and bryophyte cover value showed a strong linear correlation (Figure 3). The strongest relationship was shown between tree height and lichen cover in the undergrowth layer (Figure 3A). The cover of lichens in the forest undergrowth layer decreases with increasing tree height ($r = -0.59$). Slightly lower, the same as the previous, inverse correlations also occurred between tree height and percentage cover of *C. mitis* ($r = -0.41$) and *C. gracilis* ($r = -0.4$), the two most important habitat-forming species. In the case of *P. schreberi*, which competes strongly with lichens, a positive correlation with tree height occurred ($r = 0.43$).

Since strong relationships between tree height and percentage cover of lichens in the forest undergrowth layer were found, additionally, multiple regression analysis was performed that included data from three seasons (summer, autumn, and winter) on microhabitat parameters: average substrate temperature and average light intensity on the undergrowth surface and tree height (Table 1). Based on the multiple regression analysis performed, it was observed that the variables TH, ASTA, and ASTW (Table 1a) and TH and ALIA (Table 1b) had a large influence on the shaping of the lichen-rich undergrowth layer. The results of the next multiple regression analysis showed the relationship between the Fv/Fm of selected species (*C. mitis*, *C. uncialis*, *D. scoparium*, and *P. schreberi*) and the substrate temperature and light intensity microhabitat parameters in the aspects of daytime and seasons (Supplementary Table S1). In total, 36 models of the activity relationship of four species were made in each case for three variables, 13 of which showed statistical significance (Supplementary Table S1). For nine statistically significant models, a very high value of $r > 0.5$ of the regression coefficients was found, and for four, it was moderately strong $0.0 < r > 0.3$ (Supplementary Table S1).

In terms of species, relationships between the Fv/Fm of these species and the microhabitat parameters in terms of seasons and daytime were found to be statistically significant for the following: *C. mitis*—two out of nine results (autumn 9:00 a.m. and 12:00 a.m.); *C. uncialis*—five out of nine results (summer 9:00 a.m. and 12:00 a.m., autumn 9:00 a.m. and 3:00 p.m., and winter 9:00 a.m. and 3:00 p.m.); *D. scoparium*—two out of nine results (autumn 12:00 a.m. and winter 3:00 p.m.); and *P. schreberi*—three out of nine results (summer 9:00 a.m. and 12:00 a.m., and autumn 9:00 a.m.) (Supplementary Table S1).

3.2. Seasonal Variations in Photosynthetic Activity of Selected Lichen and Bryophyte Species

The outcomes of the two-way analysis of variance (ANOVA) demonstrated significant impacts of species, season, and their interaction on Fv/Fm values (see Table 2; depicted in Figure 4A). Across all examined species, Fv/Fm values reached their lowest during the spring and summer seasons, exhibiting an upward trend during autumn and winter (Figure 4A). Notably, during both spring and autumn, the Fv/Fm values of both lichens and *D. scoparium* demonstrated similarity and were significantly divergent from those of *P. schreberi* (Figure 4A). Moreover, during autumn, the Fv/Fm values for all species under investigation approached stabilization, resembling the levels observed during winter (Figure 4A).

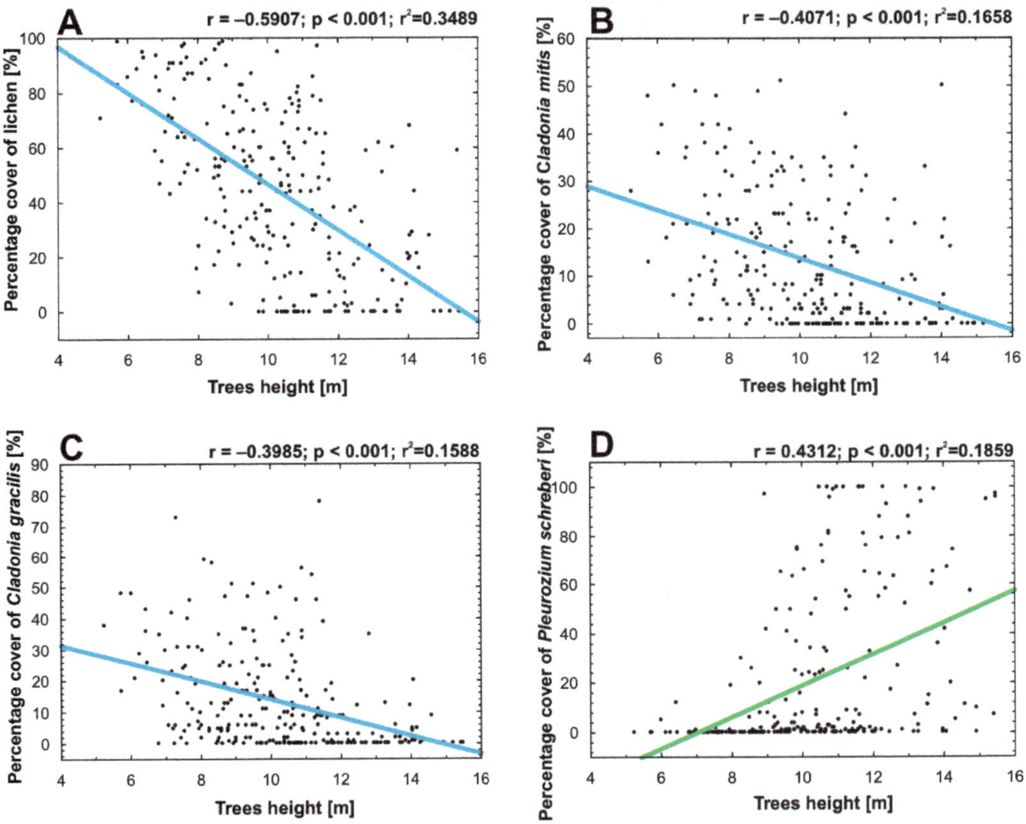

Figure 3. Linear relationship between the Scots pine trees' height in each point-framing plot and (**A**)—the percentage cover of lichens in the plots; (**B**)—the percentage cover of *Cladonia mitis* Sandst in the plots; (**C**)—the percentage cover of *Cladonia gracilis* (L.) Willd. in the plots; and (**D**)—the percentage cover of *Pleurozium schreberi* (Willd. ex Brid.) Mitt. in the plots.

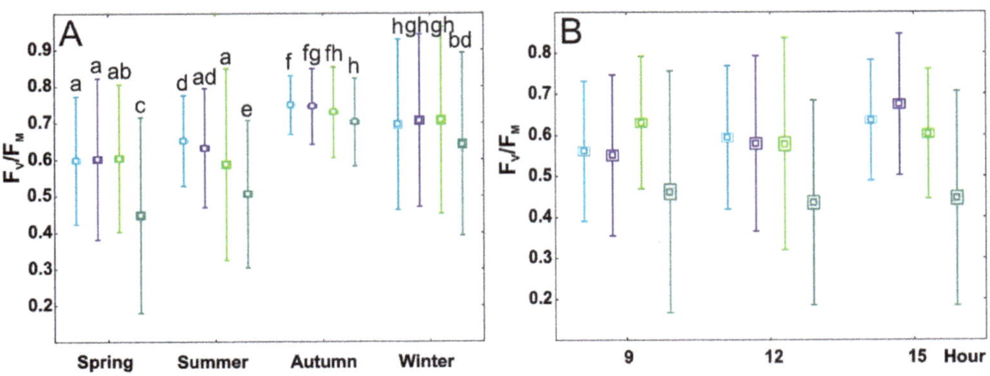

Figure 4. *Cont.*

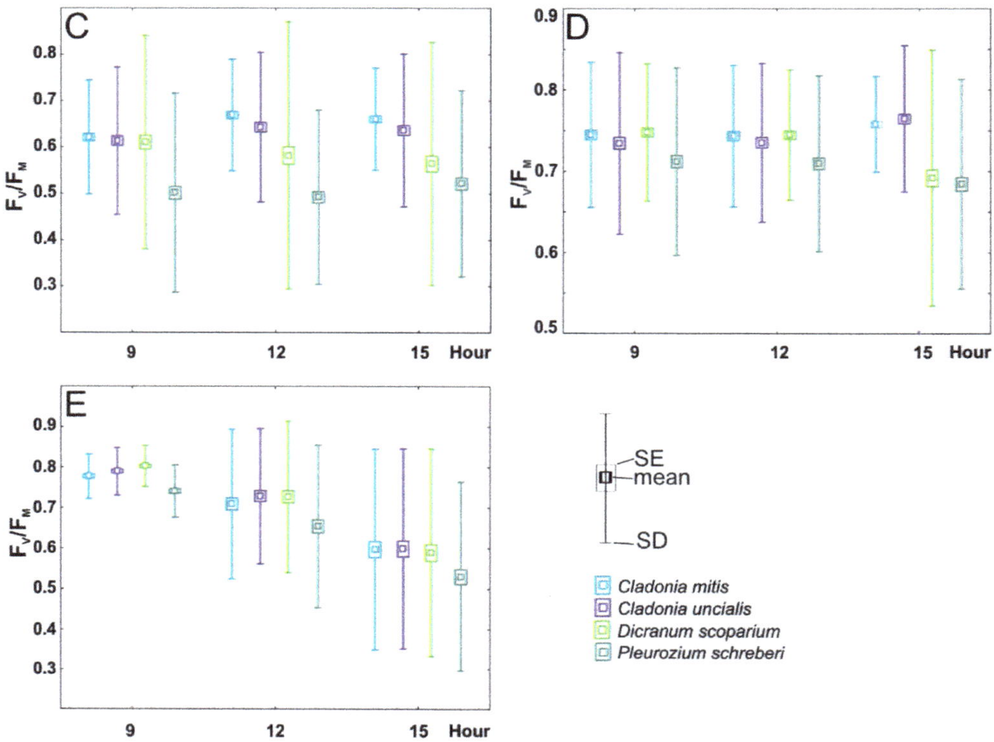

Figure 4. Variability of Fv/Fm (mean ± SE and SD) including division into species (*C. mitis* Sandst, *C. uncialis* Weber ex F.H. Wigg., *D. scoparium* (L.) Hedw., and *P. schreberi* (Willd. ex Brid.) Mitt.) in relation to the following: (**A**)—season (spring, summer, and autumn); (**B**)—daytime in spring; (**C**)—daytime in summer; (**D**)—daytime in autumn; (**E**)—daytime in winter. The lowercase letters in section A denote the statistically significant interaction among species, season, and daytime.

In terms of the impact of species and daytime, both variables exerted influence on Fv/Fm; nonetheless, their interaction was not statistically significant during the winter daytime (refer to Table 2 and Figure 4E). The values of Fv/Fm in the daytime of each season differed; however, the results of daytime in summer and autumn were not significant (Table 2, Figure 4C,D). The values of Fv/Fm for daytime in summer show a slight upward trend during the day, reaching the highest values in the afternoon (Figure 4B), while in the case of the winter, the situation was the opposite, with Fv/Fm reaching the highest values in the morning (Figure 4E). For *P. schreberi*, the Fv/Fm was significantly lower compared to that of *C. mitis*, *C. uncialis*, and *D. scoparium* (see Figure 4B,C,E).

3.3. The Relationship between Species of Lichens and Bryophytes, Forest Variables, and Microhabitat Variables

The results of the forward selection and Monte Carlo permutation tests showed that 14 of the 37 variables analyzed are statistically significant (Figure 5, Table 3). Variables were ordered according to their decreasing importance to the presence of lichen species in the study area, as shown by lambda statistics (Table 3). The CCA diagram showed the relationship between species of lichens and bryophytes, forest variables (TH, TC-2017, and CA) and microhabitat variables (ALIS, ALIA, ALIW, LIS_6, LIA_{15}, LIW_{12}, LIW_{15}, STS_6, STA_6, STS_{12}, and STA_{12}). Most of the microhabitat variables in the diagram were parallel to the second axis. This means that this axis reflects the gradient of variable microhabitat parameters on the percentage cover of lichens and bryophytes in the forest undergrowth layer. Additionally, the microhabitat variables were mostly grouped in one area, especially characterized by light intensity (Figure 5). Their ranges of individual values are shown on additional maps (Figure 6). These variables, by increasing their values, increase the percentage cover of lichen species of minor habitat formation in the forest undergrowth layer, such as the following: *C. pleurota* Hoffm., *C. pyxidata* (L.) Hoffm., *C. squamosa* Hoffm., and *C. verticillata* (Hoffm.) Ach. Two species of bryophytes, *H. splendens* (Hedw.) Schimp. and *D. spurium* Hedw., were also found in this influence area. The microhabitat variables above are also accompanied by the CA forest variable. The remaining microhabitat variables STS6, STA6, and STA12 (Figure 5) are located almost parallel to axis 2, but in the opposite direction, and range values were also presented on additional maps (Figure 7). In this case, the increase in their values influenced the habitat-forming lichens *C. portentosa* (Dufour) Coem., *C. uncialis*, *C. subulata* (L.) F.H. Wigg., *C. furcata* (Huds.) Baumg., and *C. rangiferina* (L.) Weber, and those that are insignificant to the community, *C. rei* Schaer., *C. deformis* (L.) Hoffm., *C. zopfii* Vain., and *C. phyllophora* Hoffm., as well as that of two bryophytes, *D. scoparium* and *P. piliferum* Hedw. The primary axis appears to be primarily determined by Tree Height (TH), with a significant influence from the outlier *Cetraria aculeata* exhibiting low TH values. This observation aligns with the ecological preferences of *Cetraria aculeata* (Schreb.) Fr., which thrives in open habitats. The increase in their value clearly affects the two main species of the bryophyte-rich undergrowth layer, *P. schreberi* and *D. polysetum* Sw., while their decrease affects the main species of the lichen-rich undergrowth layer, *C. mitis*, *C. arbuscula* (Wallr.) Flot., and *C. gracilis* (Figure 5).

Table 3. Results of forward selection and Monte Carlo permutation tests derived from CCA (refer to Figure 5). Environmental variables are arranged according to their model order (Lambda A). Significant variables are indicated in bold ($p < 0.05$). Abbreviations for variables were utilized in multiple regression analyses (refer to Table 1), CCA (refer to Figure 5), light intensity models (refer to Figure 6), and substrate temperature models (refer to Figure 7).

Variable	Abbreviation	Lambda A	*p*	F
Tree height	**TH**	**0.22**	**0.002**	**14.68**
Light intensity summer—6:00	**LIS6**	**0.07**	**0.002**	**4.92**
Tree cover 2017	**TC-2017**	**0.05**	**0.002**	**3.61**
Substrate temperature—autumn 12:00	**STA12**	**0.05**	**0.002**	**3.23**
Substrate temperature—summer 12:00	**STS12**	**0.05**	**0.002**	**3.67**
Crown area	**CA**	**0.04**	**0.008**	**2.3**
Light intensity—winter 12:00	**LIW12**	**0.03**	**0.006**	**2.19**
Average Light intensity—summer	**ALIS**	**0.03**	**0.004**	**2.24**
Light intensity—winter 15:00	**LIW15**	**0.03**	**0.01**	**2.04**
Average light intensity—winter	**ALIW**	**0.02**	**0.04**	**1.68**
Substrate temperature—autumn 6:00	**STA6**	**0.03**	**0.016**	**1.9**
Substrate temperature—summer 6:00	**STS6**	**0.02**	**0.028**	**1.78**
Average light intensity—autumn	**ALIA**	**0.02**	**0.038**	**1.8**
Light intensity—autumn 15:00	**LIA15**	**0.02**	**0.048**	**1.73**
Average substrate temperature—autumn	ASTA	0.03	0.11	1.5
Light intensity—winter 9:00	LIW9	0.02	0.088	1.5
Light intensity—summer 15:00	LIS15	0.02	0.222	1.22
Substrate temperature—summer 15:00	STS15	0.02	0.126	1.41
Substrate temperature—summer 9:00	STS9	0.02	0.118	1.45
Light intensity—summer 18:00	LIS18	0.01	0.128	1.33
Light intensity—winter 18:00	LIW18	0.02	0.402	1.02
Light intensity—summer 12:00	LIS12	0.01	0.364	1.08
Light intensity—autumn 18:00	LIA18	0.01	0.61	0.86

Table 3. *Cont.*

Variable	Abbreviation	Lambda A	*p*	F
Substrate temperature—autumn 18:00	STA18	0.02	0.556	0.94
Average Substrate temperature—winter	ASTW	0.01	0.348	1.11
Substrate temperature—winter 18:00	STW18	0.02	0.258	1.25
Substrate temperature—winter 6:00	STW6	0.01	0.348	1.07
Light intensity—autumn 9:00	LIA9	0.02	0.41	1.04
Light intensity—summer 9:00	LIS9	0.01	0.472	0.96
Substrate temperature—autumn 15:00	STA15	0.01	0.744	0.75
Substrate temperature—winter 12:00	STW12	0.01	0.76	0.73
Tree cover 2018	TC-2018	0.01	0.824	0.67
Light intensity—autumn 12:00	LIA12	0.01	0.772	0.74
Substrate temperature—winter 15:00	STW15	0.01	0.802	0.7
Average substrate temperature—summer	ASTS	0.01	0.894	0.55
Substrate temperature—winter 9:00	STW9	0.01	0.98	0.45
Substrate temperature—summer 18:00	STS18	0.01	0.99	0.39

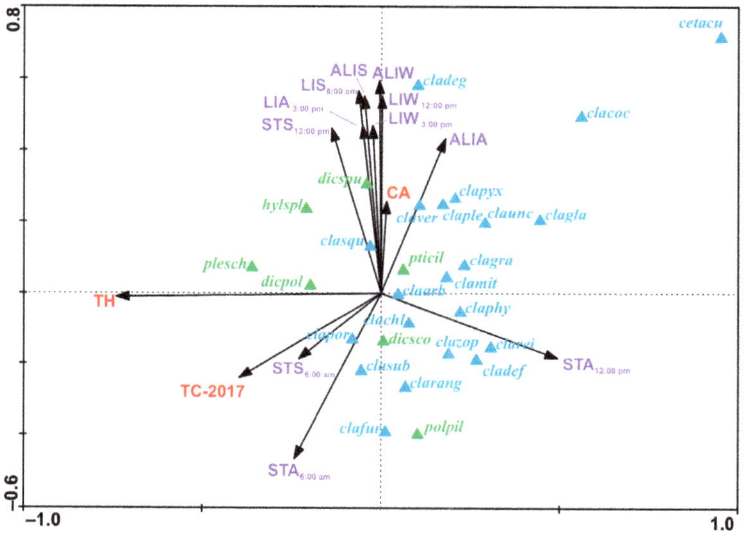

Figure 5. Canonical correspondence analysis ordination diagram showing relationship between species of lichens—blue color—and bryophyte—green color; forest variables (TH, TC-2017, and CA)—red color; and microhabitat variables (ALIS, ALIA, ALIW, LIS6, LIA15, LIW12, LIW15, STS6, STA6, STS12, and STA12)—violet color. Variables' abbreviations are explained in Table 3. The variance explained by the first canonical axis reaches 29% whereas by all canonical axes 12%. Abbreviations of lichens and bryophyte species names which were noted in all areas: *cetacu*—*Cetraria aculeata* (Schreb.) Fr.; *claarb*—*Cladonia arbuscula* (Wallr.) Flot.; *clachl*—*C. chlorophaea* (Flörke ex Sommerf.) Spreng.; *clacoc*—*C. coccifera* (L.) Willd.; *cladef*—*C. deformis* (L.) Hoffm.; *clafur*—*C. furcata* (Huds.) Baumg.; *clagla*—*C. glauca* Flörke; *clagra*—*C. gracilis* (L.) Willd.; *clamit*—*C. mitis* Sandst.; *claphy*—*C. phyllophora* Hoffm.; *claple*—*C. pleurota* Hoffm.; *clapor*—*C. portentosa* (Dufour) Coem.; *clapyx*—*C. pyxidata* (L.) Hoffm.; *clarang*—*C. rangiferina* (L.) Weber; *clarei*—*C. rei* Schaer.; *clasqu*—*C. squamosa* Hoffm.; *clasub*—*C. subulata* (L.) F.H. Wigg.; *claunc*—*C. uncialis* (L.) F.H. Wigg.; *claver*—*C. verticillata* (Hoffm.) Ach.; *clazop*—*C. zopfii* Vain.; *dicpol*—*Dicranum polysetum* Sw.; *dicsco*—*D. scoparium* Hedw.; *dicspu*—*D. spurium* Hedw.; *hylspl*—*Hylocomium splendens* (Hedw.) Schimp.; *plesch*—*Pleurozium schreberi* (Willd. ex Brid.) Mitt.; *polpil*—*Politrichum piliferum* Hedw.; *pticil*—*Ptilidium ciliare* (L.) Hampe.

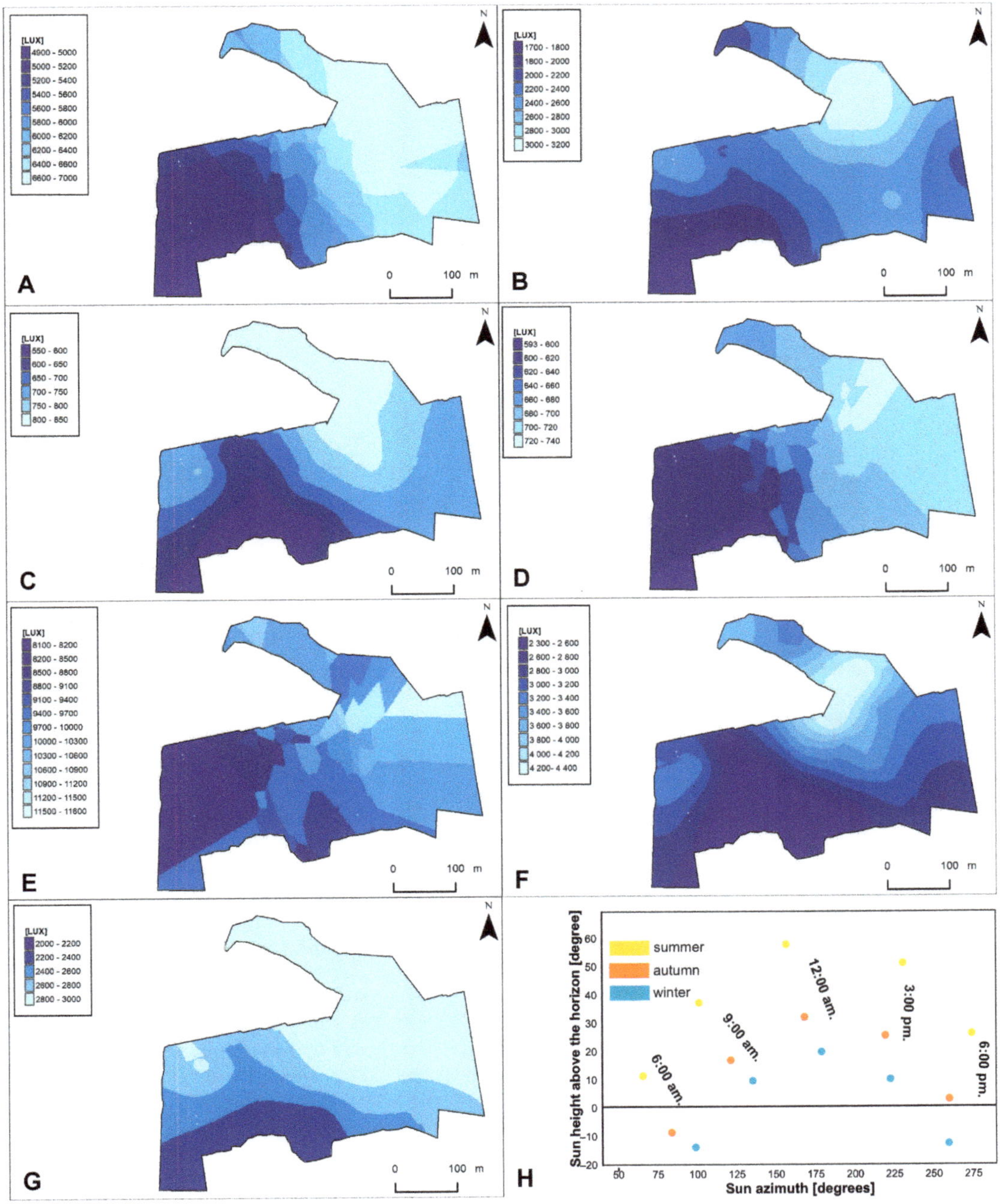

Figure 6. Light intensity models on the study area. Microhabitat variables: (**A**)—ALIS; (**B**)—ALIA; (**C**)—ALIW; (**D**)—LIS6; (**E**)—LIA15; (**F**)—LIW12; (**G**)—LIW15. Abbreviations are explained in Table 3. Relation between Sun azimuth and height above the horizon presented in part—(**H**).

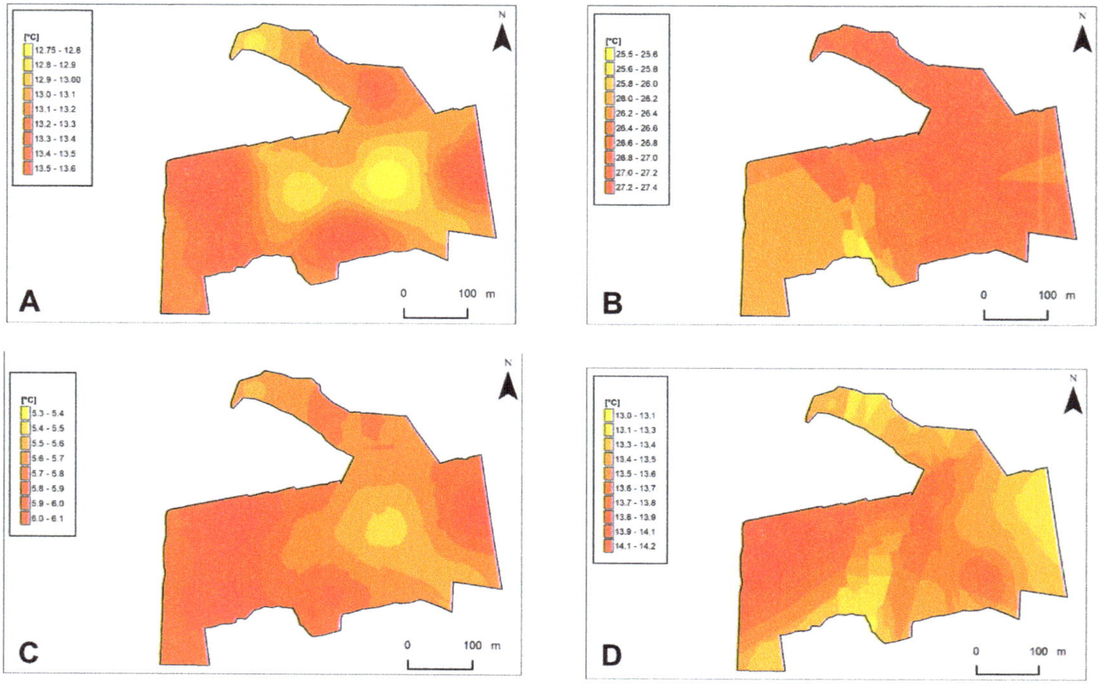

Figure 7. Substrate temperature models on the study area. Microhabitat variables: (**A**)—STS6; (**B**)—STA6; (**C**)—STS12; (**D**)—STA12. Abbreviations are explained in Table 3.

4. Discussion

4.1. Influence of Microhabitat Parameters on Photosynthetic Activity and Cover of Lichens and Bryophytes

We have already indicated in previous studies that there is a clear relationship between the height of trees and the percentage cover of lichens and bryophytes in the undergrowth layer [7]. However, those studies also considered the parameters of the undergrowth properties and were carried out on a small sample of data from several sites. The distribution of the lichen–bryophyte flaps was made based on manual mapping of the area, and the data on the energy reaching the undergrowth were calculated based on the total values of solar energy reaching the Earth's surface in that region, minus the amount of energy retained by the stand [7]. There was a great need for more detailed studies based on a wider range of data, to be completely sure of previous reports. Comparison of the model of the percentage cover of lichens in the undergrowth layer and the model of tree height in the study area (Figure 2) confirmed the previous results [7]; however, the current data allowed for more detailed insight. In a relatively low stand, up to 9 m high, the lichen undergrowth layer is very well preserved. In a tree stand of up to 13 m, the increasing share of *P. schreberi* and *D. scoparium* can be seen. Above 14 m in height, most of the undergrowth layer is completely bryophytes. These results provide a good justification as to why the lichen forest community was described as periodic and gradually disappearing as the stand ages [19–22].

Fałtynowicz [23] found in his study, which also conducted in the Tuchola Forest but in the 1980s, that there is a high correlation between the age of pine trees and the abundance of lichens and mosses, which is related to the amount of light reaching the undergrowth, as well as the amount of water available. In younger and older stands, lichens dominate [23]. The results of our research to date clearly indicate that it is not the age of the stand but the height of the trees that is of major importance in the processes of the decline of the lichen-rich undergrowth, since in the low site index V, stands have 50-year-old trees that reach a height of 9 m, and in such a habitat, the lichen undergrowth persists much longer than in the high site index I, where 50-year-old trees are 14 m high, and the lichen undergrowth begins to degenerate as they reach an age of about 20–25 years when trees exceed 9 m in height.

Numerous statistical analyses are presented in this work, including linear correlations with high correlation coefficients (Figure 3), and a more complex multiple regression analysis was performed (Table 1; Supplementary Table S1), which attempted to explain this phenomenon of the decrease in the percentage cover of lichens in the pine forest undergrowth with the increase in tree height based on the photosynthetic activity of chosen species of lichens and bryophytes. This is in line with the results of studies explaining this phenomenon of the limited availability of sunlight in increasingly mature forests [21,22] and what follows directly from the fact that lichens are photophilic [24–26] and bryophytes are photophobic organisms [27,28].

The results obtained indicate a large impact on undergrowth formation due to the relationship between forest variables: tree height and microhabitat variables; substrate temperature in autumn and winter (Table 1a); and tree height and light intensity in autumn (Table 1b). Already in previous studies [7], it was observed that substrate temperature is very variable throughout the year and that organic matter is a natural thermal insulation of the mineral layer against high temperatures in summer and low temperatures in winter [29,30]. The faster heating of the substrate in the summer months causes very rapid water loss. Furthermore, organic matter is a large reservoir of accumulated water [7].

4.2. Seasonal Variations in Photosynthetic Activity of Selected Lichen and Bryophyte Species

Current research confirms these earlier observations and shows even more clearly that the cold seasons of the year have a decisive influence on the formation of the lichen undergrowth layer. This is also confirmed by the obtained results of the seasonal variability of Fv/Fm (Figure 4). Fv/Fm was lowest in spring and summer, and it increased in, autumn, and winter when it had higher values (Figure 4A). Although lichens are pioneering organisms that occur in very extreme conditions, they prefer wetter and less sunny locations for optimal growth, as evidenced by the fact that they are called cryophilic species [31,32]. Both the results of these studies and the work of other authors confirm that the autumn and winter period is a favorable environment for lichen metabolism, due to the presence of lower temperatures, higher and constant humidity, and optimal light intensity [33]. On the other hand, other studies on bryophytes confirmed that CO_2 gas exchange takes place even at a temperature of -5 °C [34], and they contrast with the assumption that the photosynthesis process stops when cellular fluids freeze in bryophytes [35]. It is not surprising that lichens and bryophytes photosynthesize under such unfavorable conditions as autumn and winter, because it is related to their temperature preferences [10]. The temperature between 5 °C and 15 °C is the normal range of maximum apparent photosynthesis in most northern bryophytes, but activity continues at temperatures below 0° C [36]. The tolerance of lichens to freezing temperatures might be attributed to the effects of sugar alcohols [37] and antifreezing- and ice nucleation-promoting proteins [38], which occurred in lichens [39]. The ability to perform photosynthesis at low and even temperatures below 0 °C means that it does not stop even in winter. However, it is quite surprising that in the cold period, photosynthesis is at such a high level. Scientists explain this phenomenon with high tolerance to desiccation, thanks to which species of bryophytes and lichens, unlike other plants, can maintain the efficient process of photosynthesis under winter stress.

There is another important aspect to the variability of seasonal photosynthetic activity. Lichens and bryophytes can rebuild photosynthetic systems. So far, it has been found to occur faster in bryophytes that grow in light-exposed habitats and experience frequent alterations between wet and dry conditions than in woodland species subject to less frequent changes [40].

Previous pilot studies on Fv/Fm [7] showed the highest photosynthetic activity of examined lichens and bryophytes in the autumn period. The research was not carried out in winter because then the authors thought that the studied species would be in a phase of complete dormancy, and the reason for the low values of Fv/Fm [7] during spring and summer are the weather conditions, e.g., higher temperature, light intensity, and sudden and drastic fluctuations in air humidity [33,41].

Microhabitat conditions have a significant influence on water relations in the lichen thicket because they are poikilohydric species [42]. On hot and rainless days (which characterize the summer period), the lichen thallus can lose up to 97% of its water and thus enter the state of anabiosis. On the one hand, it causes the suppression of photosynthesis in them; on the other hand, it becomes a defense mechanism against solar radiation [32]. Long-term stress periods in lichens activate the production of reactive oxygen species, which negatively affect the PS II system and thus lead to cell death [43]. This mechanism is confirmed by studies conducted in warm desert areas, which have shown that high temperature causes very low or negative net photosynthesis, and in combination with water deficit, it can lead to the disappearance of the biological soil crust made of bryophytes and lichens [44]. These authors emphasize the significant role of water availability on photosynthetic activity, while bryophyte dehydration and excessive hydration inhibit the photosynthesis process [45]. A similar situation occurs for lichens, which during the natural state of physiological drought show much lower activity than after their two-hour hydration in Eppendorf [9].

Tobias and Niinemets [46] showed that with a decrease in the amount of light delivered, the level of chlorophyll concentration in *P. schreberi* increased, which resulted in better absorption of solar radiation with its lower availability. In conditions of strong sunlight, our results for *P. schreberi* showed the lowest Fv/Fm values of all the species tested. The lichen species *C. mitis* and *C. uncialis* showed similar photosynthetic efficiency throughout the seasons, which is related to their great similarity; *C. mitis* has a more intense photosynthetic process, while *C. uncialis* photosynthesizes longer under conditions of water shortage, which ultimately balances their photosynthetic activity [47]. The large values of standard deviations in the tests are the result of different atmospheric conditions during the conducted measurements. On sunny and dry days, both lichens and bryophytes reached the critical values of Fv/Fm = 0.200, while in the morning, during the presence of dew or after intense rain, these values reached the limit of Fv/Fm = 0.8. This proves that the process of photosynthesis is very quickly regenerated, and that the photosynthesis process is resumed to a maximum extent (up to an hour), with little water availability [48].

The daily cycle of Fv/Fm in lichens and bryophytes is comparable within each species in the spring, summer, and autumn seasons, and the differences are noted only in winter (Figure 4B–E). All species studied show the highest photosynthetic activity in the morning and then significantly reduce it. The explanation for this is probably a short day and an early sunset. Furthermore, the studied species form bryophyte undergrowth in tall pine forests with spreading tree crowns, which limit light and cause shading [8]. However, these are merely assumptions that require further investigation.

The results of the relationship between the photosynthetic activity of selected species (*C. mitis*, *C. uncialis*, *D. scoparium*, and *P. schreberi*) and the substrate temperature and light intensity microhabitat parameters in the aspects of daytime and seasons (Supplementary Table S1) show even more clearly how the process of photosynthesis is highly dependent on environmental factors and is specific for each species.

The results presented confirm that the TH has a large impact on the values of the substrate temperature and the intensity of light reaching the undergrowth, which in turn determines the photosynthetic activity and photosynthesis efficiency. The presented regressions state that the joint influence of temperature and light intensity determines the Fv/Fm values, which is confirmed in studies on lichens and bryophytes and in studies on vascular plants [8,49]. The differences in Fv/Fm lichens in relation to the season and daytime can be explained by the fact that these organisms are extremely resistant to unfavorable microhabitat conditions, as they have wide optimal temperature and light requirements met [50]. The results confirm the tendency that in summer and autumn, photosynthetic activity decreases with a large increase in temperature and light intensity. This phenomenon is related to dehydration of the thallus and puts it into anabiosis [51]. However, studies are known to determine the optimal temperature for the photosynthesis process that occurs in lichens in the range of 15–25 °C [52]. In the autumn period, the average temperatures in the studied area fluctuate around 14 °C, which is below the optimal values.

However, there are studies with results that differ from ours. For example, samples of *Cladonia foliacea*, *C. furcata*, *C. pyxidata*, *Diploschistes muscorum*, and *Thalloidima physaroides* had higher Fv/Fm values in spring and autumn samples compared to summer and winter [41]. The authors explain their results by stating that the environmental conditions during spring and autumn were more favorable for photosynthetic efficiency than in the other two seasons [41]. The conditions of spring and autumn, i.e., high humidity and increasing intensity of incident light, contributed to the high photosynthetic activity of the lichens. During winter and summer, the investigated species had been exposed to stress [53]. In summer, it could be explained by heat and light stress, which had a photoinhibition effect on lichen thalli [54].

Contrary to lichens, bryophytes do not like high temperatures; therefore, significant relationships in our results were only confirmed in the morning and noon hours in summer and autumn for *P. schreberi* and in autumn at noon and in winter in the afternoon for *D. scoparium*. *P. schreberi* prefers light shading, lower temperatures, and optimal light availability [55]. It is a species characteristic of wetter and shaded areas; therefore, its retreat related to the thinning of the stand, which took place in 2017 [8], is visible in the experimental plot studied.

4.3. The Relationship between Species of Lichens and Bryophytes, Forest, and Microhabitat Variables

The CCA analysis showed a clear separation of the impact of forest variables which, according to previous studies, are mainly responsible for the occurrence of lichen species characteristic of lichen Scots pine forest communities, from microhabitat variables responsible for the occurrence of less important and habitat-forming lichens and bryophytes. Some of them, STS6, STA6, and STA12, are located between the axes, which may suggest their influence on both important and less significant species for this community. Thus, CCA indicates that substrate temperature has a significant influence on the formation of the forest undergrowth layer of lichen and bryophytes. Other studies analyzing the effects of microhabitat variables for terricolous lichens showed similar results [41]. The results clearly indicate that Fv/Fm is usually higher in less arid than in arid microhabitats, which results directly from solar exposure, along with the different shading created by vascular plants, including shrubs and trees [56]. The higher soil water content provided a higher amount of humidity for rehydration/activation of terricolous lichens. Their results [41] indicate the important process of nonphotochemical quenching (NPQ), which is a mechanism employed by plants and algae to protect themselves from the adverse effects of high light intensity [57] and is usually more intensive in shaded than in exposed microhabitats of the lichen species. These observations are explained by the fact that the shaded microhabitat is usually wetter, causing more frequent irrigation of the lichen thallus, which makes them transparent and more exposed to greater amounts of solar energy supplied [58,59]. Dry lichen thalli reflect most of the sun's energy [60].

5. Conclusions

The study delves into the intricate dynamics between tree height, microhabitat factors, and the prevalence of lichens and bryophytes in the lichen Scots pine forest ecosystems. Contrary to previous notions linking stand age to undergrowth decline, our findings emphasize the pivotal role of tree height in shaping the forest undergrowth composition. Lichens exhibit photophilic tendencies, thriving under optimal humidity conditions during autumn and winter, while bryophytes' responses to environmental stress vary. Substrate temperature and light intensity emerge as key determinants of photosynthetic activity, influencing undergrowth formation. The research underscores the need for comprehensive studies to unravel the complexities of forest microhabitats and their impact on undergrowth dynamics, offering valuable insights into the resilience and adaptability of these organisms in changing environmental conditions.

Supplementary Materials: The following supporting information can be downloaded at: https://www.mdpi.com/article/10.3390/f15040675/s1, Supplementary Table S1: Result of stepwise multiple regression analysis for the effect of microhabitat parameters on the measurement of the photosynthetic activity (the maximal quantum yields of photosystem II—FV/FM) of the study cryptogams species in aspect of daytime and seasons. Variables with significant effect ($p < 0.05$) are provided in bold.

Author Contributions: Conceptualization, M.H.W. and P.F.; methodology, M.H.W. and P.F.; formal analysis, M.H.W., P.D. and P.F.; writing—original draft, M.H.W., P.D., K.W., P.W.-P. and P.F.; writing—review and editing, M.H.W., project administration, M.H.W.; investigation, M.H.W., P.D., K.W. and P.F. All authors have read and agreed to the published version of the manuscript.

Funding: The fieldwork and laboratory research leading to these results has received funding from the Forest Fund of the State Forests in Poland (no. EZ.0290.1.12.2021 and no. MZ.0290.1.12.2023).

Data Availability Statement: Data sharing is not applicable to this article.

Acknowledgments: We humbly dedicate this publication to the esteemed memory of Ludwik Lipnicki, whose guidance led us into the realm of Bory Tucholskie National Park. The authors extend heartfelt gratitude to Janusz Kochanowski of the Bory Tucholskie' National Park, for graciously permitting research within its boundaries. Special appreciation is also extended to Agnieszka Turowska of the 'Bory Tucholskie' National Park for invaluable assistance throughout project implementation and fieldwork endeavors. Our sincere thanks are owed to Piotr Wężyk for his adept execution of Li-DAR—Airborne Laser Scanning of forest stands in 2017 and 2018. Additionally, we express gratitude to the anonymous reviewers whose insightful suggestions and remarks greatly enriched the manuscript.

Conflicts of Interest: The authors declare no conflicts of interest.

References

1. Naumburg, E.; Ellsworth, D.S. Short-term light and leaf photosynthetic dynamics affect estimates of daily understory photosynthesis in four tree species. *Tree Physiol.* **2002**, *22*, 393–401. [CrossRef] [PubMed]
2. Way, D.A.; Pearcy, R.W. Sunflecks in trees and forests: From photosynthetic physiology to global change biology. *Tree Physio.* **2022**, *32*, 1066–1081. [CrossRef] [PubMed]
3. Kaiser, E.; Morales, A.; Harbinson, J.; Kromdijk, J.; Heuvelink, E.; Marcelis, L.F. Dynamic photosynthesis in different environmental conditions. *J. Exp. Bot.* **2015**, *66*, 2415–2426. [CrossRef] [PubMed]
4. Pearcy, R.W.; Way, D.A. Two decades of sunfleck research: Looking back to move forward. *Tree Physiol.* **2012**, *32*, 1059–1061. [CrossRef] [PubMed]
5. Węgrzyn, M.; Wietrzyk, P. Phytosociology of snowbed and exposed ridge vegetation of Svalbard. *Polar Bio.* **2015**, *38*, 1905–1917. [CrossRef]
6. Węgrzyn, M. New records of lichens and lichenicolous fungi from the Polish Tatra Mountains. *Pol. Bot. J.* **2008**, *53*, 163–168.
7. Węgrzyn, M.H.; Fałowska, P.; Kołodziejczyk, J.; Alzayany, K.; Wężyk, P.; Zięba-Kulawik, K.; Hawryło, P.; Turowska, A.; Grzesiak, B.; Lipnicki, L.; et al. Tree height as the main factor causing disappearance of the terricolous lichens in the lichen Scots pine forests. *Sci. Total Environ.* **2021**, *771*, 144834. [CrossRef]
8. Węgrzyn, M.H.; Kołodziejczyk, J.; Fałowska, P.; Wężyk, P.; Zięba-Kulawik, K.; Szostak, M.; Turowska, A.; Grzesiak, B.; Wietrzyk-Pełka, P. Influence of the environmental factors on the species composition of lichen Scots pine forests as a guide to maintain the community (Bory Tucholskie National Park, Poland). *Glob. Ecol. Conserv.* **2020**, *22C*, e01017. [CrossRef]
9. Węgrzyn, M.H.; Fałowska, P.; Alzayany, K.; Waszkiewicz, K.; Dziurowicz, P.; Wietrzyk-Pełka, P. Seasonal Changes in the Photosynthetic Activity of Terrestrial Lichens and Mosses in the Lichen Scots Pine Forest Habitat. *Diversity* **2021**, *13*, 642. [CrossRef]
10. Barták, M. Lichen photosynthesis. Scaling from the cellular to the organism level. In *The Structural Basis of Biological Energy Generation*; Hohmann-Marriott, M.F., Ed.; Advances in Photosynthesis and Respiration (AIPH, volume 39); Springer Dordrecht: Berlin/Heidelberg, Germany, 2014; pp. 379–400. [CrossRef]

11. Chowaniec, K.; Latkowska, E.; Rola, K. Does long-term salt stress of environmentally relevant concentrations affect the physiology of inland lichens? The importance of rainfall to restore thallus vitality. *Environ. Exp. Bot.* **2022**, *200*, 104937. [CrossRef]
12. Lange, O.L.; Büdel, B.; Heber, U.; Meyer, A.; Zellner, H.; Green, T.G.A. Temperate rainforest lichens in New Zealand: High thallus water content can severely limit photosynthetic CO_2 exchange. *Oecologia* **1993**, *95*, 303–313. [CrossRef] [PubMed]
13. Veldhuis, M.C.T.; Ananyev, G.; Dismukes, G.C. Symbiosis extended: Exchange of photosynthetic O_2 and fungal-respired CO_2 mutually power metabolism of lichen symbionts. *Photosynth. Res.* **2020**, *143*, 287–299. [CrossRef]
14. Pokorny, L.; Ho, B.C.; Frahm, J.P.; Quandt, D.; Shaw, A.J. Phylogenetic analyses of morphological evolution in the gametophyte and sporophyte generations of the bryophyte order Hookeriales (Bryopsida). *Mol. Phylogenet. Evol.* **2012**, *63*, 351–364. [CrossRef] [PubMed]
15. Wężyk, P.; Hawryło, P.; Zięba-Kulawik, K.; Szostak, M.; Kuzera, J.; Turowska, A.; Bura, M.; Wietrzyk, P.; Kołodziejczyk, J.; Fałowska, P.; et al. Using LiDAR point clouds in active protection of forest lichen communities in "Bory Tucholskie" National Park. *Arch. Fotogram. Kartogra. Rem. Sens.* **2018**, *30*, 27–41. [CrossRef]
16. Kalaji, H.M.; Rastogi, A.; Živčák, M.; Brestic, M.; Daszkowska-Golec, A.; Sitko, K.; Alsharafa, K.Y.; Lotfi, R.; Stypiński, P.; Samborska, I.A. Prompt chlorophyll fluorescence as a tool for crop phenotyping: An example of barley landraces exposed to various abiotic stress factors. *Photosynthetica* **2018**, *56*, 953–961. [CrossRef]
17. Schansker, G.; Tóth, S.Z.; Strasser, R.J. Dark-recovery of the Chl a fluorescence transient (OJIP) after light adaptation: The qT-component of non-photochemical quenching is related to an activated photosystem I acceptor side. *Biochim. Biophys. Acta* **2006**, *1757*, 787–797. [CrossRef]
18. Coe, R.A.; Lin, H.C. Light-Response Curves in Land Plants. *Photosynthesis* **2018**, *1770*, 83–94. [CrossRef]
19. Foster, D.R. Vegetation development following fire in *Picea mariana* (black spruce)—*Pleurozium* forests of south-eastern Labrador, Canada. *J. Ecol.* **1985**, *73*, 517–534. [CrossRef]
20. Ahti, T.; Oksanen, J. Epigeic lichen communities of taiga and tundra regions. *Vegetatio* **1990**, *86*, 39.e70. [CrossRef]
21. Reinecke, J.; Klemm, G.; Heinken, T. Veränderung der Vegetation nährstoffarmer Kiefernwälder im nördlichen Spreewald-Randge-biet zwischen 1965 und 2010. *Verh. Bot. Ver. Berl. Brandenbg.* **2011**, *144*, 63–97.
22. Boudreault, C.; Drapeau, P.; Bouchard, M.; St-Laurent, M.H.; Imbeau, L.; Bergeron, Y. Contrasting responses of epiphytic and terricolous lichens to variations in forest char- acteristics in northern boreal ecosystems. *Canad. J. For. Res.* **2015**, *45*, 595–606. [CrossRef]
23. Fałtynowicz, W. The dynamics and role of lichens in a managed *Cladonia*-Scotch pine forest (*Cladonio-Pinetum*). *Monogr. Bot.* **1986**, *69*, 1. [CrossRef]
24. Friedmann, E.I.; Gabun, M. Desert Algae, Lichens, and Fungi. In *Desert Biology*; Brown, G.W., Ed.; College of Fisheries Volume 2.; University of Washington Seatle: Seatle, WA, USA, 1967; pp. 165–212.
25. Bültmann, H.; Daniëls, F.J. Lichens and vegetationea case study of *Thamnolietum vermicularis*. *Bibl. Lichenol.* **2009**, *100*, 31–47.
26. Boudreault, C.; Zouaoui, S.; Drapeau, P.; Bergeron, Y.; Stevenson, S. Canopy openings created by partial cutting increase growth rates and maintain the cover of three *Cladonia* species in the Canadian boreal forest. *For. Ecol. Manag.* **2014**, *304*, 473–481. [CrossRef]
27. Valanne, N. Photosynthesis and photosynthetic products in bryophyte. In *The Experimental Biology of Bryophytes*; Dyer, A.F., Duckett, J.G., Cronshaw, J., Eds.; Academic Press: San Diego, CA, USA, 1984; pp. 257–273.
28. Marschall, M.; Proctor, M.C.F. Are bryophytes shade plants? Photosynthetic light responses and proportions of chlorophyll a, chlorophyll b and total carotenoids. *Ann. Bot.* **2004**, *94*, 593–603. [CrossRef] [PubMed]
29. Kodešová, R.; Vlasakova, M.; Fer, M.; Tepla, D.; Jakšík, O.; Neuberger, P.; Adamovský, R. Thermal properties of representative soils of the Czech Republic. *Soil Water Res.* **2013**, *8*, 141–150. [CrossRef]
30. Onwuka, B.; Mang, B. Effects of soil temperature on some soil properties and plant growth. *Adv. Plants Agric. Res.* **2018**, *8*, 34–37. [CrossRef]
31. Armstrong, R.A. The lichen symbiosis: Lichen "extremophiles" and survival on Mars. *J. Astrobiol. Space Sci. Rev.* **2019**, *1*, 378–397.
32. Kappen, L.; Schroeter, B.; Scheidegger, C.; Sommerkorn, M.; Hestmark, G. Cold resistance and metabolic activity of lichens below 0 °C. *Adv. Space Res.* **1996**, *18*, 119–128. [CrossRef]
33. Veres, K.; Csintalan, Z.; Laufer, Z.; Engel, R.; Szabó, K.; Farkas, E. Photoprotection and high-light acclimation in semi-arid grassland lichens—A cooperation between algal and fungal partners. *Symbiosis* **2022**, *86*, 33–48. [CrossRef]
34. Pannewitz, S.; Green, T.A.; Maysek, K.; Schlensog, M.; Seppelt, R.; Sancho, L.G.; Turk, R.; Schroeter, B. Photosynthetic responses of three common bryophyte from continental Antarctica. *Antarct. Sci.* **2005**, *17*, 341–352. [CrossRef]
35. Kappen, L.; Schroeter, B. Plants and Lichens in the Antarctic, Their Way of Life and Their Relevance to Soil Formation. In *Geoecology of Antarctic Ice-Free Coastal Landscapes. Ecological Studies*; Beyer, L., Bölter, M., Eds.; Springer: Berlin/Heidelberg, Germany, 2002; Volume 154. [CrossRef]
36. Kallio, P.; Valanne, N. On the Effect of Continuous Light on Photosynthesis in Mosses. In *Fennoscandian Tundra Ecosystems. Ecological Studies*; Wielgolaski, F.E., Ed.; Springer: Berlin/Heidelberg, Germany, 1975; Volume 16. [CrossRef]
37. Fontaniella, B.; Vicente, C.; Legaz, M.E. The cryoprotective role of polyols in lichens: Effects on the redistribution of RNase in *Evernia prunastri* thallus during freezing. *Plant Physiol.* **2000**, *38*, 621–627. [CrossRef]
38. Kieft, T.L.; Ruscetti, T. Molecular sizes of lichen ice nucleation sites determined by gamma radiation inactivation analysis. *Cryobiology* **1992**, *29*, 407–413. [CrossRef]

39. Doucet, C.J.; Byass, L.; Elias, L.; Worrall, D.; Smallwood, M.; Bowles, D.J. Distribution and characterization of recrystallization inhibitor activity in plant and lichen species from the UK and maritime Antarctic. *Cryobiology* **2000**, *40*, 218–227. [CrossRef] [PubMed]
40. Proctor, M.C.; Smirnoff, N. Rapid recovery of photosystems on rewetting desiccation-tolerant bryophyte: Chlorophyll fluorescence and inhibitor experiments. *J. Exp. Bot.* **2000**, *51*, 1695–1704. [CrossRef] [PubMed]
41. Veres, K.; Farkas, E.; Csintalan, Z. The bright and shaded side of duneland life: The photosynthetic response of lichens to seasonal changes is species-specific. *Mycol. Prog.* **2020**, *19*, 629–641. [CrossRef]
42. Green, T.G.A.; Lange, O.L. Photosynthesis in Poikilohydric Plants: A Comparison of Lichens and Bryophytes. In *Ecophysiology of Photosynthesis*; Schulze, E.D., Caldwell, M.M., Eds.; Springer: Berlin/Heidelberg, Germany, 1995; 1995′ Springer Study Edition; Volume 100. [CrossRef]
43. Hájek, J.; Barták, M.; Dubová, J. Inhibition of photosynthetic processes in foliose lichens induced by temperature and osmotic stress. *Biol. Plant* **2006**, *50*, 624–634. [CrossRef]
44. Grote, E.E.; Belnap, J.; Housman, D.C.; Sparks, J.P. Carbon exchange in biological soil crust communities under differential temperatures and soil water contents: Implications for global change. *Glob. Chang. Biol.* **2010**, *16*, 2763–2774. [CrossRef]
45. Perera-Castro, A.V.; Waterman, M.J.; Turnbull, J.D.; Ashcroft, M.B.; McKinley, E.; Watling, J.R.; Bramley-Alves, J.; Casanova-Katny, A.; Zunig, G.; Flexas, J.; et al. It is hot in the sun: Antarctic bryophyte have high temperature optima for photosynthesis despite cold climate. *Front. Plant Sci.* **2020**, *11*, 1178. [CrossRef]
46. Tobias, M.; Niinemets, Ü. Acclimation of photosynthetic characteristics of the bryophyte *Pleurozium schreberi* to among-habitat and within-canopy light gradients. *Plant Biol.* **2010**, *12*, 743–754. [CrossRef]
47. Lechowicz, M.J.; Adams, M.S. Ecology of *Cladonia* lichens. II. Comparative physiological ecology of *C. mitis*, *C. rangiferina*, and *C. uncialis*. *Can. J. Bot.* **1974**, *52*, 411–422. [CrossRef]
48. Tuba, Z.; Csintalan, Z.; Proctor, M.C. Photosynthetic responses of a bryophyte, *Tortula ruralis*, ssp. *ruralis*, and the lichens *Cladonia convoluta* and *C. furcata* to water deficit and short periods of desiccation, and their ecophysiological significance: A baseline study at present-day CO_2 concentration. *New Phytol.* **1996**, *133*, 353–361. [CrossRef] [PubMed]
49. Harley, P.; Guenther, A.; Zimmerman, P. 1996. Effects of light, temperature and canopy position on net photosynthesis and isoprene emission from sweetgum (*Liquidambar styraciflua*) leaves. *Tree Physiol.* **2006**, *16*, 25–32. [CrossRef] [PubMed]
50. Kershaw, K.A. Studies on lichen-dominated systems. XX. An examination of some aspects of the northern boreal lichen woodlands in Canada. *Can. J. Bot.* **1977**, *55*, 393–410. [CrossRef]
51. Sadowsky, A.; Ott, S. Symbiosis as a successful strategy in continental Antarctica: Performance and protection of *Trebouxia* photosystem II in relation to lichen pigmentation. *Polar Biol.* **2016**, *39*, 139–151. [CrossRef]
52. Tegler, B.; Kershaw, K.A. Studies on lichen-dominated systems. XXIII. The control of seasonal rates of net photosynthesis by moisture, light, and temperature in *Cladonia rangiferina*. *Can. J. Bot.* **1980**, *58*, 1851–1858. [CrossRef]
53. Maxwell, K.; Johnson, G.N. Chlorophyll fluorescence—A practical guide. *J. Exp. Bot.* **2000**, *51*, 659–668. [CrossRef] [PubMed]
54. Gauslaa, Y.; Solhaug, K.A. High-light damage in air-dry thalli of the old forest lichen *Lobaria pulmonaria*–interactions of irradiance, exposure duration and high temperature. *J. Exp. Bot.* **1999**, *50*, 697–705. [CrossRef]
55. Popov, S.Y. On the Distribution of *Pleurozium schreberi* (Bryophyta, *Hylocomiaceae*) in the East European Plain and Eastern Fennoscandia. *Mosc. Univ. Biol. Sci. Bull.* **2018**, *73*, 178–184. [CrossRef]
56. Kalapos, T.; Mázsa, K. Juniper shade enables terricolous lichens and bryophyte to maintain high photochemical efficiency in a semiarid temperate sand grassland. *Photosynthetica* **2001**, *39*, 263–268. [CrossRef]
57. Horton, P.; Ruban, A. Regulation of Photosynthesis under Stress: Molecular design of the photosystem II light-harvesting antenna: Photosynthesis and photoprotection. *J. Exp. Bot.* **2005**, *56*, 365–373. [CrossRef] [PubMed]
58. Heber, U.; Bukhov, N.G.; Shuvalov, V.A.; Kobayashi, Y.; Lange, O.L. Protection of the photosynthetic apparatus against damage by excessive illumination in homoiohydric leaves and poikilohydric bryophyte and lichens. *J. Exp. Bot.* **2001**, *52*, 1999–2006. [CrossRef] [PubMed]
59. Heber, U.; Lange, O.L.; Shuvalov, V.A. Conservation and dissipation of light energy as complementary processes: Homoiohydric and poikilohydric autotrophs. *J. Exp. Bot.* **2006**, *57*, 1211–1223. [CrossRef]
60. Ertl, L. Über die Lichtverhältnisse in Laubflechten. *Planta* **1951**, *39*, 245–270. [CrossRef]

forests

MDPI

Article

Foraging Guilds of Birds in Continuous and Fragmented Forests of Southeast China

Yuan Wang [1], Jiawen Liu [1], Shuai Gao [1], Sichun Tong [1], Zheng Wang [1,*] and Ning Li [2,*]

[1] College of Life Sciences, Nanjing Forestry University, Nanjing 210037, China; 3221700716@njfu.edu.cn (Y.W.); ljw1506182824@163.com (J.L.); njfu1517@163.com (S.G.); sctong1983@163.com (S.T.)
[2] Institute of Applied Ecology, Nanjing Xiaozhuang University, Nanjing 211171, China
* Correspondence: zhengwang@njfu.edu.cn (Z.W.); lining@njxzc.edu.cn (N.L.)

Abstract: Habitat fragmentation is one of the main factors leading to changes in bird foraging behavior. Therefore, studying bird diversity, foraging groups, and spatial utilization in fragmented habitats is of great significance for forest bird conservation. This experiment was conducted in the continuous and fragmented forests of Meihua Mountain National Nature Reserve in southeast China. We collected bird foraging behavior data using the distance sampling method and compared the composition of bird foraging groups in the two habitats between October and December 2020 and 2021. The 46 bird species observed in the fragmented habitat belonged to 3 orders and 24 families, forming a total of 8 bird foraging groups. In contrast, the continuous habitat had 45 bird species belonging to 3 orders and 19 families, forming 7 bird foraging groups. Using principal component analysis and log-linear analysis, we demonstrated significant differences in foraging location, foraging substrate, foraging height, and foraging mode between the fragmented and continuous habitats during autumn and winter. Birds in the fragmented habitat tended to pick up food from high tree crown layers, while those in the continuous habitat preferred picking up or exploring food from low shrubs and ground levels. This study revealed that the separation of foraging variables among different groups allows for efficient utilization of foraging space, and foraging behavior is influenced by habitat type. Consequently, there are notable differences in resource utilization within evergreen broad-leaved forests. These results provide valuable scientific insights into how habitats with varying degrees of fragmentation affect biodiversity and species spatial utilization.

Keywords: birds; foraging guild; utilization of foraging space; Meihua Mountain Nature Reserve

1. Introduction

A guild is a group that utilizes habitat resources at the same level of similarity [1]. Root first proposed this concept in the 1960s, and it has since been extensively studied, primarily focusing on the impact of habitat on bird foraging behavior within groups [2–4] and the species composition of bird foraging guilds [5,6]. Bird foraging guilds, as an important entry point for studying community ecology and functional diversity, have seen significant progress in multiple dimensions in recent years. Existing research indicates that the composition and structure of foraging groups are highly dependent on habitat characteristics and resource availability. For instance, Pigot et al. found that bird foraging groups in tropical forests exhibit higher functional diversity [7], while Newbold et al. demonstrated that habitat fragmentation leads to a significant reduction in carnivorous and fruit-eating groups [8]. Despite these advances, the dynamic changes of foraging groups across different geographical scales still require further exploration. This study provides theoretical foundations for avian biodiversity conservation and ecosystem management.

Habitat fragmentation is widely recognized as a major driver of global biodiversity loss, exerting profound impacts on ecological communities and species interactions [9]. Avian species, in particular, are among the most vulnerable groups due to their high sensitivity to alterations in habitat configuration and resource availability, which significantly influences their foraging behaviors and guild structures [10]. The transformation of continuous habitats into smaller, isolated patches within fragmented landscapes frequently results in substantial ecological consequences, including disrupted food web dynamics, diminished resource accessibility, and altered species composition [11]. These environmental modifications may trigger cascading effects on avian foraging guilds—ecological groups comprising species that utilize similar food resources through comparable foraging strategies—potentially leading to significant transformations in their structural organization and functional roles within ecosystems.

This study addresses the following key questions: (1) What are the characteristics of bird foraging groups in the fragmented habitat and continuous habitat? (Including similarity in resource utilization, differentiation of ecological niches, mechanisms of competition, coexistence, diversity and stability) (2) What impact does the habitat fragmentation have on the spatial utilization of bird foraging groups?

2. Materials and Methods

2.1. Study Area

Meihua Mountain Nature Reserve is located in the southwestern Fujian Province. The vegetation flora transitioned from the southern edge of the subtropical zone to the southern subtropical zone, forming typical subtropical forest vegetation with abundant vegetation types and animal and plant resources. The area of the core zone was 7041.7 km^2, the buffer zone was 2443.1 km^2, and the experimental zone was 2683.7 km^2 [12]. This study selected fragmented habitats in the Yew Ecological Garden and continuous habitats in the Tiger Park as research sites, both conducted within the experimental area.

Yew Ecological Garden as a fragmented habitat (25°16′28″–25°16′29″ N, 116°52′40″–116°52′59″ E) is in Chongtou Natural Village, Meihua Mountain Nature Reserve. The main habitat type around the tourist area is the bamboo forest. Yew Ecological Garden is dominated by evergreen broad-leaved forests with many tall trees. Overall, vegetation cover was relatively high. *Taxus. chinensis* is the most important tree species in Yew Ecological Garden. The main habitat around the Yew Ecological Garden is bamboo forests and farmland, showing a fragmented distribution. The local protection authorities have increased their efforts to protect the park; for example, the prevention of illegal fruit-picking has been enforced during the ripening period of *T. chinensis*, however, tourism activities continue to affect its growth environment. There are studies indicating that human interference leading to the continuous loss of natural habitats has an impact on the movement behavior of fruit-eating birds, which in turn affects their dispersal patterns and efficiency of southern Chinese yew seeds [13]. There are many other tall dominant trees in the evergreen broad-leaved forest habitat of Yew Ecological Garden, including *Machilus thunbergii*, *Quercus glauca*, and *Sloanea sinensis*.

The Tiger Park as a continuous habitat (25°17′16″–25°18′47″ N, 116°52′7″–116°52′58″ E) is an ecological tourism scenic spot based on forest landscape. Its location is in the southwestern part of Meihua Mountain Nature Reserve, Fujian Province, mainly distributed in evergreen broad-leaved forests that exhibit a continuous distribution pattern. The plant composition in and around the park is very complex, and the vegetation flora transitions from the southern edge of the subtropical zone to the southern subtropical zone, forming a typical subtropical forest vegetation. The main tree species in the park are *Q. glauca*, *Castanopsis eyrei*, and *S. sinensis*.

2.2. Data Collection

In the autumn and winter of October–December 2020 and October–November 2021, four transects were established using the distance sampling method in Yew Ecological Garden and Tiger Park, respectively. The transects covered evergreen broad-leaved forests in both areas, and each distance was approximately 500 m long. Five cycles of surveys were conducted on each of the two habitats, with each cycle lasting 4 days, for a total of 80 days. Observations were conducted directly using the naked eye and KOWA BD 42 × 8 XD binoculars (Kowa Company, Ltd., Nagoya, Japan) and recording the species and quantity of birds seen and heard 30 m on each side of the sample line. During each survey, we recorded both bird species and their associated foraging behaviors using a timer to record quantitative observations of bird foraging behaviors within the telescope's line of sight every 30 s. After observing bird foraging activities, the foraging location (Stay away from the trunk; Close to the trunk; Shrub; Ground), foraging substrate (Upper canopy; Middle canopy; Lower canopy; Sprig; Thick stem; Trunk; Ground; Air), foraging height (0–1 m; 1–5 m; 5–10 m; 10–15 m; >15 m), and foraging method (Glean; Hover; Sally; Probe) were recorded [10]. Before collecting the observation results, we measured the diameter at the breast height of the fruit trees.

2.3. Data Analysis

The bird classification system was divided according to the fourth edition of the "Chinese Bird Classification and Distribution List" [14]. We used the Shannon Wiener Diversity Index (H′), Pielou Uniformity Index (J), and Simpson Dominance Index (C) to evaluate the diversity of bird communities at different levels of fragmentation. The Sorensen Similarity Index (S) is used to compare the similarity of bird communities between two habitats with different degrees of fragmentation [15].

We constructed a tree diagram based on the outcomes of k-means clustering analysis, from which the aggregation level can be discerned by examining the Euclidean distance coordinates of each node. To uncover the response patterns of bird foraging strategies to the spatial structure of habitats and assess the relative impact of fragmentation types and intensities on community functional differentiation in two locations, we quantified the consistency between the clustering results of bird communities and the gradient of habitat fragmentation. Furthermore, we performed a log-linear analysis on the foraging variables of birds in the two locations using R, aiming to identify significant interactions and dependencies between habitat types and foraging variables. The percentage matrix data were used for cluster analysis and principal component analysis, and the Shapiro–Wilk test was used to compare the differences in foraging location, foraging substrate, foraging height, and foraging methods between the fragmented habitat and continuous habitat. All data were tested for normality using the Shapiro–Wilk test before further statistical testing, and data processing was performed using R 4.0.2 [16,17].

3. Results

3.1. Composition and Diversity of Foraging Birds

The fragmented habitat was used for data analysis, with 46 bird species belonging to 3 orders and 24 families, and 41 bird species belonging to the Passeriformes order, accounting for 89.13% of the total bird species. There were two species of national second-class wild protected animals: *Lophura nycthemera* and *Garrulax canorus*. There were 45 species of birds used for data analysis in the continuous habitat belonging to 3 orders and 19 families, including 38 species of Passeriformes, accounting for 84.09% of the total number of bird species. There are three species of national second-class wild protected animals: the white browed mountain partridge *Arborophila gingica*, *L. nycthemera*, and *G. canorus*. Per the Shannon diversity index and *Pielou evenness* index, the continuous habitat was more diverse than the fragmented habitat; per the Simpson dominance index, the fragmented habitat was more diverse than the continuous habitat (Table 1).

Table 1. Diversity of foraging bird communities in fragmented habitat and continuous habitat.

Site	Number of Species	Shannon Diversity Index (H′)	Pielou Uniformity Index (E′)	Simpson Dominance Index (C′)	Sorensen Similarity Index
Fragmented habitat	46	3.023	0.79	0.09	0.69
Continuous habitat	45	3.308	0.874	0.048	

3.2. Bird Foraging Guilds

We constructed the tree diagram based on the findings of k-means clustering analysis. A lower aggregation level among species suggests more similar foraging behavior patterns and indicates a greater degree of ecological niche overlap, potentially signifying competitive relationships among these species.

Based on the methodology proposed by Holmes et al., the avian community was classified into distinct foraging groups using the average Euclidean distance (đ) between species as the primary classification criterion [18]. At a threshold of đ = 12.25, our analysis revealed differential group partitioning between habitat types: the fragmented habitat was segregated into 8 distinct foraging groups, while the continuous habitat yielded 7 groups. Both habitat types shared six common foraging groups: (1) shrub and ground foraging group, (2) shrub foraging group, (3) ground foraging group, (4) lower canopy foraging group, (5) middle canopy foraging group, and (6) upper canopy foraging group. Notably, the inner trunk foraging group was exclusively identified in the fragmented habitat, representing a unique ecological adaptation to habitat patchiness (Figure 1).

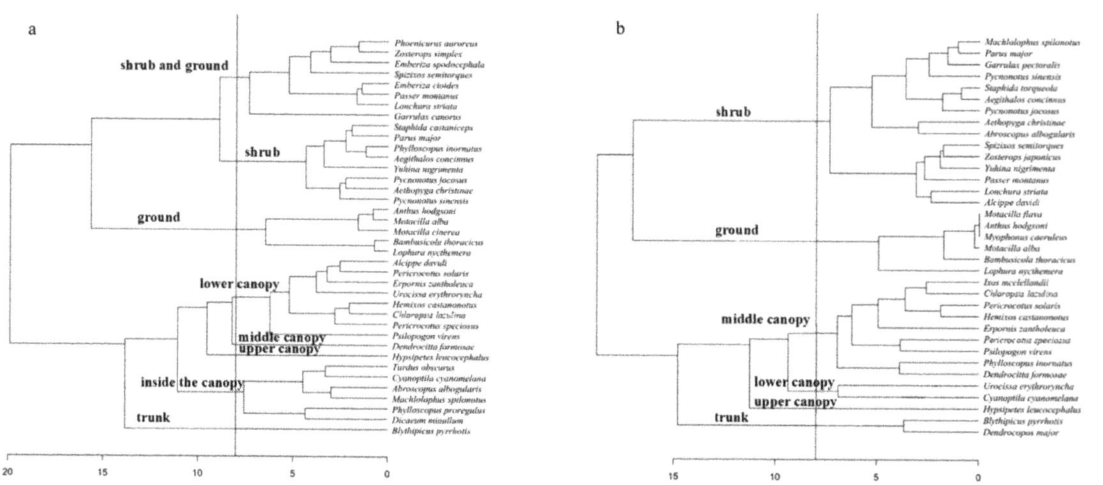

Figure 1. *Cont.*

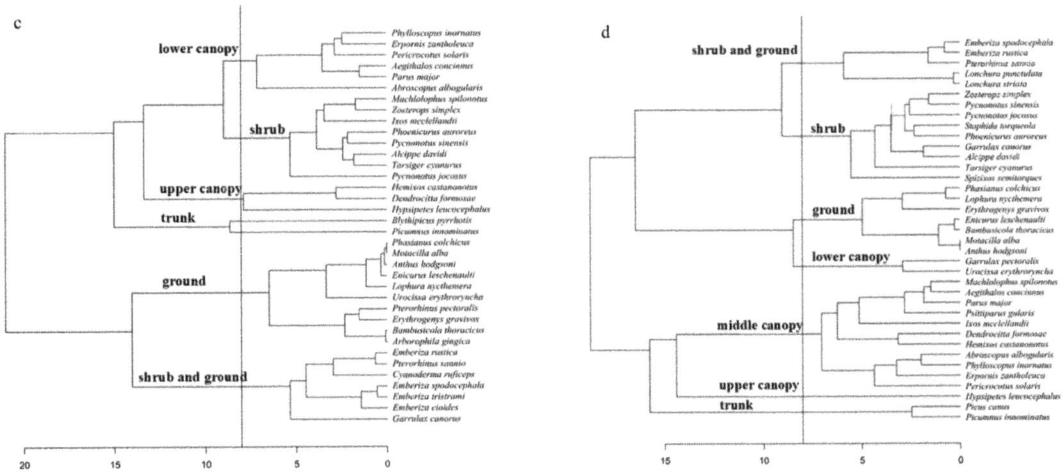

Figure 1. Cluster dendrogram of bird community foraging guild in fragmented habitat and continuous habitat. (**a**) The fragmented habitat Autumn and Winter 2020; (**b**) The fragmented habitat Autumn and Winter 2021; (**c**) The continuous habitat Autumn and Winter 2020; (**d**) The continuous habitat Autumn and Winter 2021.

3.3. *Important Variables for Dividing Bird Foraging Guilds*

The results of the principal component analysis of bird communities in the autumn and winter of the two years (2020 and 2021) showed that the characteristic values of the first two principal components in the fragmented and continuous habitat were both greater than one, with cumulative contribution rates of 70.38% and 75.86% (2020) and 70.13% and 69.64% (2021), respectively.

The principal component analysis (PCA) revealed significant differences and interannual variations in bird foraging strategies between fragmented and continuous habitats. PC1 (explaining variance 39.8%–56.4%) mainly reflects the spatial differentiation of foraging height and substrate: birds in fragmented habitats tend to prefer ground foraging in 2021 (load 0.491), while individuals in continuous habitats prefer mid to high levels (load 0.239) in 2020. PC2 (explaining a cumulative variance of 69.6%–75.9%) highlights the comparison between shrub layer and twig foraging. In 2021, the positive load of shrub foraging in patch habitat (0.431) was significantly higher than that in continuous habitat (−0.500). The interannual variation shows that the foraging strategies of birds in fragmented habitats fluctuate more (such as the reversal of ground foraging load from −0.487 to 0.491), while continuous habitats remain relatively stable. The difference in eigenvalues (with the highest PC1 eigenvalue of 2.632 in 2021) suggest that fragmentation may enhance niche differentiation. Overall, habitat types drive the functional reorganization of bird foraging behaviors by changing resource distribution; and the dynamic response of fragmented habitats is more significant (Table 2).

Table 2. Principal component analysis of bird foraging variables.

Variables of Foraging Guild		Fragmented Habitat				Continuous Habitat			
		2020		2021		2020		2021	
		PC1	PC2	PC1	PC2	PC1	PC2	PC1	PC2
Foraging location	Stay away from the trunk	0.293	−0.217	−0.217	−0.289	−0.211	0.230	−0.184	0.273
	Close to the trunk	0.195	−0.195	−0.056	−0.255	−0.108	0.249	−0.095	0.298
	Shrub	−0.001	0.519	−0.219	0.431	−0.175	−0.425	−0.207	−0.500
	Ground	−0.487	−0.107	0.491	0.113	0.494	−0.054	0.486	−0.071
Foraging substrate	Upper canopy	0.145	−0.139	−0.040	−0.123	−0.110	0.142	−0.040	0.098
	Middle canopy	0.136	−0.069	−0.123	−0.176	−0.105	0.079	−0.119	0.148
	Lower canopy	0.190	−0.117	−0.121	−0.110	−0.081	0.050	−0.110	0.165
	Sprig	0.013	0.481	−0.218	0.432	−0.171	−0.436	−0.205	−0.496
	Thick stem	−0.013	0.035	−0.001	−0.001	−0.004	0.010	−0.001	−0.004
	Trunk	0.017	−0.086	0.012	−0.134	−0.022	0.206	−0.009	0.162
	Ground	−0.488	−0.106	0.491	0.113	0.493	−0.052	0.486	−0.071
	Air	0.000	0.000	0.000	0.000	0.000	0.000	0.000	0.000
Foraging height	0–1 m	−0.030	0.295	−0.044	0.111	−0.046	−0.279	−0.040	−0.276
	1–5 m	0.136	0.305	−0.255	0.378	−0.305	0.044	−0.331	−0.041
	5–10 m	0.167	−0.160	−0.077	−0.099	−0.115	0.239	−0.099	0.329
	10–15 m	0.138	−0.220	−0.071	−0.351	−0.026	0.047	−0.015	0.058
	>15 m	0.076	−0.114	−0.045	−0.152	0.000	0.000	0.000	0.001
Foraging methods	Glean	0.065	0.196	−0.092	0.175	−0.076	−0.400	−0.082	−0.171
	Probe	−0.098	−0.162	0.102	−0.150	0.099	0.367	0.097	0.145
	Hover	0.027	−0.025	−0.008	−0.019	−0.022	0.031	−0.015	0.026
	Sally	0.005	−0.008	−0.002	−0.006	−0.001	0.002	0.000	0.000
	Eigenvalue	2.066	1.811	2.632	1.336	2.346	1.654	2.143	1.849
	Accumulated variance%	44.200	70.380	39.770	70.130	56.440	75.860	44.430	69.640

3.4. Differences in the Utilization of Foraging Space by Birds

The Mosaic plot obtained through the log-linear analysis showed that the ground foraging frequency in fragmented habitats was significantly higher than the independent model's predicted value ($p < 0.001$) at the foraging location, while it was significantly lower than the independent model's predicted value far from the tree trunk. However, the opposite was true in continuous habitats. In terms of foraging methods, the frequency of birds detecting fruits in continuous habitats was significantly higher than the independent model prediction value ($p < 0.001$), while in fragmented habitats it was significantly lower than the independent model prediction value ($p < 0.001$), and the other three foraging methods remained consistent with the expected values. On the foraging substrate, the foraging frequency of continuous habitats on the ground was significantly higher than the independent model prediction value ($p < 0.001$), while it was significantly lower than the independent model prediction value in the upper and lower canopy layers ($p < 0.001$); but the opposite was true in fragmented habitat. In terms of foraging height, the foraging frequency at ground level in continuous habitats was significantly higher than the predicted value of independent models ($p < 0.001$); in fragmented habitats, the foraging behavior of H6 (>15 m) almost disappears, which may be related to habitat loss caused by the continuous destruction of tree crowns (Figure 2).

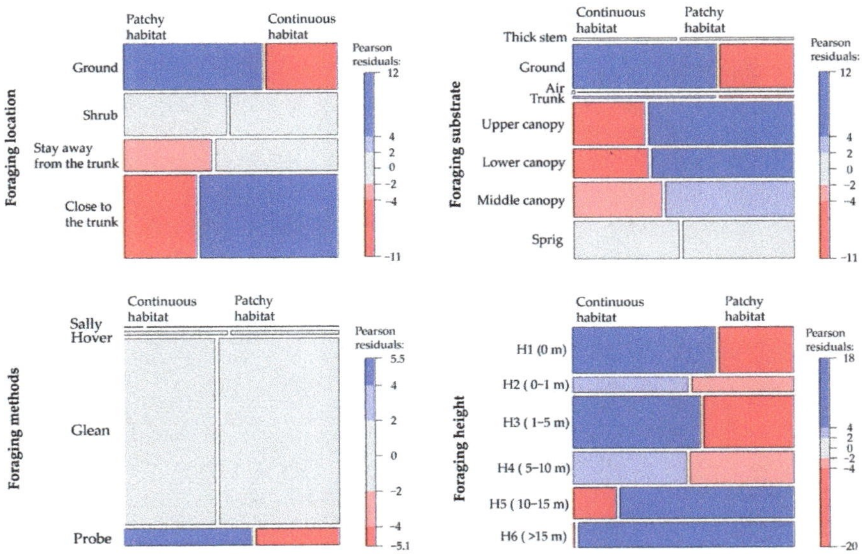

Figure 2. Log-linear analysis of habitat types and foraging variables.

4. Conclusions

The diversity and distribution of birds are strongly influenced by vegetation structure [19]. The Meihua Mountain Nature Reserve, characterized by its evergreen broad-leaved forests, supports high biodiversity with distinct vertical stratification and diverse microhabitats [20], which significantly shape avian community composition and structure [21,22]. In this study, we analyzed foraging guilds using 46 and 45 bird species in fragmented and continuous habitats, respectively. Our results reveal significant differences in foraging locations, mechanisms, vertical stratification, and foraging techniques between the two habitat types. These findings suggest that niche partitioning across multiple foraging dimensions enables optimal spatial resource utilization among avian assemblages. Moreover, the spatial distribution patterns of avian foraging activities show clear correlations with habitat characteristics.

Cluster analysis revealed that fragmented habitats formed eight foraging guilds, one more than the continuous habitats (seven guilds), with the emergence of a unique "inner trunk foraging guild." This finding contrasts the results reported by Bregman et al. in tropical forests, suggesting that subtropical bird communities exhibit distinct responses to habitat fragmentation [23]. Diversity indices showed that continuous habitats had higher Shannon diversity and evenness, supporting the "habitat heterogeneity hypothesis". In contrast, the higher Simpson dominance index in fragmented habitats implies that competitive exclusion may dominate community assembly, which aligns with findings by Karp et al. in Neotropical regions [24]. The Sørensen similarity index indicated relatively high community similarity between the two habitat types.

According to the clustering analysis, the bird community in fragmented habitats was divided into eight groups in autumn and winter, and the bird community in continuous habitats was divided into seven groups. There were significant differences in the composition of ground foraging guilds between the two locations; a reason for these differences might be that the fragmented habitat transects were distributed in patch habitats, whereas the continuous habitat transects were distributed in continuous habitats. There are scattered residential areas and roads along the edge of the patches in the fragmented habitat, and human interference is relatively large. However, the continuous habitat has a continuous evergreen broad-leaved forest habitat, which can provide more food and habitat resources for ground foraging bird groups. Within each bird group in both regions, there were birds with very low aggregation levels and very similar foraging space utilization. The potential competitive pressure was also high, which is consistent with existing research results [25,26]. However, utilizing similar resources in a similar manner within the same group does not necessarily indicate significant competition among species [27,28]. In the ground foraging guilds of the two places, the *Anthus hodgsoni* mainly fed on open surfaces, whereas *L. nycthemera* was active on forest surfaces with high vegetation coverage. Furthermore, the foraging behavior patterns of *L. nycthemera* and *Bambusicola thoracicus* were very similar, but further observation showed that *B. thoracicus* usually fed on ground clusters near shrubs. *B. thoracicus* used habitat differences to alleviate competition with *L. nycthemera* and achieve coexistence. Some studies have found that when the vegetation hierarchy is unclear, multi-layer foraging guilds may appear [26]. For example, Lu et al. conducted a bird survey in the Nonggang karst area of Guangxi and found that 16 bird species fully utilized the foraging space of each layer and were therefore classified as multi-layer foraging guilds [26]. However, this phenomenon was not observed in Meihua Mountain Nature Reserve. Although some canopy foraging guild birds, such as *Pycnonotus sinensis* and *Parus minor*, also used foraging positions below the canopy, their main foraging and habitat activities still occurred in the canopy.

The selection of foraging grounds by birds is influenced not only by their dietary preferences but also by the structure of their habitat and the availability of food resources [29]. Such selection can reveal the status, role, and relationships of birds within their ecosystems [30,31]. Our comparison of bird foraging space utilization between fragmented and continuous habitats revealed significant differences in foraging location, substrate, height, and methods between the two areas during autumn and winter. Specifically, birds in the fragmented habitat tended to pick up food from high tree crown layers, while those in the continuous habitat more frequently foraged by picking up or exploring food from low shrubs and ground levels. These differences likely arise because the fragmented habitat was dominated by tall trees, whereas the continuous habitat featured relatively lower tree heights, dense shrubs in some areas, and more complex spatial structures in the middle and lower layers. These findings further demonstrate that birds adapt their foraging behaviors to the vegetation structure of their habitat [32], enabling resource partitioning within evergreen broad-leaved forest communities and facilitating coexistence with species.

This study demonstrates significant differences in avian community composition and species diversity between fragmented and continuous forest habitats in Meihua Mountain, Fujian Province. Our results indicate that habitat fragmentation substantially influences avian foraging guild structure and foraging behaviors, leading to altered resource utilization patterns among bird species. Notably, dominant species exhibited distinct spatial adaptation strategies in response to fragmented habitats. These findings advance our ecological understanding of how habitat fragmentation affects biodiversity and species behavioral ecology. Furthermore, the study provides region-specific insights that can inform conservation strategies and management practices for forest bird populations in fragmented landscapes.

Author Contributions: Conceptualization, Z.W. and N.L.; methodology, Z.W., J.L., S.G. and N.L.; formal analysis, Y.W. and S.G.; investigation, N.L., Y.W., S.G. and S.T.; writing—original draft preparation, Y.W.; writing—review and editing, Z.W. and N.L. All authors have read and agreed to the published version of the manuscript.

Funding: This study was supported by the National Natural Science Foundation of China [grant numbers 32171528] and the Natural Science Foundation of Jiangsu Province [grant numbers BK20221180].

Data Availability Statement: The data used to support the findings of this study are available from the authors upon request.

Acknowledgments: We thank anonymous reviewers for their valuable comments.

Conflicts of Interest: The authors declare no conflicts of interest.

References

1. Root, R.B. The niche exploitation patern of the blue-gray gnatcatcher. *Ecol. Monogr.* **1967**, *37*, 317–350. [CrossRef]
2. Yilangai, M.R.; Abalaka, J.; Nsor, A.C.; Babale, A.; Karau, S.D.; Ivande, S. Effect of disturbance on bird foraging guilds in a West African dry forest. *Afr. J. Ecol.* **2023**, *61*, 461–468. [CrossRef]
3. Kavana, J.D.; Mbije, N.; Sirilo, M.T.; Bupamba, H.C. The spatio-temporal influences of habitat heterogeneity on the community structure of avian foraging guilds in human-dominated landscapes of the East African miombo woodlands. *Afr. J. Ecol.* **2023**, *62*, e13237. [CrossRef]
4. Chen, R.Y.; Li, H.Q.; Zhu, X.D.; Zhou, L.Z.; Wang, Y.Z. Analysis on the foraging guild structure and diversity of waterbirds in Caizi Lake under water level changes during overwintering. *BMC Biol.* **2024**, *41*, 77–83.
5. Ding, Z.F.; Liang, J.C.; Hu, Y.M.; Zhou, Z.X.; Sun, H.B.; Liu, L.N.; Liu, H.J.; Hu, H.J.; Si, X.F. Different responses of avian foraging guilds to spatial and environmental factors across an elevation gradient in the central Himalaya. *Ecol. Evol.* **2019**, *9*, 4116–4128. [CrossRef]
6. Burgess, E.E.; Maron, M. Does the response of bird assemblages to fire mosaic properties vary among spatial scales and foraging guilds? *Landsc. Ecol.* **2016**, *31*, 687–699. [CrossRef]
7. Pigot, A.L.; Sheard, C.; Miller, E.T.; Bregman, T.P.; Freeman, B.G.; Roll, U.; Tobias, J.A. Macroevolutionary convergence connects morphological form to ecological function in birds. *Nature* **2020**, *585*, 234–237. [CrossRef]
8. Newbold, T.; Scharlemann, J.P.W.; Butchart, S.H.M.; Şekercioğlu, Ç.H.; Alkemade, R.; Booth, H.; Purves, D.W. Ecological traits affect the response of tropical forest bird species to land-use intensity. *Biol. Conserv.* **2013**, *166*, 203–211. [CrossRef]
9. Fahrig, L. Effects of habitat fragmentation on biodiversity. *Annu. Rev. Ecol. Evol. Syst.* **2003**, *34*, 487–515. [CrossRef]
10. Sekercioglu, C.H. Bird functional diversity and ecosystem services in tropical forests, agroforests and agricultural areas. *J. Ornithol.* **2012**, *153*, 153–161. [CrossRef]
11. Wang, Z.; Chen, Q.E.; Gu, Z.Y.; Tang, N.; Li, N. Effects of landscape features on the structure and function of bird seed dispersal networks in fragmented forests. *Forest Ecol. Manag.* **2023**, *545*, 121251. [CrossRef]
12. Zeng, C. The Structural Composition, Spatial and Temporal Disparities of Wintering Avian Mixed-species Flocks in Meihuashan National Natural Reserve, Fujian Province. Master's Thesis, Xiamen University, Xiamen, China, 30 June 2021.
13. Li, N.; Wang, Z.; Lu, C.H.; Xiong, T.S.; Fu, W.Y.; Wu, J.P. Seed foraging and dispersal of Chinese yew (*Taxus chinensis* var. *mairei) by frugivorous birds within fragmented habitats. Acta Ecol. Sin.* **2014**, *34*, 1681–1689.
14. Zheng, G.M. *A Checklist on the Classification and Distribution of the Birds of China*, 4th ed.; Science Press: Beijing, China, 2023; pp. 1–512.
15. Ma, K.P.; Liu, M.Y. Methods for Measuring Biodiversity in Biological Communities Iα Methods for Measuring Alpha Diver-sity (Part 2). *Biodivers. Sci.* **1994**, *4*, 231–239. [CrossRef]
16. Rosenblad, A. Correction: Book Review of Extending the Linear Model with R: Generalized Linear, Mixed Effects and Nonparametric Regression Models by Faraway. *Psychometrika* **2024**, *89*, 1116. [CrossRef]
17. Charrad, M.; Ghazzali, N.; Boiteau, V.; Niknafs, A. NbClust: An R Package for Determining the Relevant Number of Clusters in a Data Set. *J. Stat. Softw.* **2014**, *61*, 1–36. [CrossRef]
18. Holmes, R.T.; Bonney, R.E.; Pacala, S.W. Guild structure of the Hubbard Brook bird community: A multivariate approach. *Ecology* **1979**, *60*, 512–520. [CrossRef]
19. Kornan, M.; Adamik, P. Foraging guild structure within a primaeval mixed forest bird assemblge: A comparison of two con-cepts. *Community Ecol.* **2007**, *8*, 133–149. [CrossRef]
20. Fang, J.Y.; Wang, X.P.; Shen, Z.H.; Tang, Z.Y.; He, J.S. Methods and protocols for plant community inventory. *Biodivers. Sci.* **2009**, *17*, 533–548.
21. Liu, B.; Zhou, L.Z.; Wang, W.G.; Shen, S.B.; Han, D.M. Seasonal Dynamics of the Avian Guild Structure of Mountain Second-ary Forest in Dabieshan Mountain. *Zool. Res.* **2009**, *30*, 277–287. [CrossRef]
22. Adams, B.T.; Matthews, S.N. Diverse temperate forest bird assemblages demonstrate closer correspondence to plant species composition than vegetation structure. *Ecography* **2019**, *42*, 1752–1764. [CrossRef]

23. Bregman, T.P.; Lees, A.C.; Seddon, N.; MacGregor, H.E.; Darski, B.; Aleixo, A.; Bonsall, M.B.; Tobias, J.A. Species interactions regulate the collapse of biodiversity and ecosystem function in tropical forest fragments. *Ecology* **2016**, *97*, 1782–1792. [CrossRef] [PubMed]
24. Karp, D.S.; Frishkoff, L.O.; Echeverri, A.; Zook, J.; Juárez, P.; Chan, K.M.A. Agriculture erases climate-driven β-diversity in Neotropical bird communities. *Glob. Change Biol.* **2018**, *24*, 338–349. [CrossRef] [PubMed]
25. Wells Harry, B.M.; Crego, R.D.; Opedal, Ø.H.; Khasoha, L.M.; Alston, J.M.; Reed, C.G.; Weiner, S.; Kurukura, S.; Hassan, A.A.; Namoni, M.; et al. Experimental evidence that effects of megaherbivores on mesoherbivore space use are influenced by species' traits. *J. Anim. Ecol.* **2018**, *90*, 2510–2522. [CrossRef]
26. Lu, Z.; Yang, G.; Zhao, D.D.; Wu, Y.H.; Meng, Y.J. Guild structure of forest breeding bird community in Nonggang Nature Reserve of Guangxi. *Zool. Res.* **2013**, *34*, 601–609.
27. Pérez-Crespo, M.J.; Fonseca, J.; Pineda-López, R.; Palacios, E.; Lara, C. Foraging guild structure and niche characteristics of waterbirds in an epicontinental lake in Mexico. *Zool. Stud.* **2013**, *52*, 54. [CrossRef]
28. Chatterjee, A.; Adhikari, S.; Pal, S.; Mukhopadhyay, S.K. Foraging guild structure and niche characteristics of waterbirds win-tering in selected sub-Himalayan wetlands of India. *Ecol. Indic.* **2020**, *108*, 105693. [CrossRef]
29. Huang, J.; Yuan, B.D.; Yuan, Y.F. foraging Spatial Niche and Diet of Zosterops japonicus in Spring. *Chin. J. Wildl.* **2017**, *38*, 76–80.
30. Wang, W.Y.; Guo, Y.S.; Hu, J.C.; Sun, L.H.; Zhu, L. Preliminary Study on foraging and Spatial Niche of *Pycnonotus sinensis* in Spring. *Sichuan J. Zool.* **2005**, *4*, 28–30.
31. Huan, L.; Zhang, M.X.; Yang, X.J.; Cui, L.W.; Jin, R.C. The breeding biology of Red-Whiskered Bulbul (*Pycnonotus jocosus*) in Xishuangbanna, southwest China. *Zool. Res.* **2015**, *36*, 233–240.
32. Zheng, Y.X. Research on Bird foraging Guilds in Tiantong National Forest Park, Zhejiang Province. Master's Thesis, East China Normal University, Shanghai, China, 30 June 2018.

forests

MDPI

Article

An International Perspective on the Status of Wildlife in Türkiye's Sustainable Forest Management Processes

Çağdan Uyar [1,*], Dalia Perkumienė [2,*], Mindaugas Škėma [2] and Marius Aleinikovas [2]

[1] Department of Forestry, Vocational School of Forestry, Istanbul University—Cerrahpaşa, Bahcekoy, Sarıyer, Istanbul 34473, Türkiye
[2] Institute of Forestry, Lithuanian Research Centre for Agriculture and Forestry, Akademija, 58344 Kėdainiai, Lithuania
* Correspondence: cagdanuyar@iuc.edu.tr (Ç.U.); dalia.perkumiene@lammc.lt (D.P.)

Abstract: Ensuring the sustainability of forests is among the priority measures to be taken against the decline in biodiversity, which is among the world's increasingly common concerns. This study investigated whether sustainable forest management processes are considering wildlife conservation objectives. Ten forest management processes were categorized and then analyzed for whether wildlife conservation is adequately considered. The wildlife data were grouped into four categories, with the most common being the protection of biodiversity and wildlife trade. The satisfaction level obtained according to the scoring method used was determined as the criterion of success in wildlife conservation. According to the scoring method applied, the overall success was found to be 50%. It was determined that a standard should be developed regarding the economic value of wildlife fauna and flora species and that this issue should be included in sustainable forest management strategies. Only 20 of 116 total sustainable forest management criteria considered wildlife. The African Timber Organization process, which has the most member countries, was identified as the process with the lowest number of wildlife criteria, at 2%, while the International Tropical Timber Organization process was found to have the most wildlife protection criteria at 20%. The conservation success rates for the two processes of which Türkiye is a member were also found to be quite low. It is concluded that there is a need to strengthen the place of wildlife, one of the most important living components for forests, in SFM processes both for Türkiye and internationally. The results obtained were evaluated both in terms of international criteria and practices in Türkiye. It is also recommended that future international meetings include wildlife health and diversity as a separate criterion when determining sustainable methods.

Keywords: sustainable forest management; wildlife; life on land; biodiversity protection

1. Introduction

Forests cover one-third of the land area worldwide. Approximately 10% of the world's forests are reserved for biodiversity conservation. As of 2022, Türkiye's total forest area is 29.8% (23,245,000 ha) of the country [1–3]. It is estimated that 70% of the world's plant and animal species live in forest areas [4].

For many years, timber-focused economic gain has been the main principle of forest management. Today, with the change in society's conservation interests and awareness of biodiversity loss, forest management has started to expand its objectives. A sustainable forest management (SFM) approach that takes more into account relevant issues, such as endangered species, wildlife, biodiversity, and wildlife areas, is developing. In many areas around the world, it is no longer possible to keep a single utilization target dominant in line with societal demands [5,6]. SFM is analyzed in terms of its ecological, social, and economic dimensions. In ecological terms, this means managing the forest ecosystem in a way that maintains its vegetation, animals, microorganisms, mineral substances, and hydrological and microclimatic features and the relationships between these features. The economic sustainability dimension includes the necessity for the units that carry out forest management activities to have economic viability to maintain their existence [7,8].

Another approach that is considered to be linked to processes is the economic value of wildlife. In this study, although the issue of economic valuation has not been specifically studied, we have attempted to provide complementary data in this respect and to contribute to the literature. The debate on the sustainable use of wildlife should not be limited to a conflict between advantages and disadvantages. Pitting the value of conservation against the value of development is a dead-end debate. A complementary approach allows conservation issues to meet development concerns [9]. Whereas the old-fashioned philosophy of nature and wildlife conservation was defensive, seeking to protect nature against the consequences of development, modern biodiversity conservation is a voluntary approach aimed at meeting human needs for biological resources while securing the long-term survival of the Earth's biological wealth [10]. This information is valuable for managers responsible for designing or modifying wildlife management areas. At this point, it is important to see wildlife as both a consumable and non-consumable resource [11].

For a complementary approach, recognizing the economic value of wildlife will also help to better understand its place in SFM processes. However, in addition to positive values, such as historical, educational, food, scientific research, and protection of ecological processes, wildlife can sometimes also provide negative values. Damage caused by wildlife to humans (losses), livestock (predation), agriculture (crop damage), and the natural landscape (invasive pests) are considered opposing or anti-values. However, differently minded researchers/decision-makers may consider the predation of livestock by predators as a normal and positive value [10,12]. Costs of game animals include browsing, predation, traffic accidents, and transmission of diseases. Economic valuation can also enable the development of policies in terms of property rights, taxes, etc. [13]. Furthermore, despite growing concern about wildlife management since the 1950s [14], there is a lack of studies on the economics of wildlife management [15]. There are few previous reviews on economic studies of wildlife management (reviewing methods used to calculate the costs and benefits of wildlife conservation projects) and comparisons of policies [16–20].

The increasing number of private forest owners in Europe and their intensive operations in various formats are of concern to policy makers and forestry sector representatives. Increasing demand and societal expectations for forest products are not limited to timber services. In order to manage forests effectively, forests need to be categorized according to their purpose [21–23]. Sustainability is very important in forest management, which involves long-term and multiple utilization of forest ecosystems. When planning these multiple uses, including social demands, sustainability must be included. If sustainability is missing in planning, total value generation will not be improved [24].

On the other hand, as a strategic wildlife management approach, the approach of reducing hunting-related tools, permits, or access opportunities in order to regulate or limit hunting activities is generally applied for habitat conservation, sustainability of biodiversity and prevention of human–wildlife conflicts. It is known that many countries, including Türkiye, have introduced hunting rules in this regard. For example, some traps, poaching equipment, and the use of certain types of weapons are prohibited. In addition, resources may be restricted in hunting when it conflicts with local communities, ecotourism, or sustainable land uses. Depending on management priorities, conservation methods can also be developed for populations which are declining as a result of overhunting. It is crucial that endemic wildlife species are managed without exceeding the carrying capacity of the area or without their numbers decreasing to the extent that the ecological balance is disturbed. In this regard, there are both hunting grounds and wild animal production facilities in order to contribute to the development of wildlife. In many of the wildlife development areas and nature parks in Türkiye, wild animal production is also carried out. In order to eliminate the negative effects of excessive hunting, the fact that the protection of wildlife is based on a good legal basis will provide deterrence. Effective implementation of the laws can only be realized by raising public awareness [25,26].

For example, in Türkiye, hunting-related activities are carried out under the authority of the Ministry of Agriculture and Forestry. The related works and procedures are regulated by the General Directorate of Nature Conservation and National Parks. Article 1, paragraph 2 of the Land Hunting Law No. 4915 enacted in 2003 covers hunting and wild animals and their habitats; their protection and development; hunting and wildlife management; hunting grounds; regulation of hunting, hunting tourism, and the production and trade of wild animals; raising public awareness; training of hunters; crimes and misdemeanors related to hunting and wildlife, and their follow-up and penalties. It is also states that the purpose of this law is to ensure the sustainability of hunting and wildlife [26]. Hunters use public resources. Cooperation with hunters and hunter organizations in the protection and development of wild animals can be successful. However, there is no such provision in the Turkish legislation [25].

For hunting tourism, the number and species diversity of game animals should be large enough to ensure the continuation of the species and genetic diversity. In 2013, the Hunting Management Information System (AVBIS) was put into effect in order to protect and properly use hunting and wildlife resources in Türkiye [27]. With this system, the aim is to protect wildlife more effectively with "Sustainable Hunting Management" activities by adopting planned hunting [28]. Hunting and wildlife management plans should be within a successful protection/control system for sustainable management. In addition, there should be cooperation between the Ministry and hunter associations in Türkiye.

In addition, the participation of interest groups should be ensured in decisions taken on sustainable management of wildlife. As a result, more effective contributions will be made to the protection and development of wildlife resources and biological diversity. Local people and hunters should be informed and awareness should be raised through notices, books, brochures, etc. on hunting. In doing so, public relations techniques should be utilized. In this way, local people can be more aware of the issue of wild animal poaching, and they will be encouraged to report poachers. In addition, more studies should be carried out to increase the production of wild animals in wildlife breeding stations and to release these animals into nature. In conclusion, for the sustainable management of hunting and wildlife resources, these resources should be protected, developed, and brought into the economy [25,26,28,29].

Due to the multiple use/utilization areas, flora and fauna elements that are significantly damaged in Türkiye should be included as a main subject in SFM processes for more effective conservation. The date of emergence of each SFM process, the number of members, and the number of criteria are already known [30]. However, these data are not tabulated in the literature and the scope of each criterion in terms of wildlife is unknown. This has been identified as a gap that needs to be filled. In addition, each process needs to be examined in terms of its area of coverage when developing revised SFM processes. The processes were grouped as national, regional, global, etc. in order to provide data for future research on the subject. This data is expected to have a developmental impact on SFM research.

In the Forest Europe process (Helsinki resolution/the Pan-European process), the SFM is defined as "the stewardship and use of forests and forestlands in a way, and at a rate, that maintains their biodiversity, productivity, regeneration capacity, vitality and potential to fulfill, now and in the future, relevant ecological, economic and social functions at local, national, and global levels and that does not cause damage to other ecosystems" [6,31]. The United Nations (UN) defines SFM as a dynamic and evolving concept that has the objective of conserving and enhancing the economic, social, and environmental values of all forests for the benefit of present and future generations [32,33]. According to the 2020–2030 strategy of the CIFOR (Center for International Forestry Research) and ICRAF (International Centre for Research in Agroforestry), whose work is aligned with the Sustainable Development

Goals, the Paris Agreement, and the three Rio Conventions, deforestation and biodiversity are among the five challenges of global significance that should be addressed [34,35].

In 2003, at the International Conference on the Contribution of Criteria and Indicators to the SFMs, Methods for Moving Forward, seven thematic areas were identified that form the basis of the current international (eco-regional) and regional criterion and indicator processes of SFMs. The thematic areas are as follows [30,36–39]:

- Scope and extent of forest resources (contribution to the global carbon cycle, etc.).
- Forest biodiversity.
- Health and vitality of forest ecosystems.
- Protective functions of forests: soil, water, and others.
- The production functions of forests: NTFPs (nontimber forest products), etc.
- Socioeconomic functions of forests.
- Legal, institutional, and policy framework.

According to comprehensive research, 130 of the approximately 13,200 mammalian and bird species estimated to have existed in 1600s are now extinct worldwide. Many reptiles, amphibians, fishes, and invertebrates are also endangered, and there are an estimated 20,000–25,000 endangered plants. Three-quarters of these species have become extinct or endangered as a direct result of human activities [40]. In addition, invasive alien species also have a great impact on biodiversity loss [41]. Today, in Türkiye, 481 birds, 150 mammals, and 130 reptile species are protected by the General Directorate of Nature Conservation and National Parks of the Ministry of Agriculture and Forestry [3]. As biodiversity and wildlife are closely related, it should be recognized that they are also pertinent to NTFPs involved in SFM processes.

According to target 15(7) of the United Nations' international development goals, Life on Land, urgent action is needed to end the poaching and trafficking of protected species of flora and fauna and to address both the demand and supply of illegal wildlife products [42–46]. Therefore, by taking this study into consideration, the place and importance of wildlife in development will be emphasized. It is also important to note that NTFPs are all remaining materials except timber taken out of forests for human use. These include foods, medicines, spices, essential oils, resins, gums, latex, tannins, dyes, ornamental plants, wildlife (product and live animals), fuel wood, and raw materials, notably, rattan, bamboo, small wood, and fibers [34,35,47,48]. It is also very important to involve local communities in monitoring and protection processes in order to effectively manage SFM processes.

According to the inferences obtained from the related literature, the impact of general approaches and practices on wildlife is examined, and its place in terms of Türkiye is also discussed. Conclusions for improving the situation in Türkiye with the findings obtained from the research are presented. This research investigated the effectiveness of sustainable forest management processes in taking into account wildlife conservation objectives.

2. Materials and Methods

This research investigates whether there are sufficient wildlife protection criteria for SFM processes. The Republic of Türkiye's Ministry of Environment, Urbanization and Climate Change's data on international agreements and conventions to which Türkiye is a party in the field of environment (available online: https://ab.csb.gov.tr/turkiyenin-cevre-alaninda-taraf-oldugu-uluslararasi-anlasma-ve-sozlesmeler-i-109376 accessed on 18 November 2024) and international literature from Web of Science database, Scopus database and Google Scholar articles as well as official documents and conventions that include the keywords "wildlife" and "sustainable forest management", and "wildlife" and "SFM" between 1992–2024 were used to examine the criteria and indicators of SFM processes in Türkiye. Among the analyzed data related to these criteria and indicators, it was determined that the information on wildlife conservation was quite insufficient. It was also observed that there are only local or regional studies on wildlife and SFM.

The number of member countries and the number of criteria in SFM processes were tabulated by evaluating the data in the literature. From this point of view, firstly, international SFM processes, such as global, regional, national, forest management unit, and local processes were researched and categorized into a table. Then, by analyzing the data obtained from the official website of the United Nations, the number of criteria related to wildlife was determined for each process and shown in the graph. Within 10 processes, the criteria ratios for wildlife were determined. These data were then shown as a pie chart for their percentage expression in all processes. All tables and figures in this article were compiled by the authors based on the literature review. Criteria for wildlife in the SFM processes were also identified and analyzed in terms of content. The analyzed criteria were typologically reduced into 4 groups.

Each process was analyzed separately in terms of wildlife conservation types, and the EFLD (Environmental and Forestry Law Department) scoring method was applied to each process in the 4 reduced groups. In order to examine whether a certain criterion that contributes to wildlife conservation is at a satisfactory level, the typologies were analyzed for each process and scored according to the level of protection. The process contents for each typology were explained and shown in separate groups. In addition, by evaluating which of these processes Türkiye is involved in, it was determined which processes in the SFM processes do not have a sufficient impact on wildlife despite the high number of member countries by using this scoring method [49].

A five-point scoring system, namely the EFLD method, was used for scoring (Table 1). This benchmark was previously used by environmental and forest attorneys [50–55]. The method scoreboard is as follows.

Table 1. EFLD method scoreboard.

Point	Description
0	Does not meet criterion
1	Criterion is described in general terms and is inadequate
2	Criterion is described in close terms but not enough detail
3	Criterion is partially met but there are gaps
4	Criterion is fully met

Based on this information, a study has been conducted to deepen our understanding of wildlife in terms of SFM in order to fill the gap in the subject. Considering the ongoing loss of wildlife worldwide, there is a need to evaluate SFM criteria in this respect. In addition, after all the data were obtained, document analysis was carried out again with a deductive approach in the aforementioned literature, and inductive suggestions were made together with those identified in the content analysis.

3. Results

Since the Stockholm Conference, the Brundtland Report, and the United Nations Conference on Environment and Development (Rio Summit) in 1992, significant progress has been made at regional, national, and international levels in defining and developing standards, setting criteria and indicators, and promoting and implementing the concepts of sustainability, sustainable development, and sustainable forest management. This management approach, which stems from the need for the conservation of biodiversity and other environmental concerns, has gone through 10 main stages, and some of these stages have not continued to take place within the process [56]. In this respect, it can be said that SFM criteria have a dynamic structure. Due to all the protection requirements, it is important to remember that the path to the most accurate assessments is through good research. The results of the data analysis in this regard are presented below.

3.1. Review of SFM Processes

Criteria and indicators are essential tools for the development of forest policy which provide relevant information for national forest policies, plans, and programs, and serve as the basis for cross-sectoral forest-related data collection. Sustainable forest management makes it possible to monitor, assess, and report developments at the global, regional and national levels. In this respect, one of the objectives of the implementation of the SFM criteria and indicators is to create a basis for dialogue and communication between policy makers and relevant stakeholders in forestry and other sectors for SFM and forestry policy development.

A number of initiatives have been organized under the leadership of the Food and Agriculture Organization of the United Nations to identify and develop SFM criteria and indicators at the regional and local levels. A total of ten separate regional processes involving different regions of the world have emerged. In fact, more than 150 countries have participated in these processes [57]. The SFM processes identified in this framework are as follows [47,58,59] (Table 2).

Table 2. SFM processes, number of regional member countries, and scope of the criteria.

	SFM Process	Member Country Number	Start Date of Process	Number of Criteria *
1	African Timber Organization (ATO)	13	1996	26 (R, N)
2	Dry Forests in Asia Initiative	9	1999	8 (N)
3	Dry Zone Africa Process	36	1995	7 (N)
4	International Tropical Timber Organization (ITTO)	74	1992	7 (N, F)
5	Lepaterique Process of Central America	7	1997	4 (R), 8 (N)
6	Montreal Process	12	1995	7 (N)
7	Pan-European Forest Process	41	1994	6 (R, N)
8	Tarapoto Proposal	8	1995	1 (G), 7 (N), 4 (F)
9	Near-East Process	30	1996	7 (R, N)
10	Center for International Forestry Research Process (CIFOR)	31	1998	24 (L)

* G: global, R: regional, N: national, F: forest management unit, L: local.

The general table of SFM processes shows that there are 1 G, 43 R, 83 N, 11 F, and 24 L criteria. It is important to work on addressing these processes, which are mostly national and regional, at a more global level.

3.2. Importance of Wildlife Protection

In recent years, efforts to prevent the loss of wild fauna/flora and their habitats have been prioritized. In this case, decisionmakers need to see and prioritize criteria established on this issue more frequently among the SFM framework criteria and in the reports of scientific researchers. In particular, it was evaluated that studies on linking forest biodiversity and management indicators to strengthen SFM are useful, especially for discovering important species in the area to be managed. In such studies, local people contribute to both habitat and SFM processes in terms of tracking and discovering wildlife and biodiversity [60]. In addition, it is important to examine studies on the effect of NGO participation on local people's understanding of forest management as another management tool. In such studies, it has been argued that more effective and efficient protection can be achieved if NGOs participate more actively [61]. NGOs have also played a role in helping to shape the links between government and society and have influenced the legal framework. Community-based forest management is based on the belief that communities are in the best position to manage and protect forests if they participate in decision-making on the sustainable use of forest resources. This can be seen as an enabling mechanism for the valuation of NTFPs [62]. Among the new trends, the involvement of local communities in the monitoring of forests and agricultural areas with unmanned vehicles, such as drones, to increase the effectiveness of monitoring and managing forest health, will also contribute to sustainability [63]. In addition, regular monitoring of forest health by providing access to satellite-based imagery to local communities is crucial for sustainability (i.e., publicly accessible imagery, such as Landsat, MODIS, and Sentinel data) [64].

Since forests are one of the habitats for wildlife and the conservation of forest biodiversity will affect wildlife, it is important to ensure that conservation policies in these ecosystems are effective and comprehensive. In the context of forest biodiversity (Sustainable Development Goal 15.7), pressures refer to human activities and factors that put pressure on forest ecosystems and contribute to the loss or degradation of forest biodiversity. The stressors may include deforestation, forest fragmentation, unsustainable logging practices, the introduction of invasive species, pollution, the impacts of climate change, and the illicit wildlife trade. These activities and factors directly or indirectly affect the composition, structure, and functioning of forest ecosystems, resulting in a decrease in biodiversity [65]. Among the processes in Table 1, the numbers of SFM criteria related to wildlife are as follows (Figure 1).

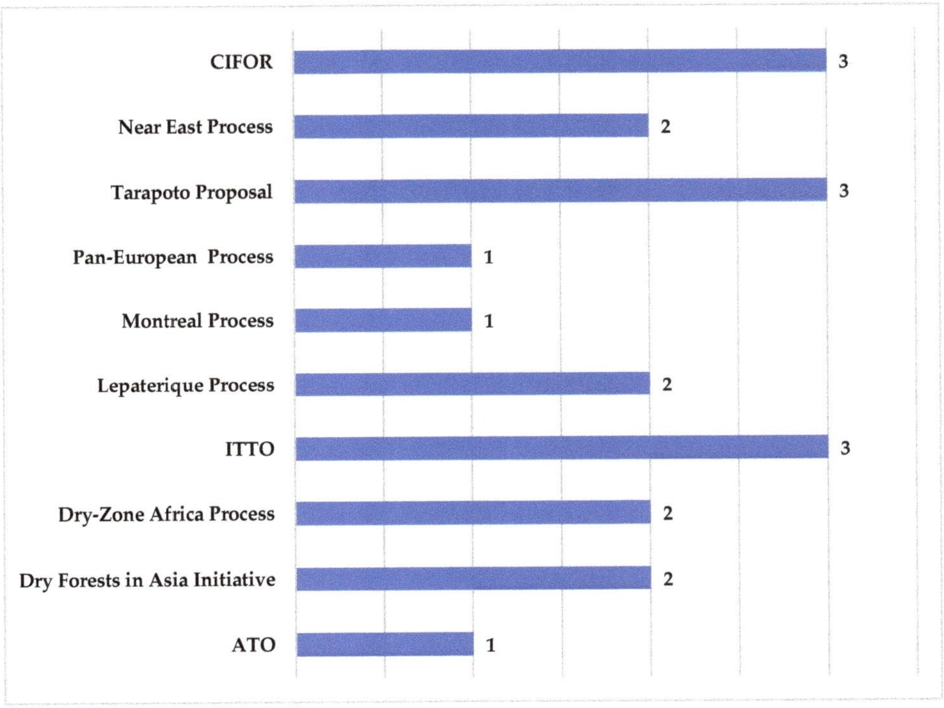

Figure 1. Criteria counts for wildlife in SFM processes.

The percentages of criteria allocated to wildlife sustainability within the SFM processes are as follows: ATO, 3.85%; Asian Dry Forest Process, 25.00%; Dry Zone Africa Process, 28.60%; ITTO, 42.80%; Lepaterique (Central America) Process, 16.70%; Montreal Process, 14.30%; Pan-European Process, 16.70%; Tarapoto Proposal, 25.00%; Near-East Process, 28.60%; and CIFOR, 12.50%. Percentages of wildlife-related criteria in the all the SFM processes are also as shown in the following pie chart (Figure 2).

The processes with the highest number of criteria for wildlife are CIFOR, the Tarapoto Proposal, and ITTO. However, the situation changes according to the pie chart that emerges when compared to the total criteria in all processes. In this case, the processes that are most amenable to wildlife in SFM processes are the ITTO, Near-East Process, and Dry Zone Africa Process. CIFOR and the Tarapoto Proposal, which have more criteria for wildlife, were found to be less relevant when calculated in general.

According to the literature review, the ITTO process includes the most criteria, and the ATO process includes the least criteria. Points that need to be improved in these processes are given in the discussion and conclusions.

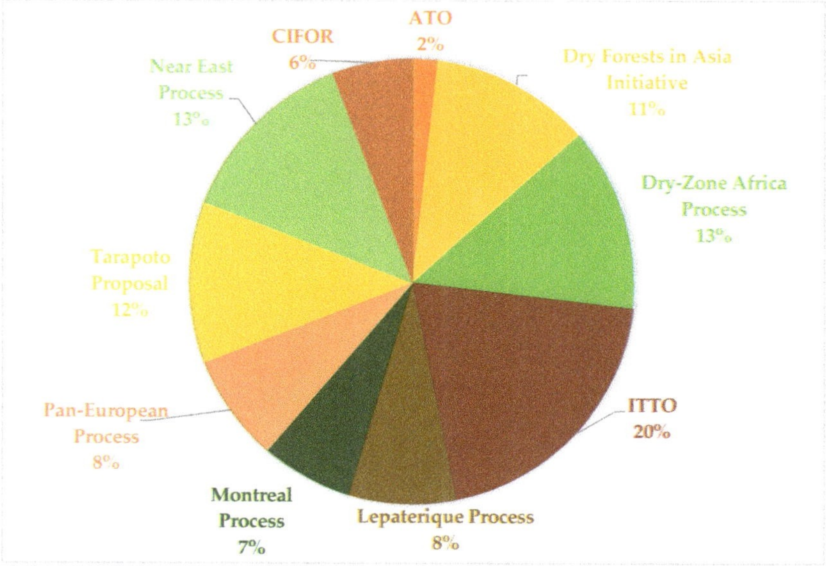

Figure 2. Percentage of criteria for wildlife sustainability.

4. Criteria for Wildlife in SFM Processes

The criteria related to wildlife in SFM processes are summarized below. The reason for including trade-related criteria is that wildlife use is directly related to the CITES Convention (Washington Convention) on the prevention of the illegal wildlife trade.

4.1. *African Timber Organization Process*

The ATO process started in January 1993. Türkiye is not a member. Among the criteria of the process, it was determined that the criterion related to wildlife is 3.2. This criterion pertains to the protection of rare and endangered species, conservation of biodiversity, and minimization of negative impacts. Eight indicators of criterion 3.2 are listed as follows [39,66]:

- Biological protection zones are established on permanent forestland where no intervention is allowed.
- The size of biological reserves has to be adapted to suit the conservation objective.
- The potential for effective conservation has to be considered in the selection of biological protection areas.
- Special provisions for the protection of sensitive areas, plains, riverbanks, and steep slopes have to be defined in the management plan.
- Forest management plans are provided for single-species or exotic-species plantations only when other types of silvicultural activities are considered by forest management experts and abandoned for justifiable reasons.
- If enrichment planting is performed in afforested forests, species that have actually been harvested in the forest will be preferred.
- Rare or endangered species have to be protected.
- High-demand nontimber forest products are the objects of conservation and domestication experiments.

4.2. *Asia Dry Forests Process*

This process started in December 1999. Türkiye is not a member. Criterion 3 pertains to the maintenance and enhancement of biodiversity. The indicators of these criterion are as follows: "scope of protected areas; number of endangered, keystone, flagship and endemic animal and plant species; list of fauna and flora; degree of nondestructive harvesting; percent cover by forest type or/and species; and presence of mechanisms for the protection of genetic resources".

Criterion 6 is related to the scope of forest resource utilization; one of the indicators of the criterion is "Import and export of wood and NTFPs" [67].

4.3. *Dry Zone Africa Process*

This process started in November 1995. Türkiye is not a member. Criterion 2 pertains to the conservation and enhancement of biodiversity in forest ecosystems for the "Permanent Interstate Committee for Drought Control in the Sahel" (CILLS) countries (CILLS Countries: Permanent Interstate Committee for Drought Control in the Sahel). The indicators of these criterion are categorized as ecosystem indicators, species indicators, and fauna/flora genetic indicators.

Criterion 6 for CILLS countries pertains to the maintenance and enhancement of socioeconomic benefits; one of its indicators is the forest sector trade balance.

Criterion 6 for SADC countries (SADC: Southern African Development Community) pertains to the maintenance and enhancement of the socioeconomic benefits of forests and other wooded areas. The criteria for CILLS and SADC include 12 indicators, 8 of which are related to the economic benefits of the forest and forest industry sectors [39,68].

4.4. *The International Tropical Timber Organization Process*

The ITTO process started in March 1992, and was revised in 1999. Türkiye is not a member. Criterion 3 pertains to forest ecosystem health and conditions. Among the indicators of the criterion, protection and conservation procedures include the "existence and implementation of quarantine and phytosanitary procedures to prevent the introduction of pests and diseases" and the "existence and implementation of procedures to prevent the introduction of potentially harmful exotic plants and animal species".

Criterion 4 pertains to the flow of forest products. This criterion concerns forest management to produce wood and NTFPs. Such production can be sustainable only in the long term if it is economically and financially feasible, environmentally friendly, and socially acceptable.

Forests allocated for production can fulfill a number of important forest functions, such as environmental protection and biodiversity conservation. These multiple roles of the forest must be maintained through concrete management practices that safeguard the potential of the full range of benefits that the forest resource can provide to society. The indicators for this criterion include indicators of resource assessment, indicators of planning procedures, and indicators of management guidelines.

Criterion 5 pertains to biodiversity. The indicators of these criteria include indicators of ecosystem diversity; indicators of diversity among endangered, rare, and threatened forest flora and fauna species; indicators of in situ and ex situ conservation of genetic diversity in commercial, endangered, rare, and threatened forest flora and fauna species; indicators of management guidelines; and indicators of monitoring and evaluation [47,69].

4.5. *The Central America Process*

This process (also called the Lepaterique Process) started in January 1997. Türkiye is not a member. Criterion 2 pertains to the protection and maintenance of environmental services provided by forest ecosystems at the regional level in Central America. Three of the thirteen indicators of this criterion are as follows: "implementation of mechanisms to regulate the capture and illegal marketing of flora and fauna species", "number of endemic, threatened and/or endangered species", and "number of forest species protected ex situ (e.g., in seed banks)".

Criterion 5 pertains to biodiversity in forest ecosystems at the national level in Central America. The indicators for this criterion are listed as follows: "percentage and area of forest species in various categories of protected areas", "number of endemic, threatened and/or endangered species", "estimates of wildlife species dependent on forest habitats", "area and length of biological corridors per forest ecosystem", "area and percentage of primary and secondary forests and plantations", and "number of species protected ex situ (e.g., in seed banks)" [70].

4.6. Montreal Process

This process started in June 1992. Türkiye is not a member. Member countries agreed on a legally non-binding dataset for national implementation of the SFM and agreed to review and consider possible elements for criteria and indicators at the forest management unit level. Criterion 1 pertains to biodiversity conservation. The indicators of this criterion are as follows [71]:

- ecosystem diversity,
- interspecies diversity and genetic diversity of the species, and their subheadings.

4.7. The Pan-European Process

The Pan-European process (also called the Helsinki resolution) started in June 1993 and is also known as the Helsinki Process. Türkiye is a member. The process is supervised by the Ministerial Conferences on Forest Protection in Europe (Forest Europe). The topics that form the basis of forestry practices in Europe include "Domestic policies that are effective in sustainable forest management and development", "Certification", "Basic forestry measures-monitoring", "Rural development", "Biodiversity", "Climate change and carbon management", "Forest industry", "Forest production materials and phytosanitary", "International forestry policy processes", "Cooperation development", and "International trade and foreign relations" [72].

It is possible to establish a direct relationship between these European criteria and wildlife through the issues of "biodiversity", "forest production materials and phytosanitary" and "international trade". Among the criteria identified in this process, the criterion related to wildlife is stated as follows: Criterion 4 pertains to the maintenance, conservation, and enhancement of biodiversity in forest ecosystems [47].

4.8. The Tarapoto Proposal

This process started in February 1995. Türkiye is not a member. Criterion 4 is at the national level and pertains to the protection of the forest and biodiversity. Two of the eight indicators of this national-level criterion are listed as "measures for the in situ conservation of endangered species" and "measures for the conservation of genetic resources".

Criterion 10 is at the management unit level and pertains to the protection of forest ecosystems. One of the six indicators of this management unit level criterion is "measures for the conservation, recovery and sustainable use of wild populations of endangered species".

Criterion 12 is at the global level, and pertains to the economic, social, and environmental services performed in the Amazon forest. Two of the seven indicators of this global criterion are as follows: "contribution to meeting global demand for sustainably produced wood and NTFPs" and "contribution to biodiversity conservation" [39,47].

4.9. The Near-East Process

This process started in October 1996. Türkiye is a member. Criterion 2 pertains to the conservation of biodiversity in forest areas. The indicators of this criterion are listed under the headings "ecosystem indicators", "fauna and flora species indicators", and "genetic indicators".

Criterion 6 pertains to the maintenance and improvement of socioeconomic functions and conditions. Among the indicators of this criterion, the headings of the indicators of economic conditions are as follows: "value of wood products", "value of nontimber forest products", "value of recreation", "value of hunting", "share of forest industry in

GNP/GDP", "value obtained from secondary forest industries", "value obtained from biomass energy", "forest trade balance", and "level of investment in the forest industry and other industries based on forest products" [47].

4.10. Center for International Forestry Research

CIFOR criteria and indicators were first used in September 1998. Türkiye is not a member. One of CIFOR's six main principles is ecology. The criteria for "maintenance of ecosystem integrity", which is linked to ecology, are "preservation of processes that maintain biodiversity in managed forests", "preservation of ecosystem function", and "preservation of processes that maintain genetic diversity". CIFOR includes 24 criteria and 98 indicators which are used as a "starting platform" for countries to develop their own local sets of criteria and indicators. These criteria include indicators and validators to protect biodiversity and its processes [47,48].

5. Types of Criteria for Wildlife Conservation

From the analysis of SFM processes, four main types of wildlife conservation were identified. These are wildlife trade, biodiversity conservation, NTFPs, and wildlife health. Among the 10 processes, 6 regulate the criteria for wildlife trade, 6 are for biodiversity conservation, 5 are for NTFPs, and 3 are for wildlife health. It would also be useful to proportionally represent all the criteria in these SFM processes in terms of the criteria headings related to wildlife (Figure 3).

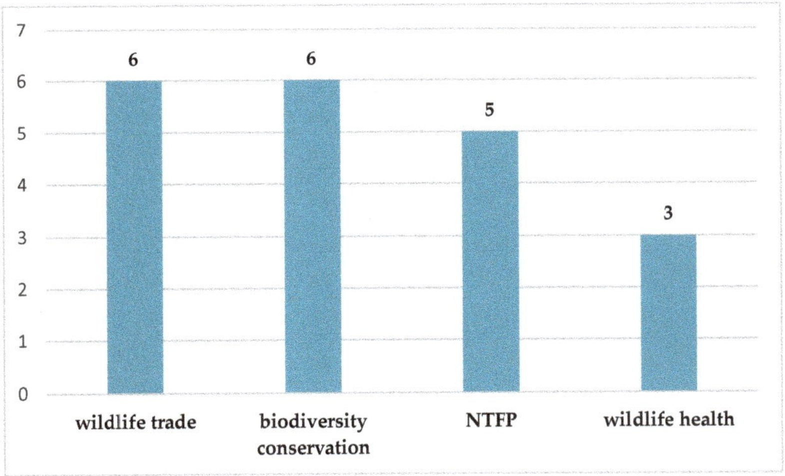

Figure 3. Criteria counts pertaining to wildlife protection.

When each process was analyzed in terms of the four protection types that it combines in terms of wildlife, the following results were obtained:

1. The contents of the processes with provisions on wildlife trade are based on "import and export, product demand, biodiversity conservation in commercial, illegal marketing of flora and fauna species, international trade and foreign relations, investment in forest industry, share of forest industry in GNP/GDP, value of products and forest trade balance".
2. The contents of the processes with provisions on biodiversity conservation are based on "rare or endangered species protection, biological reserves and biological protection zones, maintenance and enhancement of biodiversity, inventory works, genetic resources, works on forest species, habitat and forest ecosystem, reclamation on areas,

in situ ex situ protection works, genetic diversity, measures for the conservation, resilience and sustainable use of wildlife, ecosystem functions integrity".

3. The contents of the processes with provisions on NTFPs are based on "export and import of non-wood forest products, forest management for the production of timber and NTFP's, contribution to meeting global demand for sustainably produced nontimber forest products, value of nontimber forest products, maintenance of ecosystem integrity".
4. The contents of the processes with provisions on wildlife health are based on "measures for the in situ conservation of endangered species, existence and implementation of procedures to prevent the introduction of potentially harmful exotic plant and animal species, resource assessment, planning procedures and indicators of management guidelines, management practices that preserve the potential for the full range of benefits that the forest resource can provide to society, biodiversity conservation works such as management guidelines on species and forest conservation, measures for the conservation of genetic resources, maintenance and preservation of biodiversity processes".

When the 10 processes are analyzed in terms of wildlife conservation, the most related processes are ITTO, Tarapoto, and CIFOR. Other processes were seen to have a more general approach. When the other processes are examined, they are evaluated as weak even though they contain wildlife protection criteria in two aspects.

The wildlife criteria analyzed according to the EFLD method in terms of all processes were scored for the effectiveness of the conservation types in the processes. The score distribution in this regard is as shown in Table 3.

Table 3. Wildlife conservation type points of each process.

Process	Wildlife Conservation Type	Point
ATO	Biodiversity conservation	1
Asia Dry Forests Process	Wildlife trade, NTFPs	2
Dry Zone Africa Process	Wildlife trade, biodiversity conservation	2
ITTO	Wildlife trade, NTFPs, wildlife health	3
Lepaterique Process	Wildlife trade, biodiversity conservation	2
Montreal Process	Biodiversity conservation	1
Helsinki Process	Wildlife trade	1
The Tarapoto Proposal	Biodiversity conservation, NTFPs, wildlife health	3
The Near-East Process	Wildlife trade, NTFPs	2
CIFOR	Biodiversity conservation, NTFPs, wildlife health	3

According to the results obtained by calculating the score in the table, the overall success of all processes in terms of wildlife sustainability was found to be 50%. It is important to discuss the data on this subject and to develop them in line with the conclusions.

6. Discussion

The analysis of 10 processes revealed that the SFM criteria and indicators provide an international reference for policy makers in formulating national policies and programs. However, in many countries, there are still insufficient data to assess flora and fauna, as well as gaps in monitoring [49,73]. Therefore, there are no detailed data on wildlife in the international meetings on SFM processes.

A report of the United Nations Intergovernmental Science-Policy Platform on Biodiversity and Ecosystem Services (IPBES 2019) revealed that nearly 1 million plant and animal species are threatened with extinction due to human activities [74]. The Living Planet Report (2020), published by the WWF, reminds us that vertebrate species populations have declined by 68% in the nearly 50 years between 1970 and 2016. In addition to these findings, research has shown that protected areas are among the most important tools available for slowing the loss of biodiversity [75].

When the processes are analyzed, several criteria for wildlife reveal what needs to be improved and renewed. A healthy diversity of wildlife is a key element of biodiversity. If biodiversity suffers, so does wildlife. It is very important to protect biodiversity in terms of both habitat and species diversity. In the processes that have been implemented for several years and developed on the basis of conservation measures, wildlife is included in very few criteria. When this situation is analyzed in percentages, it is seen that timber-related processes predominantly exclude wildlife. However, it is illogical to consider these elements separately in wood quality and ecosystem integrity.

According to the result of wildlife conservation type points (Table 3), the processes that need to be developed in terms of the criteria reduced to four protection types have to be evaluated in terms of increasing the success of international sustainable forestry in the future. Wildlife protection types, which are one of the most important components for the sustainability of forests, should be developed throughout the SFM, and there should be more than four protection types. It is seen that Türkiye is a member of the Helsinki and Near-East processes. It has been seen that there is a need to bring new updates on wildlife health to these two processes. Compared to other processes, the lack of effort on biodiversity conservation and wildlife health in both processes was also seen as a common deficiency. This situation is explained in the SFM (2020) publication as well as in many reports which indicate that there is not enough flora and fauna inventory to evaluate and consider [2].

Regarding the Near-East process, the forest trade balance is expressed as an indicator. Under the criterion "conservation of biodiversity in forest areas", there is no mention of the protective use of wildlife. According to the other criterion, poaching of wild animals was cited as one of the reasons for the loss of natural forest areas. Therefore, although there are no detailed data on wildlife in the SFM processes in which Türkiye has been involved, it is known that there is a commercial wildlife supply and demand situation that continues today. Although wildlife has been included more frequently in the agenda of the recent SFM (2020) report, effective conservation success has not yet been achieved. In many scientific meetings, the need to improve legal infrastructure, increase the population of wildlife, and take measures to protect genetic purity have been emphasized [2]. These meetings have also emphasized how state ownership under economic pressure in Türkiye can lead to forest loss, and the need for a more adaptive policy approach [76]. There are also studies showing that 75% of visitors give negative comments when they see park management that does not take into account wildlife and their habitats [77].

Human activities are the most important driving force affecting wildlife populations through direct and indirect interactions [78,79]. Over the years, strict conservation aims were formulated and humans and wildlife were separated in order to reduce the negative effects of human practices, including agriculture, leisure, and hunting [80]. However, the positive effects of practices, such as hunting, should not be ignored. As opposed to overexploitation and poaching, controlled hunting is very useful for ecosystem sustainability. For example, hunting is a very necessary control mechanism in terms of combating invasive alien species, has economic benefits, and helps to make the population suitable for the carrying capacity of the area, etc. [29,41].

Hunting income is also a source of livelihood for many local communities. Approved practices, such as sport hunting, are also widely used as a preventive or curative measure. A separate source of income is also obtained for ecotourism or sport hunting practices in the areas. In addition, the products obtained after planned hunting, such as trophy hunting, contribute to the society and local economy through both direct consumption and trade [79].

Wildlife management aligns with each of the above criteria. In particular, sustainability, biodiversity, climate change, policies, and international trade are the main objectives of protecting wildlife habitats, flora, and fauna. Therefore, it is not possible to assess wildlife independently of the SFM criteria. When the criteria counts pertaining to wildlife protection are analyzed, the weakest of the four main types of protection are NTFPs and wildlife health. It is thought that international conventions on these topics and the biodiversity conservation agenda can also be taken into consideration to determine the improvements to be prioritized in the studies on the subject. Healthy and diverse wildlife requires a healthy and sustainable forest. Hopefully, taking this study into consideration will contribute to creating a new perspective on preventing the loss of biodiversity on land.

The importance of NTFPs and forest data (including flora and fauna specific studies) for strengthening the bioeconomy in Europe is emphasized [73,81]. Good knowledge and valuation of forest products (living and non-living) can both diversify forest incomes and enhance ecological integrity. Scientific research on transparency to increase access to forest data is also in progress. There are also studies investigating contradictions in forest utilization licenses and management rights. The main purpose of these studies is to contribute to sustainability [82,83].

Today, forestry faces a complex management situation. The understanding of SFM has moved far beyond the original meaning of ensuring a sustainable timber yield. SFM strategies must fulfill ecological, economic, and social functions without harming other ecosystems. In this understanding, forest management actions cannot be viewed as isolated or monocausal [5]. In the planning and revision of forests, local people and stakeholders making decisions together with protected area management, i.e., ensuring participation in sustainable conservation, will prevent the unclear division of responsibilities among authorities and encourage cooperation and success [84].

When the framework criteria are determined and continue to be developed at the International Conference on Criteria and Indicators (CICI) according to the current FAO data, it can be concluded that national and international institutions should continue to conduct research on difficult-to-measure criteria and indicators. These may be focused on particular issues, such as biodiversity, NTFPs, nonmarket values, soil and water conservation, carbon sequestration, and sociocultural aspects [36–38,47,59].

It is also very important to protect and restore the natural structure of existing habitats in line with the ecological demands of wild animals. In addition, bio-restoration plans should be regularly implemented to ensure that the carrying capacity of habitats and the target species are optimally maintained and improved [85]. In forests, which are the priority ecosystem for wildlife, forest cover needs to be improved primarily. To achieve this goal, afforestation, sub-flora maintenance, control of grazing, restoration of plant species preferred by wildlife, reproduction of shrubs, trees, and bushes in the areas, laying of trimmed branches, combating erosion in the area, maintenance and development of water resources, and combating the direct and indirect effects of forest fires should be monitored regularly [86]. Environmental arrangements in areas where wildlife populations are present should be carried out with the highest care, and factors that threaten the needs and survival of wild animals should be considered. Animals, which are the active element of wildlife, have demands, such as food, shelter, dust bathing, roaming, and water [86–88].

Proper management of hunting also involves the rational organization of hunting range capacity. It is known that both good monitoring of population dynamics and the correct implementation of hunting will help to improve habitat health and increase wildlife potential. It is clearly seen that the main elements of wildlife management, such as inventory, education, monitoring, and supervision, need to be managed, especially within the scope of SFM.

7. Conclusions

It is not possible to evaluate wildlife separately from SFM criteria. A forest without wildlife and a humanity without forests are unimaginable. The planning of forest products according to socioeconomic factors and other functional purposes requires that wildlife be included in these plans. Economic planning in forests without considering wildlife diversity will have consequences that threaten the existence of forests. It is important to calculate the economic value of wildlife, which is counted among NTFPs. A standard should be developed for determining the economic value of wildlife fauna and flora species, and this issue should be included in the SFM criteria. This is because there are many criteria and indicators that point to this value. While many issues, such as the illegal wildlife trade, NTFPs, and invasive alien species are being studied, it has been determined as a deficiency in terms of developing this standard. The economic valuation of wildlife is considered to be within the scope of NTFPs, one of the four prominent wildlife protection typologies. It is clear that more work needs to be carried out to develop a standardization for the economic valuation of wildlife, which is known to have a place in all areas of SFM. As seen in many international definitions of SFM, it is an economic policy, and it is critical to include the living and non-living parts of wildlife.

In Türkiye, the economic value of the entire hunting and wildlife sector has not yet been calculated even for the largest cities. Products obtained from the skins, feathers, etc., of wild animals also make a significant contribution to the economy. If the ecological, cultural, and social values of hunting and wildlife are also taken into consideration, the total economic value will increase much more and the contribution of the sector to GDP will increase at the same rate. In this respect, it is important to carry out the necessary scientific research in order to calculate the total economic value of hunting and wildlife mathematically. It would be useful to examine the examples of countries that have achieved success in this regard. In addition, a database in which the records likely to be used for this purpose are kept in a healthy way should also be established.

The SFM criteria and indicators provide an international reference for policy makers for formulating national policies and programs. Despite these limitations, in many countries, there is still a lack of sufficient data for the assessment of flora and fauna, and monitoring deficiencies exist. It is a priority for Türkiye and many other countries to develop standards and to pay more attention to inventory studies.

Social participation in decision-making, the research/development dimension, and the resource utilization dimension of forest villagers, as well as the integration of wildlife in decision-making processes, will ensure full protection. Considering the social participation in decision-making, the research and development (R&D) perspective, and the resource utilization perspective of forest villagers, the inclusion of wildlife in decision-making processes will ensure their full protection.

The majority of the SFM criteria are directly related to wildlife. While there are many criteria related to wildlife, there are still none that directly address wildlife. It is observed that wildlife is not sufficiently mentioned even among the prominent indicators in a criterion as broadly defined as biological diversity. This will always be a weakness in maintaining forest unity. Therefore, wildlife health and diversity should be added as separate criteria in addition to the studies and opinions that set the criteria for SFM. In addition, in future international meetings, it is crucial that the issue of wildlife be brought to the agenda as a main heading rather than a subheading to make sustainable decisions regarding wildlife.

Sustainability in forests and wildlife are not issues to be planned independently of one other. Forests are the primary habitat of wildlife. In this case, while there are many criteria related to wildlife, the fact that a criterion directly related to wildlife is still not regulated is an issue that will always be missing in the protection of forest unity.

According to criteria that are broadly defined as relating to biological diversity, wildlife is not mentioned enough even among the leading indicators. It should also be emphasized that wildlife management and the legal infrastructure cannot be considered separately. This legal infrastructure should be in accordance with current international law, effective, and applicable. Scientific researchers at Türkiye should classify these important criteria as priorities in their implementation policies.

From a general framework of SFM processes, it is seen that among the criteria for identifying wildlife, the most common criteria are biodiversity conservation and the wildlife trade. When the results are analyzed in general, only 20 criteria point to wildlife among a total of 116 SFM criteria, and this is considered to be insufficient. This general conclusion is expected to contribute to international conventions and other meetings to be organized on biodiversity conservation.

The fact that only 17% of the criteria of the 10 processes include wildlife is a major deficiency. Within the existing criteria, which are typologically divided into four headings, wildlife health should be developed as a priority. The fact that wildlife criteria, of which there are very few in the overall SFM criteria, achieve only 50% success even when grouped together, shows the importance of developing this subject and adding it as a main criterion in international SFM processes.

Author Contributions: Conceptualization, Ç.U., D.P., M.Š. and M.A.; methodology, Ç.U. and D.P.; software, Ç.U.; validation, Ç.U., D.P. and M.A.; formal analysis, Ç.U., M.Š. and M.A.; investigation, Ç.U. and D.P.; resources, Ç.U., D.P. and M.Š.; data curation, Ç.U. and M.A.; writing—original draft preparation, Ç.U. and D.P.; writing—review and editing, Ç.U., D.P., M.Š. and M.A.; visualization, Ç.U. and D.P.; supervision, M.Š. and M.A.; project administration, Ç.U. and D.P.; funding acquisition, D.P., M.Š. and M.A. All authors have read and agreed to the published version of the manuscript.

Funding: This research received no external funding.

Data Availability Statement: The data presented in this study are available on request from the corresponding authors.

Conflicts of Interest: The authors declare no conflicts of interest.

References

1. Forest Europe. State of Europe's Forests 2020, Food and Agriculture Organization of the United Nations. Ministerial Conference on the Protection of Forests in Europe—FOREST EUROPE, Liaison Unit Bratislava. 2020. Available online: https://foresteurope.org/publications_type/state-of-europes-forests-2020/ (accessed on 5 May 2024).
2. SFM. Sürdürülebilir Orman Yönetimi Kriter ve Göstergeleri Türkiye Raporu 2019, T.C. Tarım ve Orman Bakanlığı, Strateji Geliştirme Dairesi Başkanlığı. (Turkish Report on Sustainable Forest Management Criteria and Indicators 2019). 2020. Available online: https://www.ogm.gov.tr/tr/e-kutuphane/kitaplik/ulusal-duzey-surdurulebilir-orman-yonetimi (accessed on 12 June 2024). (In Turkish)
3. Ministry of Environment, Urbanisation and Climate Change (MEUCC). Çevresel Göstergeler, 2023. Çevresel Etki Değerlendirmesi, İzin ve Denetim Genel Müdürlüğü. Ankara. Available online: https://cevreselgostergeler.csb.gov.tr/cevresel-gostergeler-2023-kitapcigi-haber-288092 (accessed on 14 May 2024).
4. Gündoğan, A.C.; Turhan, D.E.; Aydın, C.İ.; Berke, M.Ö. 100 Maddede Sürdürülebilirlik Rehberi. (The 100 Bullet Points Guide to Sustainability). 2016. Available online: https://www.skdturkiye.org/files/yayin/100-maddede-surdurulebilirlik-rehberi.pdf (accessed on 9 December 2024). (In Turkish)
5. Wolfslehner, B.; Vacik, H. Evaluating sustainable forest management strategies with the Analytic Network Process in a Pressure-State-Response framework. *J. Environ. Manag.* **2008**, *88*, 1–10. [CrossRef]
6. Rodriguez Franco, C.; Conje, J. The Evolution of the Dialogue and Perspectives on Sustainable Forest Management with Special Emphasis on the United States of America. *J. Sustain. For.* **2022**, *42*, 747–791. [CrossRef]
7. Beriozovas, O.; Perkumienė, D.; Škėma, M.; Saoualih, A.; Safaa, L.; Aleinikovas, M. Research Advancement in Forest Property Rights: A Thematic Review over Half a Decade Using Natural Language Processing. *Sustainability* **2024**, *16*, 8280. [CrossRef]
8. Makutėnienė, D.; Perkumienė, D.; Makutėnas, V. Logarithmic Mean Divisia Index Decomposition Based on Kaya Identity of GHG Emissions from Agricultural Sector in Baltic States. *Energies* **2022**, *15*, 1195. [CrossRef]
9. Pranskuniene, R.; Perkumienė, D. Public Perceptions on City Landscaping during the Outbreak of Coronavirus Disease: The Case of Vilnius Pop-Up Beach, Lithuania. *Land* **2021**, *10*, 32. [CrossRef]
10. Williamson, L. Evolution of a landmark law. In *Restoring American Wildlife*; Kallman, H., Ed.; United States Department of the Interior; Fish and Wildlife Service: Washington, DC, USA, 1987.
11. Hay, M.J.; McConnell, K.E. Harvesting and non consumptive wildlife recreational decisions. *Land Econ.* **1984**, *60*, 388–396. [CrossRef]
12. Chardonnet, P.; Clers, B.D.; Fischer, J.; Gerhold, R.; Jori, F.; Lamarque, F. The value of wildlife. *Rev. Sci. Tech.-Off. Int. Épizooties* **2002**, *21*, 15–52. [CrossRef]
13. Conover, M.R.; Pitt, W.C.; Kessler, K.K.; Dubow, T.J.; Sandborn, W.J. Review of human injuries, illnesses, and economic losses caused by wildlife in the United States. *Wildl. Soc. Bull.* **1995**, *23*, 407–414.
14. Gordon, I.J.; Hester, A.J.; Festa-Bianchet, M. Review: The management of wild large herbivores to meet economic, conservation and environmental objectives. *J. Appl. Ecol.* **2004**, *41*, 1021–1031. [CrossRef]
15. Gren, I.M.; Häggmark-Svensson, T.; Elofsson, K.; Engelmann, M. Economics of wildlife management—An overview. *Eur. J. Wildl. Res.* **2018**, *64*, 22. [CrossRef]

16. Rasker, R.; Martin, M.; Johnson, R. Economics: Theory versus practice in wildlife management. *Conserv Biol.* **1992**, *6*, 338–349. [CrossRef]
17. Mawdsley, J.; O'Malley, R.; Ojima, D. A review of climate-change adaptation strategies for wildlife management and biodiversity conservation. *Conserv. Biol.* **2009**, *29*, 1080–1089. [CrossRef] [PubMed]
18. Pack, S.; Golden, R.; Walker, A. *Comparison of National Wildlife Management Strategies: What Works Where, and Why*; Wildlife Consulting Associates: Santa Rosa, CA, USA, 2013.
19. Shwiff, S.A.; Anderson, A.; Cullen, R.; White, P.C.L.; Shwiff, S.S. Assignment of measurable costs and benefit to wildlife conservation projects. *Wildl. Res.* **2013**, *40*, 134–141. [CrossRef]
20. Loomis, J.; Donnelly, D.; Sorg-Swanson, C. Comparing the economic value of forage on public lands for wildlife and livestock. *Rangel. Ecol. Manag. J. Range Manag. Arch.* **1989**, *42*, 134–138. [CrossRef]
21. Roitsch, D.; Abruscato, S.; Lovrić, M.; Lindner, M.; Orazio, C.; Winkel, G. Close-to-nature forestry and intensive forestry–Two response patterns of forestry professionals towards climate change adaptation. *For. Policy Econ.* **2023**, *154*, 103035. [CrossRef]
22. Niedziałkowski, K.; Chmielewski, P. Challenging the dominant path of forest policy? Bottom-up, citizen forest management initiatives in a top-down governance context in Poland. *For. Policy Econ.* **2023**, *154*, 103009. [CrossRef]
23. Škėma, M.; Doftartė, A.; Perkumienė, D.; Aleinikovas, M.; Perkumas, A.; Sousa, H.F.P.E.; Pimenta Dinis, M.A.; Beriozovas, O. Development of a Methodology for the Monitoring of Socio-Economic Indicators of Private Forest Owners towards Sustainable Forest Management: The Case of Lithuania. *Forests* **2024**, *15*, 1657. [CrossRef]
24. Başkent, E.Z. Ormancılık Sektörünün Lokomotifi; Planlama. *Anadolu Orman Araştırmaları Derg.* **2015**, *1*, 1–7.
25. Perkumienė, D.; Doftartė, A.; Škėma, M.; Aleinikovas, M.; Elvan, O.D. The Need to Establish a Social and Economic Database of Private Forest Owners: The Case of Lithuania. *Forests* **2023**, *14*, 476. [CrossRef]
26. Mevzuat. Land Hunting Law. 2003. Law No. 4915. Available online: https://www.mevzuat.gov.tr/mevzuatmetin/1.5.4915.pdf (accessed on 2 December 2024).
27. AVBIS. 2013. The Hunting Management Information System. Available online: https://www.tarimorman.gov.tr/DKMP/Link/25/Avbis (accessed on 2 December 2024).
28. Aydar, U. Coğrafi Bilgi Sistemleri kullanılarak yaban hayatı çeşitliliği ve avlak alanları risk haritalarının ü retilmesi. *Coğrafi Bilgi Sist. Derg.* **2020**, *2*, 49–56.
29. Pehlivan, Ş.Y.; Izyajen, S.A.D. İstanbul İlinde Avlanma ve Yaban Hayati Açisindan Elde Edilen Gelirler Üzerine İncelemeler. *Turk. J. For. Sci.* **2024**, *8*, 230–249. [CrossRef]
30. FAO. Sustainable Forest Management. 2024. Available online: https://www.fao.org/sustainable-forests-management/en/ (accessed on 22 June 2024).
31. Forest Europe. Second Ministerial Conference on the Protection of Forests in Europe. Helsinki/Finland. 1993. Available online: https://www.foresteurope.org/docs/MC/MC_helsinki_resolutionH1.pdf (accessed on 20 August 2024).
32. Putz, F.E.; Thompson, I.D. Defining sustainable forest management (SFM) in the tropics. *Achiev. Sustain. Manag. Trop. For.* **2020**, *10*, 25. [CrossRef]
33. Gordeeva, E.M. Evolution of the International Forest Regulation. *Kutafin Law Rev.* **2021**, *8*, 155–198. [CrossRef]
34. CIFOR. Evidence-Based Forestry. 2020. Available online: https://www.cifor-icraf.org/knowledge/publication/4342/ (accessed on 13 November 2024).
35. CIFOR-ICRAF. CIFOR-ICRAF Strategy 2020–2030. Bogor, Indonesia: Center for International Forestry Research (CIFOR) and Nairobi, Kenya: World Agroforestry (ICRAF). 2021. Available online: https://www.cifor-icraf.org/publications/pdf/CIFOR-ICRAF-Strategy.pdf (accessed on 24 November 2024).
36. CICI. Report of the International Conference on the Contribution of Criteria and Indicators for Sustainable Forest Management: The Way Forward. In Proceedings of the CICI-2003, Guatemala City, Guatemala, 3–7 February 2003; Volume 2. Available online: http://www.fao.org/docrep/005/J0077E/J0077E00.htm#TopOfPage (accessed on 23 May 2024).
37. Akyol, A.; Tolunay, A. Sürdürülebilir orman yönetimi ölçüt ve göstergelerinin Türkiye için modellenmesi. *SDÜ Orman Fakültesi Derg.* **2014**, *15*, 21–32. (In Turkish) [CrossRef]
38. Innes, J.L. Criteria and indicators of sustainable forest management. In *Sustainable Forest Management*; Routledge: London, UK, 2016; pp. 53–63.
39. Castañeda, F. Criteria and indicators for sustainable forest management: International processes, current status and the way ahead. *Unasylva-FAO* **2000**, *51*, 34–40.
40. Inskipp, T.; Wells, S. *International Trade in Wildlife*; Routledge: London, UK, 2019; pp. 1967–1970. [CrossRef]
41. Tolunay, D. Türkiye'de ekosistem tahribat faktörü olarak habitat ve arazi kullanim değişiklikleri. *Memleket Siyaset Yönetim* **2021**, *16*, 279–304.
42. Lemaître, S.; Hervé-Fournereau, N. Fighting wildlife trafficking: An overview of the EU's implementation of its action plan against wildlife trafficking. *J. Int. Wildl. Law Policy* **2020**, *23*, 62–81. [CrossRef]
43. Tisdell, C.A. Biodiversity and the UN's Sustainable Development Goals. *Sustain. Life Land* **2021**, 25.
44. Bandeira, V.; Fonseca, C. Wildlife and Sustainable Development. In *Life on Land*; Springer: Cham, Switzerland, 2021; pp. 1104–1117. [CrossRef]

45. Yuliantiningsih, A.; Jati, B.K.H.; Parluhutan, D.J.; Nordin, R. Nexus Between Illegal Wildlife Trade and Financial Crime: How to Counter It? A Case Study in Southeast Asia. In *Financial Crimes: A Guide to Financial Exploitation in a Digital Age*; Springer International Publishing: Cham, Switzerland, 2023; pp. 19–33. [CrossRef]
46. Nijman, V.; Abdullah, A.; Adinda, E.; Ardiansyah, A.; Campera, M.; Chavez, J.; Dewi, T.; Hedger, K.; Imron, M.A.; Shepherd, C.R.; et al. Indonesia's sustainable development goals in relation to curbing and monitoring the illegal wildlife trade. *Sustain. Dev.* **2024**, *32*, 5393–5403. [CrossRef]
47. Castañeda, F.; Palmberg-Lerche, C.; Vuorinen, P. *Criteria and Indicators for Sustainable Forest Management: A Compendium*; Forest Management Working Paper; FAO: Rome, Italy, 2001.
48. CIFOR. *The CIFOR Criteria and Indicators Generic Template*; The Criteria & Indicators Toolbox Series; Center for International Forestry Research (CIFOR): Jakarta, Indonesia, 1999; ISBN 979-8764-29-3. [CrossRef]
49. Uyar, Ç.; Elvan, O.D. Legal analysis of the CITES convention in terms of Turkish administrative and judicial processes. *Int. Environ. Agreem.* **2024**, *24*, 515–538. [CrossRef]
50. Aydin Coskun, A.; Gencay, G. Kyoto Protocol and deforestation: A legal analysis on Turkish environment and forest legislation. *For. Policy Econ.* **2011**, *13*, 366–377. [CrossRef]
51. Elvan, O.D. The legal environmental risk analysis (LERA) sample of mining and the environment in Turkish legislation. *Resour. Policy* **2013**, *38*, 252–257. [CrossRef]
52. Elvan, O.D.; Turker, Y.O. Analysis of Turkish groundwater legislation and policy regarding international principles and conventions. *Water Sci. Technol.* **2014**, *69*, 2155–2165. [CrossRef]
53. Elvan, O.D.; Birben, Ü.; Unal, H.E. The effectiveness of the Bern Convention on wildlife legislation and judicial decisions in Turkey. *Int. Environ. Agreem. Politics Law Econ.* **2021**, *21*, 305–321. [CrossRef]
54. Elvan, O.D.; Birben, U.; Ozkan, U.Y.; Yildirim, H.T.; Turker, Y.O. Forest fire and law: An analysis of Turkish forest fire legislation based on Food and Agriculture Organization criteria. *Fire Ecol.* **2021**, *17*, 12. [CrossRef]
55. Turker, Y.O.; Aydin, A. How ready is the Turkish Legislation for the green deal? *Energy Clim. Change* **2022**, *3*, 100084. [CrossRef]
56. Hakverdi, A.E. Türkiye'de sürdürülebilir orman yönetimi kriter ve göstergelerinin değerlendirilmesi. *Turk. J. For.* **2020**, *21*, 332–343. [CrossRef]
57. Nyyssönen, A.; Ahti, A. (Eds.) *Proceedings of FAO Expert Consultation on Global Forest Resources Assessment 2000 in Cooperation with ECE and UNEP with the Support of the Government of Finland (Kotka III), Held in Kotka, Finland, 10–14 June 1996.* p. 369. Available online: https://jukuri.luke.fi/handle/10024/521249 (accessed on 10 June 2024).
58. Durusoy, İ. Türkiye Ormancılığı İçin Sürdürülebilir Orman Yönetimi Ölçütlerinin Belirlenmesi. *Orman. Dergis* **2012**, *8*, 41–49. (In Turkish)
59. Uyar, Ç. Sürdürülebilir Ormancılık Açısından Uluslararası CITES Sözleşmesinin Hukuki Analizi. Ph.D. Thesis, İstanbul University-Cerrahpaşa, Istanbul, Türkiye, 2023. (In Turkish)
60. Oettel, J.; Lapin, K. Linking forest management and biodiversity indicators to strengthen sustainable forest management in Europe. *Ecol. Indic.* **2021**, *122*, 107275. [CrossRef]
61. Ito, K.; Oura, Y.; Takeya, H.; Hattori, S.; Kitagawa, K.; Paudel, D.; Paudel, G. The influence of NGO involvement on local people's perception of forest management: A case study of community forestry in Nepal. *J. For. Res.* **2005**, *10*, 453–463. [CrossRef]
62. Duthy, S.; Bolo-Duthy, B. Empowering people's organizations in community-based forest management in the Philippines: The community organizing role of NGOs. *Ann. Trop. Res.* **2003**, *25*, 13–28.
63. Mohan, M.; Richardson, G.; Gopan, G.; Aghai, M.M.; Bajaj, S.; Galgamuwa, G.P.; Vastaranta, M.; Arachchige, P.S.P.; Amorós, L.; Corte, A.P.D.; et al. UAV-supported forest regeneration: Current trends, challenges and implications. *Remote Sens.* **2021**, *13*, 2596. [CrossRef]
64. Torres, P.; Rodes-Blanco, M.; Viana-Soto, A.; Nieto, H.; García, M. The role of remote sensing for the assessment and monitoring of forest health: A systematic evidence synthesis. *Forests* **2021**, *12*, 1134. [CrossRef]
65. Linser, S. *Potential for Revision of Forest Europe Indicators Under Criterion 4 "Maintenance, Conservation and Appropriate Enhancement of Biological Diversity in Forest Ecosystems"*; Forest Europe: Bonn, Germany, 2024; 32p.
66. ATO. *The African Timber Organisation (ATO) Initiative on Principles, Criteria and Indicators for Sustainable Management of African Forests*; General Presentation; Organisation Africana de la Madera: Liberville, République Gabonaise, 1998.
67. Dry Forest. Regional Initiative for the Development and Implementation of National Level Criteria and Indicators for the Sustainable Management of Dry Forests in Asia, 1999. "Workshop on National-Level Criteria and Indicators for Sustainable Management of Dry Forests in Aia/Suth Aia" Organised by FAO/UNEP/ITTO Indian Institute of Forest Management, Bhopal, India. 30 November–3 December 1999. Available online: https://www.fao.org/3/ac135e/ac135e04.htm (accessed on 14 October 2023).
68. Dry Zone. Criteria and Indicators for Sustainable Forest Management in SADC countries Within the Framework of the Dry-Zone Africa Process. In Proceedings of the UNDP/FAO/SADC Meeting, Lilongwe, Malawi, 18 December 1999.
69. ITTO. *Doğal Tropikal Ormanların Sürdürülebilir Yönetimi için Kriterler ve Göstergeler. Uluslararası Tropikal Kereste Organizasyonu (ITTO)*; Politika Geliştirme Serisi 7; ITTO: Yokohama, Japan, 1998. (In Turkish)
70. Lepaterique. Results of the FAO-CCAB-AP Experts' Meeting on Criteria and Indicators for Sustainable Forest Management in Central America. 1997. Tegucigalpa, Honduras, 20–24/01/97. Available online: https://www.fao.org/4/ac135e/ac135e07.htm (accessed on 9 December 2024).

71. Process, M. *Criteria and Indicators for the Conservation and Sustainable Management of Temperate and Boreal Forests. The Montreal Process*; Canadian Forest Service: Hull, QC, Canada, 1995. [CrossRef]
72. Şener, F.N.; Tolunay, A. Sürdürülebilir orman yönetimi süreçlerinde Türkiye'nin konum analizi. *Bartın Orman Fakültesi Derg.* **2016**, *18*, 146–157. (In Turkish) [CrossRef]
73. Lahteenmaki-Uutela, A.; Rantala, S.; Swallow, B.; Lehtiniemi, H.J.; Pohjola, T.; Paloniemi, R. Increasing Access to Forest Data for Enhancing Forest Benefits to All. *Silva Fenn.* **2023**, *11*, 231–461. [CrossRef]
74. IPBES. *Summary for Policymakers of the IPBES Global Assessment Report on Biodiversity and Ecosystem Services*; Intergovernmental Science-Policy Platform on Biodiversity and Ecosystem Services (IPBES): Bonn, Germany, 2019; ISBN 978-3-947851-13-3. [CrossRef]
75. Almond, R.E.; Grooten, M.; Peterson, T. *Living Planet Report 2020-Bending the Curve of Biodiversity Loss*; World Wildlife Fund: Gland, Switzerland, 2020; ISBN 978-2-940529-99-5. [CrossRef]
76. Birben, Ü. State Ownership of Forests from Different Angles: Policy, Economics, and Law. *Environ. Monit. Assess.* **2019**, *191*, 502. [CrossRef]
77. Prakash, S.L.; Perera, P.; Newsome, D.; Kusuminda, T.; Walker, O. Reasons for Visitor Dissatisfaction with Wildlife Tourism Experiences at Highly Visited National Parks in Sri Lanka. *J. Outdoor Recreat. Tour.* **2019**, *25*, 102–112. [CrossRef]
78. Woodroffe, R.; Thirgood, S.; Rabinowitz, A. (Eds.) *People and Wildlife, Conflict or Co-Existence?* Cambridge University Press: Cambridge, UK, 2005; Volume 9.
79. Nyhus, P.J. Human–wildlife conflict and coexistence. *Annu. Rev. Environ. Resour.* **2016**, *41*, 143–171. [CrossRef]
80. Crespin, S.J.; Simonetti, J.A. Reconciling farming and wild nature: Integrating human–wildlife coexistence into the land-sharing and land-sparing framework. *Ambio* **2019**, *48*, 131–138. [CrossRef]
81. Huber, P.; Kurttila, M.; Hujala, T.; Wolfslehner, B.; Sanchez-Gonzalez, M.; Pasalodos-Tato, M.; de-Miguel, S.; Bonet, J.A.; Marques, M.; Borges, J.G. Expert-Based Assessment of the Potential of Non-Wood Forest Products to Diversify Forest Bioeconomy in Six European Regions. *Forests* **2023**, *14*, 420. [CrossRef]
82. Krott, M. *Forest Policy Analysis*; Springer Science & Business Media: Berlin, Germany, 2005.
83. Rochmayanto, Y.; Nurrochmat, D.R.; Locke, C.; Casse, T.; Nugroho, B.; Darusman, D. Evaluating the "Village Forests" in Indonesia: Property Rights and Sustainability Perspectives. *Small-Scale For.* **2022**, *21*, 461–481. [CrossRef]
84. Hovik, S.; Sandström, C.; Zachrisson, A. Management of protected areas in Norway and Sweden: Challenges in combining central governance and local participation. *J. Environ. Policy Plan.* **2010**, *12*, 159–177. [CrossRef]
85. Oğurlu, İ.; Ünal, Y.; Aksan, Ş. Yaban Hayatında Biyorestorasyon. In Proceedings of the III. Ulusal Karadeniz Ormancılık Kongresi, Artvin, Turkey, 20–22 May 2010; pp. 1225–1232.
86. Mol, T. *Yaban Hayatı (Wildlife)*; İstanbul Üniversitesi Publishing: Fatih, Istanbul, 2006; ISBN 975-404-766-9. (In İstanbul)
87. Vincent, C.H.; Hanson, L.A.; Argueta, C.N. *Federal Land Ownership: Overview and Data*; CRS Report R42346; Congressional Research Service: Washington, DC, USA, 2017. Available online: https://fas.org/sgp/crs/misc/R42346.pdf (accessed on 9 December 2024)
88. Ye, J.; Xiao, C.; Feng, Z.; Qiao, T. A review of global wilderness area identification since the 21st century. *J. Environ. Manag.* **2024**, *358*, 120946. [CrossRef]

forests

MDPI

Review

Roe Deer, Lithuania's Smallest and Most Abundant Cervid

Linas Balčiauskas

Nature Research Centre, Akademijos 2, 08412 Vilnius, Lithuania; linas.balciauskas@gamtc.lt; Tel.: +370-685-34141

Abstract: A review of 113 scientific and other publications on the smallest and most abundant deer in Lithuania, the roe deer (*Capreolus capreolus*), is presented, along with an analysis based on compound annual population growth rates of population numbers, hunting bags, and roadkill. This review covers the species' history in the country from the last glaciation onwards, the changes in numbers from 1934 to 2023, hunting bag changes from 1965 to 2022, roadkill numbers from 2002 to 2022, data on roe deer reproduction, habitat selection, genetic diversity, pathogens, and damage to forest stands. It also provides an overview of species management and selection for trophies. Despite the exponential increase in roe deer numbers since 1990 and the dominance of this species in roadkill, even on urban roads, the number and density of animals and the damage they cause to forest stands are relatively low compared to other European countries. Within the observed period, drops in numbers were related to harsh winters in 1969/1970 and 1995/1996. Poaching, especially coinciding with weakened enforcement during the period of political–economic transition in the early 1990s, has also had a negative impact on population numbers, as have recent increases in the number of large predators. Population growth over the recent period does not correlate with hunting mortality, which has remained stable at an average of 16.9% since 2002.

Keywords: *Capreolus capreolus*; abundance; hunting; roadkill; population management; genetics; pathogens; Lithuania

1. Introduction

With a vast range in the Palearctic region, the roe deer (*Capreolus capreolus*) inhabits most of Europe, excluding Ireland, southern islands such as Sicily, Corsica, Sardinia, Cyprus, and the far northern latitudes [1–3]. The species has been present in Europe for over half a million years, enduring through glacial and interglacial periods, and it has been one of the most abundant mammals in the Late Quaternary [4].

Randi et al. highlighted that genetically, European roe deer can be categorized into three haplogroups: one in the west (mainly Iberia), another in the east (Balkans and Greece), and a third in the central part of the continent (spanning across Europe, including the eastern region); however, this study did not include samples from Lithuania [5]. Subsequently, Lorenzini and Lovari [6] classified Lithuanian roe deer into a central–northern group, alongside animals from Romania, Turkey, Greece, and Poland. Roe deer populations in eastern and central parts of Europe exhibit the highest genetic diversity, likely attributable to introgression from the Siberian roe deer (*Capreolus pygargus*) [7].

The roe deer is a generalist browser, but they also eat grasses and sedges. Species occupies "deciduous wooded areas with glades and clearings, as well as in fields interspersed with wood patches, mostly in temperate climates" [1]. However, in areas with small forest fragments and intensive agriculture, the so-called "field roe deer" ecotype evolved [8]. Its biology differs from that of woodland roe deer in several respects [9]. Field roe deer foraging in agricultural areas outside forests can have substantial impacts on crops [10]. The field roe deer ecotype is also widespread in Lithuania [11–13].

Roe deer are the most abundant deer in the mid-latitudes [1,14], and management of this species is therefore linked to ecological, economic, and other aspects [15]. Recently, the management of the species has faced a number of challenges [16,17] due to the overabundance of ungulates in Europe [18], including human safety issues due to the increasing number of wildlife and vehicle accidents, even in urban areas [19].

Roe deer are important as browsers, playing a significant role in forest regeneration [20]. Furthermore, they can serve as a model species for studying the impacts of climate change [21] and landscape change [22]. In heterogeneous European landscapes, roe deer could be utilized as an indicator species [23]. Despite any disagreement with the authors' position, roe deer have been recognized as ecosystem engineers [24].

Roe deer have been a good indicator of political and economic change in the post-1990 post-socialist countries of Eastern and Central Europe [25,26]. Currently, the management of roe deer populations in European National Parks has shown the need for an integrated management approach [27].

However, many aspects of roe deer ecology and management in Lithuania are either unexplored or inaccessible to non-native speakers. The aim of this review is therefore to provide a comprehensive overview of roe deer in Lithuania, including the history of the species, its current status and abundance, ecology, and other aspects of the species' management, based on published sources. Changes in the management of the species in relation to legislation are assessed. In the absence of publications on population dynamics, we have supplemented the review with trend analysis. Management options have been analyzed in the light of the political and governance changes that have occurred in the country since the 1960s.

2. Materials and Methods

2.1. Scope of Literature Sources and Filtering of References

Given that many of the references on roe deer are in the national language and those previously published are in Russian, we carried out several literature searches. The two main terms to be included in the search were "roe deer" as a species and "Lithuania" as a geographical reference. In some references, Lithuania was not included in the search terms because all Baltic countries were included.

Further on, we used search terms including biology, ecology, abundance, diet, reproduction, hunting, management, and roadkill of roe deer (Table 1).

Table 1. Key terms used in the literature search.

Key Term	Definition	Notes on Usage
Roe deer	*Capreolus capreolus*, member of the Cervidae family	We also used "stirna" (Lithuanian) and "Evropejskaja kosulia" (Russian)
Lithuania	Territorial scope of the search	We also used "Lietuva", "Baltic countries" and "Baltic states", as well as "Baltijos šalys"
History	Period between last glaciation and world War II	Includes archaeological findings and archives
Management	Includes hunting, survey, and monitoring	The hunting yield and the abundance of species were associated with alterations in the legal framework

Three search engines were used to search for published information sources: the Web of Knowledge, Google Scholar, and Google. The latter was needed for books published in Lithuanian. Our search period covered the period from 1900 to 2024. Boolean combinations of terms were used, such as

$$\text{(roe deer OR } \textit{Capreolus capreolus}\text{) AND (history OR survey OR hunting OR roadkill OR monitoring OR management) AND Lithuania,} \quad (1)$$

also using the same terms in Lithuanian and Russian languages.

Two problems with the key term "roe deer" in Lithuanian were the fact that it has the same species name in Latvian and that it is popular as a surname or company name. For this reason, the resulting reference lists were manually sorted to exclude sources not related to the animal.

Student works, such as Master's and Bachelor's theses, were critically analyzed and singled out because they did not contain original information on the species. The literature sources that did not provide original information and relied on other sources were also excluded. Thus, the list of sources to be reviewed included 114 sources. The remainder of the cited sources are related to other countries.

2.2. Study Site and Sources of Empirical Data

The study area with its forest cover, main cities, and the main roads is shown in Figure 1. Data on the recent forest cover and forest cover dynamics are taken from the Official Statistics Portal [28,29]. Data on regulations concerning roe deer management were collected from official sources [30–34]. The hunting periods for roe deer are taken from the hunting calendar [35].

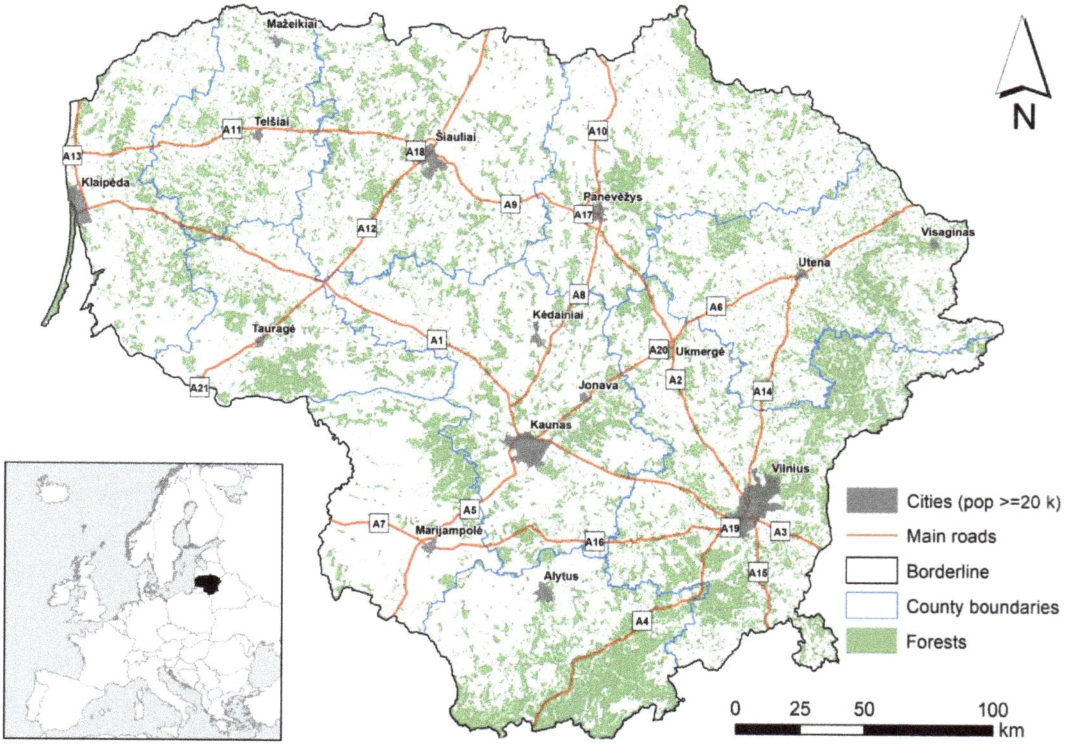

Figure 1. Map of Lithuania showing forests, cities, and main roads in 2021. The inset indicates the position of the country in Europe. Note: Between 1934 and 1940, the territorial boundaries of the country were altered in the western and southeastern regions due to political factors.

The size of the national roe deer herd (number of animals surveyed) and the number of hunting bags (number of animals hunted per season) were used to describe population dynamics and trends. We used official data on the number of surveyed animals [36] and data on the hunting bag [37], previously available in the archives of the Ministry of the

Environment of the Republic of Lithuania, with data for the period 2015–2023 only being available on the internet.

The data on roadkill of roe deer were provided by the Lithuanian Police Traffic Supervision Service, comprising the sum of reported roadkill documented in officers' reports. In the past decade, the verification of roadkill has been conducted by the nearest hunting clubs. Additionally, unregistered roadkill numbers of roe deer were compiled by professional biologists from the Nature Research Centre. These figures represent the sum of animals found roadkilled by scientists but not documented in the reports of police officers.

Until 1990, the main survey method was snow-tracking, adjusted according to the number of animals recorded visually. Game counts in the snow by foresters and hunters were carried out in a coordinated manner on the same day throughout the country, and the results obtained were aggregated at several levels. Subsequently, after the 1990s, the population size was taken as the sum of the animals present in all hunting areas in the country, as reported by hunting clubs. In the last decade, wildlife cameras were used by nearly all clubs, however, there are no publications on the roe deer numbers. After 2018, the snow-track index [38] has again been used to analyze population trends, but not the number.

We understand that official figures may be biased, but this is the only source of game numbers. These data have been used in publications covering Europe and the Baltic States [14,16,26,27,39], Lithuania [11], and at the local level [40].

Data on roe deer damage were taken from the annual reports [41].

2.3. Data Analyses

We analyzed trends in population numbers, hunting bag numbers, and roadkill numbers using the compound annual growth rate (CAGR) and regression analysis. Various time periods were selected to represent different population phases.

CAGR was expressed as the number required for the population or hunting bag to change from the beginning value to an end value. We used the following expression:

$$\mathrm{CAGR} = \left(\frac{\mathrm{EV}}{\mathrm{IV}}\right)^{\frac{1}{n}} - 1, \quad (2)$$

where IV is the initial value, EV is the end or final value, and n is the number of years [42]. Larger CAGR values indicate faster growth, while negative CAGR values indicate a reduction in numbers over the specified time period.

We employed basic statistics, including Pearson correlation and the standard error of the mean, for our analysis. Regression analysis was conducted to determine the best fit to the raw data based on the least squares criteria, utilizing either linear, power, or exponential regression models. The best regression model was selected using the coefficient of determination, R^2.

All calculations were done in MS Excel, ver. 16.0.5387.1000 and Statistica for Windows, ver. 6.0 [43].

3. Results

3.1. Historic Data on Presence of Roe Deer in Lithuania

As pointed out by Sommer et al. [44], "the European roe deer is a typical faunal element of the Holocene", and it was present in Europe for over 600,000 years, surviving several glaciations in refugia as were other ungulate species [45]. During the last glaciation period, the distribution range of roe deer was at its minimum [44]. Central and northern Europe were subsequently recolonized after the glaciation, as fossils of the species older than 11,000–12,000 years are not present [6]. The current distribution of the species is limited by permafrost, leading to the absence of roe deer beyond the Arctic Circle in Scandinavia [5]. Furthermore, forest re-establishment after deglaciation was followed by the expansion of the species' range.

While the teeth of ungulates were commonly found in Mesolithic burials across Europe [46], those of roe deer were not discovered in Lithuania [47]. From the Early Neolithic to the Early Bronze Age, the number of findings of roe deer bones and teeth was considerably lower compared to those of moose (*Alces alces*) and red deer (*Cervus elaphus*) [48]. According to these authors, roe deer bones accounted for only 1.54% of the animal remains in the Early Neolithic, 2.99% in the Middle Neolithic, 2.32–6.66% in the Late Neolithic periods, and 3.93% in the Early Bronze Age. Furthermore, based on stable isotopes, there is no evidence of roe deer presence in the diet of humans during that period [49].

From the Late Paleolithic to the Middle Bronze Age, hunting was the main economic activity in Lithuania, but roe deer were not often hunted—hunting was focused on the largest ungulates such as reindeer (*Rangifer tarandus*), moose, and red deer [50]. As the landscape changed and forests expanded from the river valleys into the rest of the country, the composition of game also changed, but roe deer were again in the minority.

As the Iron Age progressed, the importance of hunting gradually diminished with the appearance of other sources of meat, such as domestic animals [50]. M. Vitkūnas [51] demonstrated that the proportion of game animal bones in archaeological findings decreased over time: it was up to 35.4% in the last millennium BC, then decreased to 27.5% in the first 400 years AD, further dropping to up to 7% before the end of the first millennium AD, and finally to 3.1% until 1400 AD. The proportion of roe deer bones in the latter case ranged from 3% to 3.7% of all bones at several sites. This suggests that either roe deer were not numerous or larger game were the main target of hunting [52]. Additionally, roe deer were not numerous in later periods, as evidenced by materials from the 14th to 17th centuries excavated in the castle of Vilnius, the capital city, where roe deer bones were absent [53].

In the 15th and 16th centuries, roe deer were a common and widespread animal in Lithuania [54–56]. In the 16th and 19th centuries, their numbers declined due to the colder climate [57].

Towards the 20th century, roe deer distribution declined and became fragmented due to uncontrolled hunting [11] and the influence of predators [58]. In Lithuania, a decline in roe deer numbers due to both of these causes was reported at the end of the 18th century [59]. In the second half of the 19th century, roe deer in Lithuania became rare or endangered [11,60].

3.2. Numbers of Roe Deer in Lithuania from the 20th Century to the 1990s

At the European level, the roe deer population started to increase again in the 20th century [1]. However, in Lithuania, the trends were different. Poaching and wolf population growth led to a critical decline in roe deer numbers during the First World War [11]. It was not until 1928 that it was concluded that roe deer were no longer threatened with extinction due to their rapid reproduction and ability to adapt to their environment [61].

In the 1930s, the number of roe deer increased (Figure 2), from 13,930 in 1934 to 35,030 in 1949 [11]. The compound growth rate of the roe deer population during this period was CAGR = 0.202. However, during the Second World War and the post-war years, the number of roe deer decreased drastically due to poaching and preying by wolves. In 1948, only 8500 roe deer remained, CAGR = −0.146. Wolf numbers at the end of the 1940s were at about 1700 [59].

The next decade, 1948–1958, was characterized by a slow increase in roe deer numbers (CAGR = 0.084); despite the extermination, numbers of wolves remained high, and poaching still had a substantial influence. Subsequently, numbers increased more rapidly (CAGR = 0.138 during 1959–1968), reaching 70,800 individuals in 1968. The growth of the population in some regions halted even without roe deer being hunted [62], followed by a reduction in breeding intensity [63].

The harsh and deep snow winter of 1969/1970 resulted in a significant decrease (CAGR = −0.20) in roe deer numbers, declining from 68,500 in 1969 to 44,000 in 1971 [11].

Subsequently, roe deer numbers remained at the same level (CAGR = −0.002) until 1991, when the population was 44,400 roe deer.

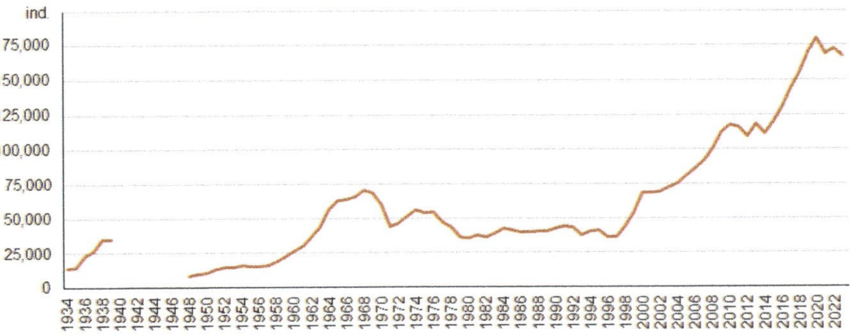

Figure 2. Changes of the roe deer numbers in Lithuania, 1934–2023, based on [11,36,59]. Data for 1940–1942 and 1944–1946 are not available, while data for 1943 are not reliable.

3.3. Numbers of Roe Deer in Lithuania during 1990–2023

In the last decade of the 20th century, roe deer numbers remained nearly stable (CAGR = 0.003), at just over 40,000 individuals [11]. A reduction in roe deer numbers in 1996–1997 was attributed to intensive hunting (see below), an increase in wolf numbers, and increased poaching associated with weakened controls during the political–economic transition [26,64,65]. Additionally, the harsh and deep snow winter of 1995/1996 resulted in a decrease in roe deer numbers from 41,000 to 36,500 in the country. However, locally, the loss ranged from 20% to 50%, and in some forests, it amounted to up to 80% of the roe deer population [66].

In general, after 1998, the increase in the roe deer herd in Lithuania lasted until 2020 (see Figure 2). The growth rate was not very steep (CAGR = 0.059), with the fastest growth occurring during the years 2014–2020 (CAGR = 0.084). The roe deer population growth after 1990 is best described as exponential (R^2 = 0.94), as indicated by the equation

$$roe\ deer\ number = 33,782e^{0.0518x} \tag{3}$$

where x is the year.

3.4. Roe Deer Hunting

The first period of legal hunting of roe deer in the 20th century was 1936–1939, during which time the number of individuals hunted increased from 486 to 868 and their share in the population was 1.96%–2.48%. Following the subsequent decline in the population, roe deer hunting only resumed in 1965 (Figure 3).

From 1965 to 1996, the numbers of roe deer bags remained stable (CAGR = 0.010), averaging about 6500 ± 500 annually, with a minimum of 910 in 1970 and a maximum of 10,844 in 1976, followed by a second drop in 1979. Meanwhile, the proportion of the population that hunted fluctuated widely (see Figure 4), averaging 14.5% ± 1.3%. However, it reached a minimum of 1.5% in 1970, dropped to 5.1% in 1979, and peaked at 28.3% in 1993. From 1993 to 1997, the proportion of hunted roe deer sharply decreased (CAGR = −0.882). In 1997, under a hunting ban for roe deer, only two individuals were reported to have been hunted.

From 1999 to 2019, the increase in the roe deer hunting bag was steady (CAGR = 0.115), with only a small drop in 2011 (see Figure 3). During this period, the hunting bag size increased from 3558 to 31,564 individuals, with the latter number representing the absolute maximum. The increase in the hunting bag (CAGR = 0.053) from 1999 to 2019 was consistent with the population increase. Therefore, the proportion of the population that hunted remained stable from 2002 onwards, averaging 16.9%. There was a drop in roe deer hunting bag size and proportion of the population hunted from 2020 to 2022 (see Figures 3 and 4); however, there are no data on changes in hunting effort.

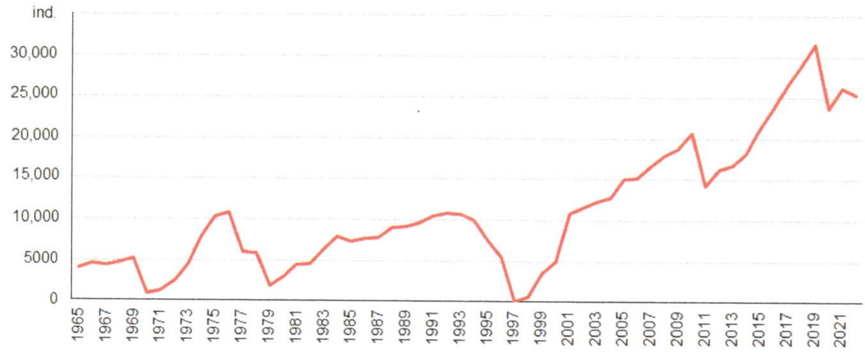

Figure 3. Changes of the roe deer hunting bag in Lithuania, 1965–2022, based on [11,37,59].

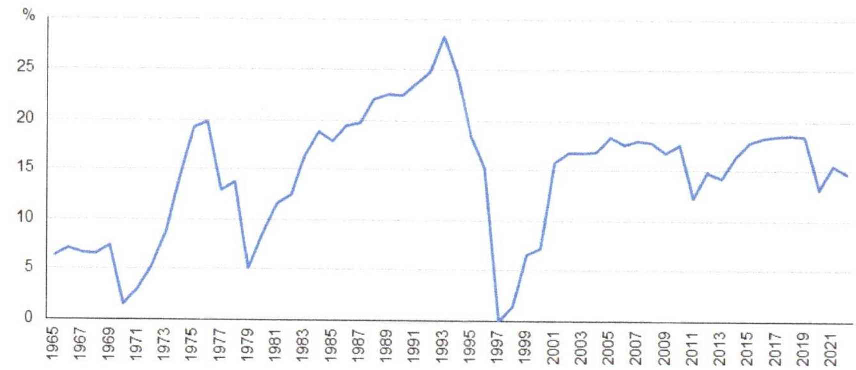

Figure 4. Proportion of the roe deer hunting bag from the population numbers in Lithuania, 1965–2022. Based on [11,36,37,59].

3.5. Roe Deer Roadkill in 2002–2022

Roe deer are the most abundant species among ungulate roadkill incidents [67]. Out of 50,679 animals killed on the roads of Lithuania during 2002–2022, 32,944 were ungulates, including 32,707 cervids. Among these, 29,199 collisions were registered as cases of roe deer roadkill, while 57 incidents went unreported to the Lithuanian Police Traffic Supervision Service. The correlation between roadkill incidents and population numbers is very high ($r = 0.94$, $p < 0.001$). Extrapolating from unregistered roe deer roadkill numbers based on the sampling effort, it is estimated that approximately 9200 individuals were not recorded by the Service during 2002–2022 [67].

The changes in roadkilled roe deer numbers on main, national, regional, and urban roads, based on [19], are presented in Table 2. An exponential increase in registered roe deer roadkill numbers, from 150 in 2002 to over 3500 individuals in 2019, 2021, and 2022, was significant ($R^2 = 0.97$) and best expressed on urban roads.

The decrease in roadkill incidents in 2020, observed on main, national, and regional roads, can be attributed to COVID-19-related human and vehicular mobility restrictions [68]. However, on urban roads, the number of roe deer roadkill incidents increased during the COVID-19 restrictions [19]. This increase was one of the most notable consequences of the so-called "anthropause" [69].

Table 2. Roe deer roadkill on the main, national, regional, and urban roads in 2002–2022, and roe deer proportion of the roadkilled mammals: TM, N—the total number of roadkilled mammals; RD main—number of roadkilled roe deer on the main, RD nat—national, RD reg—regional, RD urb—urban roads; RD, n—total of roadkilled roe deer; RD, %—percent of roadkilled roe deer from all roadkilled mammals, RDF, %—percent of roe deer roadkilled in the forest habitat.

Year	TM, N	RD Main	RD Nat	RD Reg	RD Urb	RD, n	RD, %	RDF, %
2002	373	42	93	15		150	40.2	29.3
2003	447	52	92	23		167	37.4	22.9
2004	571	104	128	33		265	46.4	26.2
2005	641	102	152	41		295	46.0	29.6
2006	781	105	201	49		355	45.5	28.8
2007	1625	158	257	79	41	535	32.9	30.8
2008	1534	145	258	84	36	523	34.1	23.2
2009	2774	159	255	73	44	531	19.1	22.8
2010	1350	135	308	95	69	607	45.0	19.5
2011	1265	129	244	94	78	545	43.1	23.6
2012	1379	175	326	87	83	671	48.7	25.0
2013	2120	180	408	126	111	825	38.9	22.3
2014	2810	283	525	229	161	1198	42.6	23.4
2015	2824	336	679	295	176	1486	52.6	21.7
2016	3327	403	988	402	282	2075	62.4	22.1
2017	2554	341	735	295	223	1593	62.4	21.9
2018	4065	548	1283	576	463	2870	70.6	19.6
2019	5027	733	1495	724	700	3652	72.6	18.2
2020	4717	505	1060	503	1325	3393	71.9	18.5
2021	5281	513	1357	696	1281	3847	72.8	17.8
2022	4902	573	1255	615	1230	3673	74.9	16.8

Two patterns of roe deer roadkill were observed: an increase in the proportion of roe deer in the total roadkill (Table 2) and the habitat distribution of roadkill, which was not related to forest habitats (Figure 5). The proportion of roe deer in the total roadkill exceeded 50% after 2014, rising to nearly 75% in 2022. Decreases in this proportion in 2008, 2009, and 2013 were related to intensive roadkill studies conducted at the Nature Research Centre, which focused on registering the number of smaller mammals [70].

Among other cervids, roe deer roadkill incidents were the least related to forest habitats [67]. With the increase in roe deer numbers roadkilled in urban areas [19], the proportion of roadkill in forested areas decreased, particularly in recent years. From 2018 to 2022, the proportion of roe deer roadkilled in forest habitats was consistently less than 20% (see Table 2).

This pattern is most evident in the southwestern, western, and central–northern parts of the country, where forest cover is below average [67], creating conditions conducive to the existence of the field roe deer ecotype [11,71]. These animals exhibit morphometric differences [12] and genetic distinctions [72] from forest-dwelling roe deer. Field roe deer are also common in northeast Poland [8,9]. The formation of this ecotype depended on scarce forest cover and the abundance of forage in agricultural areas [10]. Landscape changes in post-Soviet countries [16,25,26] have favored the formation or support of the field ecotype of roe deer.

The influence of cervid roadkill on population management in Lithuania has been studied for nearly two decades [73]. While roadkill incidents were very important for moose during years of low population density and limited hunting [74], this was not the case for red deer, as roadkill numbers are very limited [75]. Roadkill incidents involving roe deer are the most notable in proportion to the hunting bag of all cervids, averaging 7.7%. However, this proportion increased to 16.5%–16.6% in 2020–2021 [67].

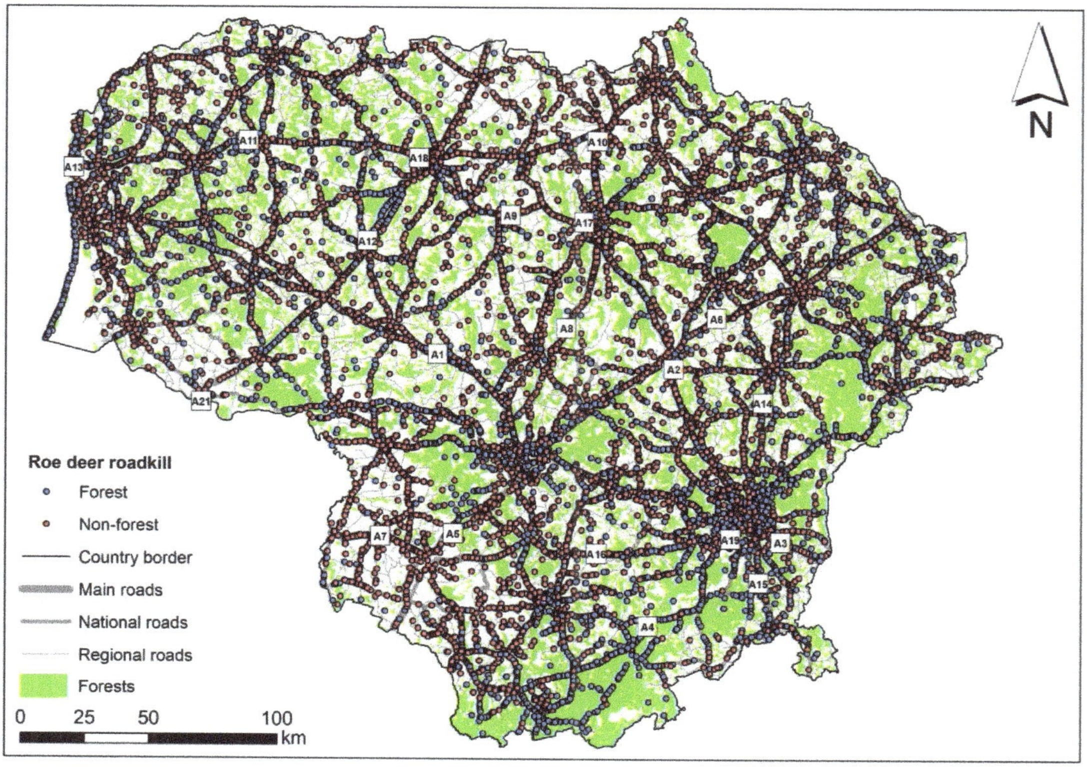

Figure 5. Roe deer roadkill in respect to forest habitat, 2002–2022 [67]. One dot represents one roadkill.

There is no single solution to the problem of roadkill; therefore, in Lithuania, several approaches were investigated, including the following:

- The use of chemical repellent to deter ungulates from entering main roads through gaps in wildlife fencing was investigated as a quick solution before the fence was repaired [76]. It was demonstrated that the reduction in ungulate entry, particularly in roe deer, exceeded 40% when using Wam Porocol®, Witasek PflanzenSchutz GmbH, Feldkirchen, Austria.
- Understanding the temporal patterns of roadkill incidents [77–79] reveals that the majority of casualties occur during dusk in May and September, primarily between 8:00 p.m. and 10:00 p.m., and on weekends (Friday to Sunday) [80]. However, roe deer are not the main contributors to human casualties.
- Analysis of the structure of surrounding habitats [81–83] has demonstrated that the movement of ungulates, including roe deer, is influenced by the distant habitat structure, leading to the identification of key habitat features important for the formation of roadkill hotspots [83,84].
- Testing the effectiveness of wildlife fencing as a mitigation tool to prevent ungulates from entering roads has been conducted [84].
- Identifying intersections between urban corridors and wildlife serves as a framework for planning wildlife-vehicle collision mitigation measures [85].
- Given limitations in the use of wildlife fences, there is a need to improve driver awareness regarding hotspot locations and the most dangerous times for collisions [84].

Fencing on main roads can lead to the redistribution of ungulate roadkill toward roads characterized by lower speed and traffic intensity.

3.6. Environmental Parameters and Roe Deer Ecology

The increase in roe deer numbers and harvest has been observed across much of Europe, particularly in the western part [14]. The synchrony of harvested numbers across the continent suggests common factors affecting abundance. Hagen et al. [86] found "a significant positive correlation between the North Atlantic Oscillation winter index (December–March) and annual changes in roe deer bag records for Czech Republic, Germany, Latvia, Lithuania, Norway, and Sweden". They concluded that periodic fluctuations in population size and harvest are induced by the interplay of environmental factors and human-induced changes.

In Lithuania, roe deer ecology is inadequately understood, with much of the available data being over three decades old [11]. Previous studies concluded that roe deer densities are directly proportional to land fertility and the proportion of deciduous trees in forest stands, while inversely proportional to woodland cover. Land fertility and afforestation are indirect factors influencing the protective qualities and fodder basis of roe deer habitats [87,88]. This suggests that the increased anthropogenic nature of the landscape contributes to higher ungulate densities in Lithuania [89], a phenomenon reflected in their territorial movements [90].

Predation, particularly by wolves, and mortality during severe winters have the most substantial negative impact on roe deer abundance [64–66,91,92] The winters of 1969/1970 and 1995/1996 serve as notable examples of such negative consequences for roe deer [11]. Snow cover exceeding 40 cm was considered detrimental to the species survival [93].

The influence of environmental parameters, particularly the composition of forest stands, on roe deer abundance has been studied since the late 1960s [94], with a focus on the food resources available to roe deer and their impact on forest stands. Subsequent analyses of the roe deer diet in relation to forest cover [95,96] not only determined the species composition of plants eaten (including 20 species of trees and shrubs and 95 herbaceous plant species), but also assessed the importance of agricultural habitats [97,98]. Supplementary food sources from agricultural fields were acknowledged as the main factor contributing to increased cervid abundance in the country after the 1980s [89,90]. Another important source of food for roe deer was the remnants of forest cutting [99], which were also processed into special pellets [100].

In the 1990s, the distribution of roe deer across various landscape types and forest stands was assessed based on winter pellet counts [101], revealing the highest importance of forest stands over 20 years old, as well as young pine stands on coastal and sandy plains [102]. Additionally, the spatial distribution of roe deer was found to be dependent on the presence of undergrowth and dwarf shrubs [103].

Roe deer have been observed as city-dwelling animals [104]. Subsequently, the various effects of urbanization on cervids, including roe deer, were analyzed by Brazaitis et al. [105]. It is reasonable to assume that roe deer numbers in urban areas have been increasing over the last decade, with a significant turning point being the limitations on human mobility during the COVID-19 pandemic [19,68].

Data on the reproduction of roe deer in Lithuania are outdated, primarily referencing the 1970s [11]. Calves are typically born in May–June [59]. The average number of embryos per pregnant doe was recorded at 1.9, with 96.6% of adult females breeding [63,106]. Regarding the presence of yellow bodies in roe deer ovaries, 12.6% of females had one, 61.8% had two, 22.3% had three, and 3.3% had four. Older roe deer (6+ years old) tended to have more yellow bodies, averaging 2.5, compared to middle-aged (3–5 years old) and young (1–2 years old) individuals, with 2.2 and 1.9 yellow bodies, respectively [11]. In other countries, large litters (>3 offspring) in roe deer are exceptions, but they might have the potential to further increase populations if monitored more frequently [107].

There are few publications on the field roe deer ecotype in Lithuania, although its formation in Southwest Lithuania began in 1965–1967 [12]. This region is characterized by very rich soil and minor forest cover, comprising only 9% of the area. Field roe deer are distinguished by their larger body mass, ranging from 20.1 to 20.4 kg in juvenile females and males, to 27.7–31.0 kg in adult animals, respectively. In contrast, adult forest roe deer do have a body mass of 26.1 kg, and bucks 29.1 kg [88], resulting in a difference of 1.6–1.8 kg. In Poland, field roe deer were reported to be even heavier than forest ones, by 2.2 kg for males and 3.2 kg for females [108]. Additionally, the cranium and trophy value of field roe deer were reported to be larger [109].

3.7. Genetic Diversity of Lithuanian Roe Deer

Genetic analysis conducted at various spatial scales across Central and West Europe revealed the presence of mtDNA from the Siberian roe deer. In the Pannonian region of Hungary, this species was represented in nearly 3% of samples [110]. Conversely, in northeastern Poland, mtDNA from the Siberian roe deer was found in half of the investigated roe deer [111]. These findings suggest either natural hybridization in the past or the translocation of Siberian roe deer by humans. Interestingly, no genetic differences were observed between field and forest roe deer by these authors, indicating that the species' ability to thrive in such diverse environments relies on phenotypic plasticity.

Three studies conducted on a continental scale have assessed genetic diversity patterns across the range of roe deer, incorporating samples from Lithuania. In the Eastern European deme, which spans Belarus, Lithuania, Latvia, and the European part of Russia, the highest number of haplotypes of the Siberian roe deer was identified [7]. It was observed that the introgression of the Siberian lineage decreased from eastern to central Europe [112]. These patterns likely depend on the survival of roe deer during the last glacial maximum across a wide range, including two northern refugia. During recolonization, the Baltic Sea has been regarded as a barrier to gene flow [113]. Consequently, the genetic makeup of the species in Lithuania holds interest. However, research on this subject has been notably limited thus far.

Using RAPD (random amplified polymorphic DNA) and enzyme studies, researchers identified nine polymorphic loci and confirmed differences in genetic variability among local roe deer populations [114]. Subsequently, analysis of data from the 457 bp mtDNA control region sequences revealed widespread introgression of Siberian roe deer, with three out of six identified haplotypes being specific to roe deer from Lithuania [115]. Further analysis of microsatellite and D-loop sequences demonstrated a high level of molecular genetic variation in roe deer populations in Lithuania [116].

Genetic differentiation between the field and forest ecotypes, as assessed using mtDNA and AMOVA, was found to be not significant. This includes differences in genetic diversity parameters such as heterozygosity and allelic richness [71]. The authors concluded that both field and forest ecotypes of roe deer in Lithuania are interbreeding and share a recent common gene pool.

3.8. Pathogens of Lithuanian Roe Deer

Roe deer can suffer from very dangerous diseases such as foot-and-mouth disease, brucellosis, tuberculosis, and rabies [58], and they have a variety of endoparasites [117]. Infectious diseases in roe deer have not been studied [11].

Pathogens of roe deer were analyzed during three periods. In the first period, the focus was on coccidian and helminth pathogens. Seven coccidian species were identified, including *Eimeria lituanica* n.sp., discovered in roe deer [118]. The most frequently encountered coccidians in roe deer were *Eimeria capreoli* and *E. ponderosa* [119]. In the internal organs of roe deer, 27 species of helminths were identified, including 3 flukes, 3 tapeworms, and 22 roundworms [120–123]. However, this represented only a fraction of the helminth fauna known to infect roe deer at that time, with 68 species identified in the Soviet Union [58]. The most frequently found roundworms were *Chabertia ovina* (prevalence—80% of investigated roe deer), *Bunostomum trigonocephalum* (25%), and *Trichocephalum capreoli* (14%). *Ch. ovina* and *Varestrongylus capreoli* were the most pathogenic, particularly in cases of intense infection [123]. The helminth fauna of roe deer is influenced by various factors, including the age of the animals, the availability of food, the prevalence of helminths in domestic livestock, and the density of the roe deer herd.

In the second period, beginning in the late 1990s, investigations into the muscle parasites of roe deer were primarily conducted. Four species of *Sarcocystis* were identified from the cysts, including *S. capreolicanis*, *S.* cf. *hofmanni*, and *S. gracilis* [124]. The prevalence of *Sarcocystis* infection among roe deer was 87.5% in central, northern, and eastern Lithuania. The number of cases with intense infection in roe deer was higher than in other ungulates, with a median intensity of 26.8% [125]. The main limitation of this period was the conventional method of light microscopy, which was not supported by molecular methods.

The prevalence of *Sarcocystis* infection, ranging from 94.4% to 100.0% in various muscle groups of roe deer, was confirmed in the third period of investigations [126]. This study marked the last investigation solely based on light microscopy. Molecular methods confirmed the presence of six *Sarcocystis* species in roe deer from Lithuania: *S. capreolicanis*, *S. gracilis*, *S. linearis*, *S. oviformis*, *S. silva*, and *S. entzerothi* [127], with the latter described as a new species [128].

In the last decade, molecular methods have been instrumental in identifying pathogens associated with cervids. Two *Babesia* species, *B. microti* and *B. venatorum*, were identified in ticks, *Ixodes ricinus* and *Dermacentor reticulatus*, collected from cervids [129]. Other pathogens identified in roe deer include *Anaplasma phagocytophilum*, with a prevalence of 47%, and *Bartonella* spp., with a prevalence of 15.7% [130]. Using nested PCR targeting different parts of the 18S rRNA gene of *Babesia* spp., these pathogens were detected in roe deer with a prevalence of 70.8%, identifying *Babesia capreoli*, *B. odocoilei*, *B. venatorum*, and *B. divergens* [131]. According to the 16S–23S rRNA ITS region, samples of *Babesia* from wild animals, including roe deer, were similar to *B. shoenbuchensis*, *B. chomelii*, and *B. birtlesi* [132].

Roe deer have been shown to be susceptible to hepatitis E virus (HEV), with an overall prevalence of IgG antibodies being 1.20%. Such a low prevalence suggests that roe deer are accidental hosts [133]. However, a later finding revealed that 22.58% of the roe deer samples tested positive for HEV RNA using primers targeting the ORF1 fragment. This underscores the need for further studies to understand HEV transmission between animals and humans in the Baltic countries [134].

3.9. Roe Deer Damage to Forest Stands

The most substantial damage caused by roe deer to forest saplings and undergrowth has been reported in deciduous forests, particularly affecting oak, ash, maple, and, to some extent, spruce. Studies from the 1960s and 1970s indicated that intact oak trees accounted for only 0.5%–2.0% of the total [135]. In addition to the damaged area (Figure 1), it was reported that 36%–39% of trees were damaged in 1969, 32% in 1975, 30% in 1980, and 39% in 1981. Later, the intensity of damages was not presented in the annual reports [41].

In mixed forests, roe deer damage is relatively low, while pure pine forests are generally unaffected. Damage to agriculture is considered insignificant, even in cases where the animals graze in fields throughout the year [11].

In general, roe deer damage is significantly lower compared to that caused by moose or red deer. To mitigate damage to forests, various protective measures have been developed and tested in the country [136]: different methods for maintaining the most valuable tree species, fencing of critical areas, implementation of chemical and biotechnical measures, and control of deer populations.

Despite increasing roe deer numbers, their damage decreased significantly (Figure 6), from 12,114 ha in 1967 to 489 ha in 1997 (CAGR = −0.101). After 1998, deer species are no longer analyzed separately (State Forest Service 2024). It should be noted that, during the same period, damage caused by red deer and moose increased [74,75].

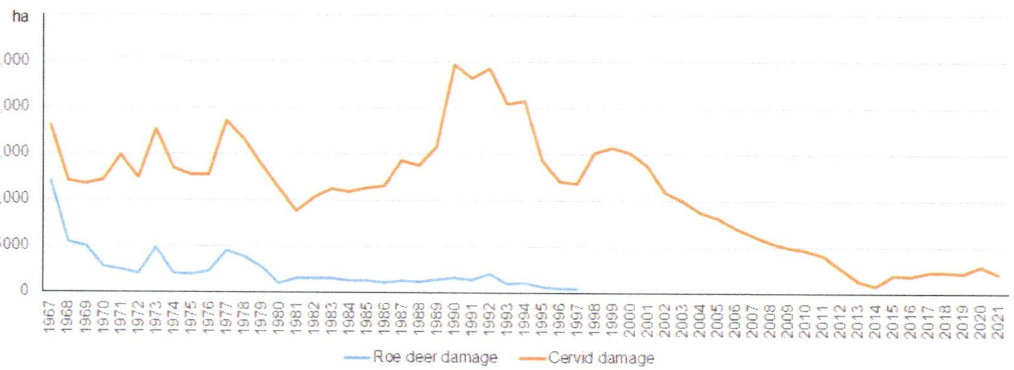

Figure 6. Dynamics of roe deer-induced forest damage in Lithuania from 1967 to 1997, measured by the extent of damaged areas (ha) [41].

Damage caused by all cervid species reached its maximum in 1990–1992, covering 23–24,000 ha of forest, later gradually decreasing to 2–3000 ha after 2014 (Figure 6). Scientific investigations of roe deer damage in the last 40 years are absent. Sabalinkiene et al. found the negative influence of ungulates on natural forest regeneration and afforestation even under cervid densities not exceeding the highest optimal numbers [137].

3.10. Roe Deer Population Management

Roe deer management in Lithuania is based on regulating their numbers through legal and administrative measures, which underwent several changes during the analyzed period [11,74,75]. The main management method is hunting [16]. However, roe deer are not the main commercial target on hunting grounds [40,138].

During the first two periods (Table 3), roe deer hunting was either forbidden or hunters were not actually interested in increasing the hunting bag. Later, after introducing the so-called "norms" or allowable maximum density of cervids (Table 4), overhunting of roe deer as well as of moose and red deer [74,75] occurred due to concerns about the perceived threat of cervids to forest regeneration and quality.

Table 3. Roe deer population management in Lithuania by time periods.

Years	Period	Short Description	References
1948–1964	Hunting ban	During the period of the post-war hunting ban, the roe deer population increased 6.7 times (see Figure 2).	[11,36,59]
1965–1988	Central regulation	The hunted roe deer were delivered to state-owned meat factories; therefore, hunter interest was kept to minimum.	[139]
1986–1994	Overhunting	Based on the opinion that deer overabundance is threatening forests, the hunting bag was set at over 20% of the winter population (Figure 4). This period ended with a decrease in the population and a temporary ban on roe deer hunting in 1997.	[140–142]
1990–1994	Hunting reorganization	In the reorganization period following Lithuania's independence, administrative and legal changes negatively affected roe deer numbers. Overhunting was driven by norms indicating maximum density limits for deer in forests.	[30,143]
1995	Hunt regulations	The hunting of roe deer in the Republic of Lithuania is regulated by the Adopted Regulation for Hunting and the Rules of Hunting. Roe deer hunting quotas are set annually.	[34]

Table 3. *Cont.*

Years	Period	Short Description	References
2002	Law on Hunting	Concepts of hunting, hunting plot units, and selectional hunting established. Hunting grounds rent terms set for no shorter than 10 years.	[144]
2020	Limits lifted	From 2020 onwards, roe deer can be hunted without limits.	[34]
2021	Selective deer hunting lifted	There are no mandatory regulations specifically addressing selective deer hunting; the only restriction pertains to the length of the hunting season. Currently, roe deer bucks can be hunted from 1 May to 31 October, while does and calves can be hunted from 1 October to 31 January.	[34]

Table 4. Norms of allowable maximum density of roe deer, red deer, and moose in the forests of Lithuania (1995), which become Norms of recommended maximum density in 2005 [30].

Forest Type	Number of Animals (Per 1000 ha of Forest)		
	Moose	Red Deer	Roe Deer
Deciduous forests with admixture of coniferous trees	4	15	55
Mixed coniferous forests with admixture of deciduous stands	3	12	45
Mixed coniferous forests with scarce deciduous admixture	2	12	34
Pure pine forests	1	6	14

Institutional and legal changes also influenced roe deer population management. In 1989, wildlife management was transferred to the Ministry of Forestry, leading to a period where the highest proportion of the roe deer population was hunted, particularly during the early years of independence. Norms of allowable density of ungulates, incorporated into legal acts [30,145], mandated hunters to regulate cervid populations. According to these norms, the density of one species of cervids could be increased if the numbers of other species were reduced. For instance, one moose was considered equivalent to three red deer, and one red deer was equivalent to four roe deer. However, the allowable density of roe deer (Table 4) was multiplied by two in forests smaller than 300 ha.

The last three periods were pivotal for the dynamics of roe deer numbers. The implementation of the Law on Hunting in 2002, which allowed the renting of hunting lands for extended periods, enabled hunter organizations to pursue long-term wildlife management strategies [144]. Consequently, the numbers of other deer species besides roe deer also increased [74,75]. The lifting of limits on roe deer hunting and the removal of binding restrictions on selective roe deer hunting, with the only remaining constraint being the length of the hunting season, stabilized roe deer numbers in recent years at approximately 170 thousand.

4. Discussion

We do not dispute the official roe deer numbers, as in many countries, official statistics and bag records are the primary sources of data [16,17,26]. According to Burbaitė and Csányi, the actual roe deer population size in Europe may be around 1.5 times larger than the official data suggest [14]. In the 1980s and 1990s, the real number of roe deer in Lithuania could have been 1.3–1.6 times higher than officially reported, depending on hunters' reluctance to hand over hunted animals to the state [11]. While roe deer hunting was restricted, from the 1990s onwards, the reported number of roe deer may have been increased to achieve a higher limit. However, these expert opinions have not been thoroughly verified at the country level.

The variation in roe deer population density across Europe is considerable, ranging from 11 to 5380 individuals per 100 km^2 [146]. In Lithuania, the density falls towards the lower end of this spectrum, averaging around 347 individuals per 100 km^2, with observed limits between 100 and 886 individuals per 100 km^2 of the forest area [11]. Specifically, in the continental pine forests of Lithuania, roe deer densities are comparatively low, ranging from 30 to 100 individuals per 100 km^2 [93], while in the mixed forests of southern Lithuania, densities are higher, averaging about 150–160 individuals per 100 km^2 [40].

There are no publications on the density of the population in the country, taking into account the changes in the area of the country between 1923 and 1945. In 1937, the total forest area in Lithuania was 9064.82 km^2, while the country area in 1939 was 55,265.82 km^2 (https://osp.stat.gov.lt/statistiniu-rodikliu-analize#/, accessed on 12 January 2024). After 1945, the area of Lithuania remained stable at 65,300 km^2. Thus, in 1937–1939, the density of roe deer was 48–63 ind. per 100 km^2 of the country, or 295–386 ind. per 100 km^2 of the forest area. After the minimum densities found in 1948, 13 ind. per 100 km^2 of the country, or 66 ind. per 100 km^2 of the forest area, later roe deer densities increased. In 1961, these indices were 46 and 195; in 1983, 60 and 214; in 2002, 106 and 345; in 2013, 181 and 555, respectively. Maximum densities reached 276 ind. per 100 km^2 of the country, or 837 ind. per 100 km^2 of the forest area in 2020. Therefore, roe deer has consistently been the most abundant cervid in the country [74,75], outnumbering moose 8–12 times and red deer 2–4 times in 2012–2022. In 2017, the size of the roe deer bag in Lithuania was second to Poland, with 26,592 individuals harvested, surpassing Latvia and Estonia, which reported 22,135 and 6264 individuals, respectively [147]. Therefore, we can conclude that the roe deer population in Lithuania is robust.

Studies conducted in eastern Poland, where field roe deer populations are abundant, much like in the southwestern part of Lithuania, have revealed that these populations are particularly vulnerable to poaching, which is identified as the primary source of mortality [148]. Similarly, in the intensively managed cultural landscape of the Czech Republic, predation by red foxes (*Vulpes vulpes*) and the use of harvesting machines have been identified as important factors contributing to fawn mortality [149]. Although poaching specifically targeting field roe deer was not mentioned as a specific threat [98], guidance notes on the protection of game fauna during agricultural operations were issued in Lithuania [150].

One of the distinctive features of ungulate population management in the country is the selective hunting of ungulate males, a practice that has been in place for over 40 years [151–153]. Selective hunting and the criteria for selecting deer were established based on the age variability of cervid antlers [154], and were later formalized into legal requirements [145], including the mandate to pass an exam for permission to hunt adult animals [155]. The requirement for selective hunting was only lifted in 2021 (see Table 3).

In the case of roe deer, a similar strategy of enhancing the value of trophy morphometric parameters was not employed in Latvia or Estonia, the other Baltic countries with similar game fauna and environments [156]. As a result, the largest roe deer trophies are found in Lithuania, and the trophy values of younger roe deer bucks in Lithuania are comparable to the best mature individuals in Estonia and Latvia, despite the lower availability of good quality food.

One potential direction for further research on roe deer ecology and population management could be to assess the relationship between the abundance of several deer species, which has only been considered in terms of permissible densities in Lithuania thus far. A study conducted in northern Poland suggests that red deer may outcompete roe deer, and high densities of red deer could have a negative impact on roe deer numbers and spatial distribution [157]. While red deer numbers in Lithuania are increasing [75], this relationship has not yet been analyzed at various spatial scales.

5. Conclusions

The review concludes that the methodologies used in the studies analyzed had a number of shortcomings, such as:

1. The survey of roe deer numbers may not provide an accurate picture, even with broadly coordinated snow or visual surveys, due to potential overlap in roe deer herds between surveyed areas.

2. Aggregated data on roe deer poaching in the country are lacking, leading to incomplete hunting bag counts. It is presumed that poaching levels peaked in the 1960s and early 1990s, coinciding with weakened enforcement during periods of political–economic change [11,143].
3. Determining habitat preferences from pellet counts may be biased by differences in pellet decomposition rates and detectability in open and forest habitats.
4. Attribution of deer damage to specific deer species lacks sufficient documentation in the cited sources. Methods such as animal tracks and feces in damaged stands, as well as measuring the thickness of browsed shoots, still leave space for bias due to the presence and numbers of different deer species in the area.
5. The review highlights the scarcity of research on roe deer biology and ecology in recent decades, except in areas related to road ecology and pathogens.

Funding: This research was conducted under long-term research program of Nature Research Centre with no external funding.

Data Availability Statement: No new data were created or analyzed in this study. Data sharing is not applicable to this article.

Acknowledgments: We thank Jolanta Urbelionytė, from the Ministry of the Environment of the Republic of Lithuania, and Virgilijus Vasiliauskas, from the State Forest Service, for their help with roe deer survey and damage data. We acknowledge the help of Deivis Kulvietis and Artūras Kajėnas from the Lithuanian Police Traffic Supervision Service for their help with deer roadkill. Andrius Kučas provided maps of roe deer roadkill.

Conflicts of Interest: The author declares no conflicts of interest.

References

1. Andersen, R.; Duncan, P.; Linnell, J.D. (Eds.) *The European Roe Deer: The Biology of Success*; Scandinavian University Press: Oslo, Norway, 1998; pp. 1–376.
2. Wilson, D.E.; Reeder, D.M. (Eds.) *Mammal Species of the World: A Taxonomic and Geographic Reference*, 3rd ed.; JHU Press: Baltimore, MD, USA, 2005; pp. 1–2147.
3. Lorenzini, R.; Hewison, M.; Gaillard, J.M.; Garofalo, L.; Rossi, L.; Morellet, N.; Verheyden, H.; Lovari, S.; Lister, A.M.; Mattioli, S. European roe deer *Capreolus capreolus* (Linnaeus, 1758). In *Handbook of the Mammals of Europe*; Springer International Publishing: Cham, Switzerland, 2022; pp. 165–195.
4. Sommer, R.S.; Fahlke, J.M.; Schmölcke, U.; Benecke, N.; Zachos, F.E. Quaternary history of the European roe deer *Capreolus capreolus*. *Mamm. Rev.* **2009**, *39*, 1–16. [CrossRef]
5. Randi, E.; Alves, P.C.; Carranza, J.; Milošević-Zlatanović, S.; Sfougaris, A.; Mucci, N. Phylogeography of roe deer (*Capreolus capreolus*) populations: The effects of historical genetic subdivisions and recent nonequilibrium dynamics. *Mol. Ecol.* **2004**, *13*, 3071–3083. [CrossRef] [PubMed]
6. Lorenzini, R.; Lovari, S. Genetic diversity and phylogeography of the European roe deer: The refuge area theory revisited. *Biol. J. Linn. Soc.* **2006**, *88*, 85–100. [CrossRef]
7. Plis, K.; Niedziałkowska, M.; Borowik, T.; Lang, J.; Heddergott, M.; Tiainen, J.; Bunevich, A.; Šprem, N.; Paule, L.; Danilkin, A.; et al. Mitochondrial DNA diversity and the population genetic structure of contemporary roe deer (*Capreolus capreolus*) in Europe. *Mamm. Biol.* **2022**, *102*, 1743–1754. [CrossRef]
8. Kałuziński, J. Dynamics and structure of a field roe deer population. *Acta theriol.* **1982**, *27*, 385–408. [CrossRef]
9. Pielowski, Z. Some aspects of population structure and longevity of field roe deer. *Acta Theriol.* **1984**, *29*, 17–33. [CrossRef]
10. Putman, R.J. Foraging by roe deer in agricultural areas and impact on arable crops. *J. Appl. Ecol.* **1986**, *23*, 91–99. [CrossRef]
11. Baleišis, R.; Bluzma, P.; Balčiauskas, L. *Lietuvos kanopiniai žvėrys [Ungulates of Lithuania]*, 3rd ed.; Akstis: Vilnius, Lithuania, 2003; pp. 1–217. (In Lithuanian)
12. Pėtelis, K.; Brazaitis, G. Morphometric data on the field ecotype roe deer in Southwest Lithuania. *Acta Zool. Litu.* **2003**, *13*, 61–64. [CrossRef]
13. Sabalinkiene, G.; Danusevicius, D.; Manton, M.; Brazaitis, G.; Simkevicius, K. Differentiation of European roe deer populations and ecotypes in Lithuania based on DNA markers, cranium and antler morphometry. *Silva Fenn.* **2017**, *51*, 1743. [CrossRef]
14. Burbaitė, L.; Csányi, S. Roe deer population and harvest changes in Europe. *Est. J. Ecol.* **2009**, *58*, 169–180. [CrossRef]
15. Cederlund, G.; Bergqvist, J.; Kjellander, P.; Gill, R.; Gaillard, J.M.; Boisaubert, B.; Ballon, P.; Duncan, P. Managing roe deer and their impact on the environment: Maximising the net benefits to society. In *The European Roe Deer: The Biology of Success*; Scandinavian University Press: Oslo, Norway, 1998; pp. 337–372.

16. Apollonio, M.; Andersen, R.; Putman, R. (Eds.) *European Ungulates and Their Management in the 21st Century*; Cambridge University Press: New York, NY, USA, 2010; pp. 578–604.
17. Carpio, A.J.; Apollonio, M.; Acevedo, P. Wild ungulate overabundance in Europe: Contexts, causes, monitoring and management recommendations. *Mammal Rev.* **2021**, *51*, 95–108. [CrossRef]
18. Valente, A.M.; Acevedo, P.; Figueiredo, A.M.; Fonseca, C.; Torres, R.T. Overabundant wild ungulate populations in Europe: Management with consideration of socio-ecological consequences. *Mammal Rev.* **2020**, *50*, 353–366. [CrossRef]
19. Balčiauskas, L.; Kučas, A.; Balčiauskienė, L. Mammal Roadkills in Lithuanian Urban Areas: A 15-Year Study. *Animals* **2023**, *13*, 3272. [CrossRef] [PubMed]
20. Partl, E.; Szinovatz, V.; Reimoser, F.; Schweiger-Adler, J. Forest restoration and browsing impact by roe deer. *Forest Ecol. Manag.* **2002**, *159*, 87–100. [CrossRef]
21. Davis, M.L.; Stephens, P.A.; Kjellander, P. Beyond climate envelope projections: Roe deer survival and environmental change. *J. Wildl. Manag.* **2016**, *80*, 452–464. [CrossRef]
22. Maublanc, M.L.; Bideau, E.; Launay, C.; Monthuir, B.; Gerard, J.F. Indicators of ecological change (IEC) as efficient tools for managing roe deer populations: A case study. *Eur. J. Wildl. Res.* **2016**, *62*, 189–197. [CrossRef]
23. Breyne, P.; Mergeay, J.; Casaer, J. Roe deer population structure in a highly fragmented landscape. *Eur. J. Wildl. Res.* **2014**, *60*, 909–917. [CrossRef]
24. Martin, J.; Vourc'h, G.; Bonnot, N.; Cargnelutti, B.; Chaval, Y.; Lourtet, B.; Goulard, M.; Hoch, T.; Plantard, O.; Hewison, A.M.; et al. Temporal shifts in landscape connectivity for an ecosystem engineer, the roe deer, across a multiple-use landscape. *Landscape Ecol.* **2018**, *33*, 937–954. [CrossRef]
25. Bragina, E.V.; Ives, A.R.; Pidgeon, A.M.; Kuemmerle, T.; Baskin, L.M.; Gubar, Y.P.; Piquer-Rodríguez, M.; Keuler, N.S.; Petrosyan, V.G.; Radeloff, V.C. Rapid declines of large mammal populations after the collapse of the Soviet Union. *Conserv. Biol.* **2015**, *29*, 844–853. [CrossRef] [PubMed]
26. Bragina, E.V.; Ives, A.R.; Pidgeon, A.M.; Balčiauskas, L.; Csányi, S.; Khoyetskyy, P.; Kysucká, K.; Lieskovsky, J.; Ozolins, J.; Randveer, T.; et al. Wildlife population changes across Eastern Europe after the collapse of socialism. *Front. Ecol. Environ.* **2018**, *16*, 77–81. [CrossRef]
27. Calkoen, S.T.S.V.; Muhlbauer, L.; Andren, H.; Apollonio, M.; Balčiauskas, L.; Belotti, E.; Carranza, J.; Cottam, J.; Filli, F.; Gatiso, T.T.; et al. Ungulate management in European national parks: Why a more integrated European policy is needed. *J. Environ. Manag.* **2020**, *260*, 110068. [CrossRef]
28. Official Statistics Portal. Forests. Available online: https://osp.stat.gov.lt/lietuvos-aplinka-zemes-ukis-ir-energetika-2021/aplinka/miskai (accessed on 10 January 2024).
29. Official Statistics Portal. Lietuvos Miškingumas Pagal Apskritis [Lithuania's Forest Cover by County]. Available online: https://osp.stat.gov.lt/infografikas-lietuvos-miskingumas-pagal-apskritis (accessed on 10 January 2024).
30. LR Aplinkos Ministro 2005 m. Kovo 21 d. Įsakymas nr. D1-162. Dėl Medžioklėtvarkos Projektų Rengimo, Derinimo, Teikimo Tvirtinti ir Tvirtinimo Tvarkos Aprašo Patvirtinimo. [Regarding the Preparation, Coordination, Submission for Approval of Hunting Regulation Projects and Approval of the Description of the Approval Procedure]. Available online: https://e-seimas.lrs.lt/portal/legalAct/lt/TAD/TAIS.252979?jfwid=9iufxps7h (accessed on 10 January 2024).
31. Lietuvos Respublikos Aplinkos Ministras Įsakymas dėl Lietuvos Respublikos Aplinkos Ministro 2000 m Birželio 27, D. Įsakymo nr. 258 "Dėl Medžioklės Lietuvos Respublikos Teritorijoje Taisyklių Patvirtinimo" Pakeitimo 2018 m. Spalio 18 d. Nr. D1-892 [Regarding Approval of Rules for Hunting in the Territory of the Republic of Lithuania]. Available online: https://e-seimas.lrs.lt/portal/legalAct/lt/TAD/38fc75e0d2d011e8a82fc67610e51066 (accessed on 10 January 2024).
32. Lietuvos Respublikos Aplinkos Ministras. 2018 m. Gegužės 7 d. Įsakymas Nr. D1-365 "Dėl Lietuvos Respublikos Aplinkos Ministro 2000 m. Birželio 27 d. Įsakymo Nr. 258 "Dėl Medžioklės Lietuvos Respublikos Teritorijoje Taisyklių Patvirtinimo" Pakeitimo". [Change Regarding Approval of Rules for Hunting in the Territory of the Republic of Lithuania]. Available online: https://www.e-tar.lt/portal/legalAct.html?documentId=9379426051dd11e884cbc4327e55f3ca (accessed on 10 January 2024).
33. Lietuvos Respublikos Aplinkos Ministras. 2020 m. Gegužės 8 d. Įsakymas Nr. D1-268 "Dėl Lietuvos Respublikos Aplinkos Ministro 2000 m. Birželio 27 d. Įsakymo Nr. 258 "Dėl Medžioklės Lietuvos Respublikos Teritorijoje Taisyklių Patvirtinimo" pakeitimo". [Change Regarding Approval of Rules for Hunting in the Territory of the Republic of Lithuania]. Available online: https://www.e-tar.lt/portal/legalAct.html?documentId=c4cf1e70910a11ea9515f752ff221ec9 (accessed on 10 January 2024).
34. Lietuvos Respublikos Aplinkos Ministras. 2020 m. Spalio 14 d. Įsakymas Nr. D1-631 "Dėl Medžioklės Lietuvos Respublikos Teritorijoje Taisyklių Patvirtinimo" [Regarding the Approval of Hunting Rules in the Territory of the Republic of Lithuania]. Available online: https://www.e-tar.lt/portal/legalAct.html?documentId=348629a00e1211ebb74de75171d26d52 (accessed on 11 February 2024).
35. Medžioklės Kalendorius 2024 m. Available online: https://medziokleskalendorius.lt/ (accessed on 10 January 2024).
36. Medžiojamųjų Žvėrių Apskaita [Game Survey]. Available online: https://am.lrv.lt/lt/veiklos-sritys-1/gamtos-apsauga/medziokle/medziojamuju-zveriu-apskaita (accessed on 1 February 2024).
37. Sumedžioti Žvėrys ir Paukščiai [Hunted Animals and Birds]. Available online: https://am.lrv.lt/lt/veiklos-sritys-1/gamtos-apsauga/medziokle/sumedzioti-zverys-ir-paukstiai/ (accessed on 1 February 2024).

38. Medžiojamųjų Žvėrių Apskaita Pagal Pėdsakus Sniege [Game Snow-Track Counts]. Available online: https://vstt.lrv.lt/lt/veiklos-sritys/biologines-ivairoves-apsauga/zveriu-apskaita/medziojamuju-zveriu-apskaita-pagal-pedsakus-sniege/ (accessed on 1 February 2024).
39. Kawata, Y.; Ozolins, J.; Andersone-Lilley, Z. An analysis of the game animal population data from Latvia. *Baltic For.* **2008**, *14*, 75–86.
40. Barkauskas, T.; Belova, O.; Griciuvienė, L. Differences in ungulate population use in different hunting ground units in Lithuania. *Baltic For.* **2020**, *26*, 514. [CrossRef]
41. Valstybinė Miškų Tarnyba. Miškų Sanitarinė Būklė [Sanitary Condition of Forests]. Available online: https://amvmt.lrv.lt/lt/veiklos-sritys/misko-sanitarine-apsauga/misku-sanitarine-bukle (accessed on 5 January 2024).
42. Compound Annual Growth Rate—CAGR. Available online: https://www.investopedia.com/terms/c/cagr.asp#what-is-compound-annual-growth-rate--cagr (accessed on 20 December 2023).
43. TIBCO Software Inc. Data Science Textbook. Available online: https://docs.tibco.com/data-science/textbook (accessed on 15 January 2024).
44. Sommer, R.S.; Zachos, F.E. Fossil evidence and phylogeography of temperate species: 'glacial refugia' and post-glacial recolonization. *J. Biogeogr.* **2009**, *36*, 2013–2020. [CrossRef]
45. Niedziałkowska, M.; Jędrzejewska, B.; Honnen, A.C.; Otto, T.; Sidorovich, V.E.; Perzanowski, K.; Skog, A.; Hartl, G.B.; Borowik, T.; Bunevich, A.N.; et al. Molecular biogeography of red deer *Cervus elaphus* from eastern Europe: Insights from mitochondrial DNA sequences. *Acta Theriol.* **2011**, *56*, 1–12. [CrossRef] [PubMed]
46. Grünberg, J.M. Animals in Mesolithic burials in Europe. *Anthropozoologica* **2013**, *48*, 231–253. [CrossRef]
47. Rimkus, T. In search of Lithuania coastal Mesolithic: Review of current data and the aims of an ongoing research project. *Fornvännen* **2019**, *114*, 1–11.
48. Daugnora, L.; Girininkas, A. *Osteoarcheologija Lietuvoje*; Savastis: Vilnius, Lithuania, 1996; pp. 1–198.
49. Simčenka, E.; Kozakaitė, J.; Piličiauskienė, G.; Gaižauskas, L.; Piličiauskas, G. Human Diet During the Stone Age and Early Metal Period (7000–1 cal BC) in Lithuania: An update. *Radiocarbon* **2022**, *64*, 1171–1189. [CrossRef]
50. Girininkas, A.; Daugnora, L. Hunting in the territory of Lithuania from the Late Palaeolithic to the Middle Ages. In *Hunting in Northern Europe until 1500 AD: Old Traditions and Regional Developments, Continental Sources and Continental Influences*; Grimm, O., Schmölcke, U., Eds.; Neumünster: Wachholtz, Germany, 2013; pp. 567–595.
51. Vitkūnas, M. Medžioklė XIII–XIV a. (tyrinėjimų Pietryčių Lietuvoje duomenimis) [Hunting in the 13th–14th Centuries (on the research data in Southeastern Lithuania)]. *Istorija. Liet. Aukštųjų Mokykl. Moksl. Darb.* **2009**, *76*, 3–17. (In Lithuanian)
52. Volkaitė-Kulikauskienė, R. *Lietuviai IX–XII Amžiais*; Mintis: Vilnius, Lietuva, 1970; pp. 59–60.
53. Daugnora, L.; Piličiauskienė, G. XIV–XVII a. osteologinės medžiagos, 1988–1990 m. iškastos Vilniaus žemutinės pilies teritorijoje, analizė [The analysis of the osteological material of the 14th–17th c. excavated in 1988–1990 in the territory of the Vilnius Lower Castle]. *Liet. Archeol.* **2005**, *28*, 207–216. (In Lithuanian)
54. Rzączyński, G. *Historia Naturalis Curiosa Regni Poloniae, Magni Ducatus Lithvaniae*; Typis Collegii Soc. Jesu: Sandomiria, Poland, 1721; pp. 1–489.
55. Ładowski, S. *Hystoria Naturalna Królestwa Polskiego*; W Drukarni Uprzywilejowaney Ignacego Grebla: Kraków, Poland, 1783; pp. 1–206.
56. Eichwald, E. *Naturhistorische Skizze von Lithauen, Volhynien und Podolien*; Zawadzki: Wilna, Lithuania, 1830; pp. 1–256.
57. Paaver, K.L. *Formirovanie Teriofauny i Izmenchivost'mlekopitayushchikh Pribaltiki v Golotsene (Formation of the Theriofauna and the Variability of Baltic Mammals in the Holocene)*; Akad. Nauk ESSR: Tartu, Estonia, 1965; pp. 1–494.
58. Heptner, V.G.; Nasimovič, A.A.; Bannikov, A.G.E. *Mammals of the Soviet Union: Volume 1, Ungulates*; Model Press: New Delhi, India, 1989; pp. 243–301.
59. Prūsaitė, J. (Ed.) . *Lietuvos Fauna. Žinduoliai [Fauna of Lithuania. Mammals]*; Mokslas: Vilnius, Lithuania, 1988; pp. 214–218. (In Lithuanian)
60. Plater, A. *Spis Zwierząt Ssących, Ptaków i Ryb Krajowych*; Zawadzki: Wilno, Lithuania, 1852; pp. 1–160.
61. Ivanauskas, T. Mūsų medžiojamieji gyvuliai. Stirna. *Mūsų rytojus* **1928**, *6*. Available online: https://dokumen.pub/tadas-ivanauskas-gyvenimas-ir-veikla.html (accessed on 15 January 2024).
62. Prusaite, J.; Blazys, A.; Mickus, A.; Bluzma, P. Number dynamics and structure of the unhunted roe deer population. *Lith. Acad. Sci. Ser. C* **1973**, *62*, 115–125.
63. Prusaite, J.A.; Blazys, A.S.; Baleisis, R.M. Intensity of breeding and fecundity of the European roe deer in North Lithuania. *Liet. TSR Moksl. Akad. Darb. Ser. C* **1977**, *3*, 105–110.
64. Bluzma, P. Estimation of the state of lynx and wolf populations in Lithuania. *Acta Zool. Litu.* **1999**, *9*, 35–41. [CrossRef]
65. Bluzma, P. Lietuvos medžiojamieji žinduoliai: Populiacijų dinamika ir dabartinė būklė. *Theriol. Litu.* **2001**, *1*, 12–21.
66. Bluzma, P.; Baleišis, R. The cervid (Cervidae) monitoring in Lithuania (1993–1998): Abundance, dynamics and distribution. *Acta Zool. Litu.* **1999**, *9*, 61–70. [CrossRef]
67. Balčiauskas, L.; Kučas, A.; Balčiauskienė, L. The Impact of Roadkill on Cervid Populations in Lithuania. *Forests* **2023**, *14*, 1224. [CrossRef]
68. Balčiauskas, L.; Stratford, J.; Kučas, A.; Balčiauskienė, L. Lockdown's Silver Lining? Different Levels of Roadkill during the COVID-19 Times in Lithuania. *Animals* **2023**, *13*, 2918. [CrossRef]

69. Perkins, S.E.; Shilling, F.; Collinson, W. Anthropause Opportunities: Experimental Perturbation of Road Traffic and the Potential Effects on Wildlife. *Front. Ecol. Evol.* **2022**, *10*, 833129. [CrossRef]
70. Balčiauskas, L.; Stratford, J.; Balčiauskienė, L.; Kučas, A. Importance of professional roadkill data in assessing diversity of mammal roadkills. *Transp. Res. D Transp. Environ.* **2020**, *87*, 102493. [CrossRef]
71. Narauskaitė, G.; Pėtelis, K.; Maksvytis, M. Šilutė region seacoast roe deer *Capreolus capreolus* L. population quality. *Acta Biol. Univ. Daugavp.* **2011**, *11*, 29–34.
72. Narauskaitė, G.; Danusevičius, D.; Østbye, K.; Pėtelis, K. Genetic Differentiation of Field and Forest Ecotypes of Roe Deer (*Capreolus capreolus* L.) in Lithuania Based on DNA Markers. *Rural. Dev.* **2013**, *6*, 376–381.
73. Balčiauskas, L. The influence of roadkill on protected species and other wildlife in Lithuania. In Proceedings of the 2011 International Conference on Ecology and Transportation (ICOET 2011), Seattle, WA, USA, 21–25 August 2011; pp. 647–655. Available online: https://trid.trb.org/view/1347982 (accessed on 20 December 2023).
74. Balčiauskas, L.; Kawata, Y.; Balčiauskienė, L. Moose Management Strategies under Changing Legal and Institutional Frameworks. *Sustainability* **2020**, *12*, 8482. [CrossRef]
75. Balčiauskas, L.; Kawata, Y. Red Deer in Lithuania: History, Status and Management. *Sustainability* **2022**, *14*, 14091. [CrossRef]
76. Balčiauskas, L.; Jasiulionis, M. Reducing the incidence of mammals on public highways using chemical repellent. *Balt. J. Road Bridge E* **2012**, *7*, 92–97. [CrossRef]
77. Ignatavicius, G.; Valskys, V. The influence of time factors on the dynamics of roe deer collisions with vehicles. *Landsc. Ecol. Eng.* **2018**, *14*, 221–229. [CrossRef]
78. Kučas, A.; Balčiauskas, L. Temporal patterns of ungulate-vehicle collisions in Lithuania. *J. Environ. Manag.* **2020**, *273*, 111172. [CrossRef] [PubMed]
79. Galinskaitė, L.; Ignatavičius, G. Dynamics of Roe Deer (*Capreolus capreolus*) Vehicle Collisions in Lithuania: Influence of the Time Factors. *Int. J. Transp. Veh. Eng.* **2022**, *16*, 108–112.
80. Balčiauskas, L.; Kučas, A.; Balčiauskienė, L. Roadkill Patterns on Workdays, Weekends and Long Weekends: Anticipating the Implications of a Four-Day Work Week. *Diversity* **2024**, *16*, 84. [CrossRef]
81. Balčiauskas, L.; Wierzchowski, J.; Kučas, A.; Balčiauskienė, L. Habitat Suitability Based Models for Ungulate Roadkill Prognosis. *Animals* **2020**, *10*, 1345. [CrossRef] [PubMed]
82. Galinskaitė, L.; Ulevičius, A.; Valskys, V.; Samas, A.; Busher, P.E.; Ignatavičius, G. The Influence of Landscape Structure on Wildlife–Vehicle Collisions: Geostatistical Analysis on Hot Spot and Habitat Proximity Relations. *ISPRS Int. J. Geo-Inf.* **2022**, *11*, 63. [CrossRef]
83. Kučas, A.; Balčiauskas, L. Roadkill-Data-Based Identification and Ranking of Mammal Habitats. *Land* **2021**, *10*, 477. [CrossRef]
84. Kučas, A.; Balčiauskas, L. Impact of Road Fencing on Ungulate-Vehicle Collisions and Hotspot Patterns. *Land* **2021**, *10*, 338. [CrossRef]
85. Kučas, A.; Balčiauskas, L.; Lavalle, C. Identification of Urban and Wildlife Terrestrial Corridor Intersections for Planning of Wildlife-Vehicle Collision Mitigation Measures. *Land* **2023**, *12*, 758. [CrossRef]
86. Hagen, R.; Heurich, M.; Storch, I.; Hanewinkel, M.; Kramer-Schadt, S. Linking annual variations of roe deer bag records to large-scale winter conditions: Spatio-temporal development in Europe between 1961 and 2013. *Eur. J. Wildl. Res.* **2017**, *63*, 97. [CrossRef]
87. Marma, B.; Padaiga, V. Stirnų mitybos fiziologijos klausimu. *Girios* **1975**, *5*, 24–25.
88. Bluzma, P. Lietuvos Stirnos. *Mūsų Gamt.* **1976**, *7–9*, 12.
89. Balčiauskas, L. Антропогенность ландшафта Литвы как фактор высокой численности копытных [Anthropogenicity of the landscape of Lithuania as a factor of high abundance of ungulates]. *Весци Ан Бсср Сер. Биол.* **1988**, *2*, 105–107. (In Russian)
90. Balčiauskas, L. Сеть троп копытных как отражение перемещений животных и использования ими обжитой территории [Network of the paths of ungulates as a reflection of the animal movements and use of the inhabited territory]. In *Млекопитающие в Культурном Ландшафте Литвы*; Ekologijos Institutas: Vilnius, Lithuania, 1990; pp. 131–137. (In Russian)
91. Marma, B. Stirnų kritimų priežastys ir pasekmės. *Girios* **1970**, *9*, 17–19.
92. Падаига, В. факторы, определяющие зимние падежы косуль. *Зоологичеслий Журнал* **1971**, *10*, 1546–1552.
93. Belova, O. The impact of herbivorous mammals on woody vegetation in the different stages of forest succession. *Baltic For.* **2012**, *18*, 100–110.
94. Padaiga, V. Stirnos ir kietieji lapuočiai. *Girios* **1967**, *1*, 10–11.
95. Балейшис, Р.М.; Прусайте, Я.А. Питание европейской косули в небольшом лиственном лесу Северной Литвы в 1975–1976 г.г. *Труды Академии Наук Литовской Сср Серия В* **1980**, *1*, 85–91.
96. Прусаите, Я.; Балеишис, Р.; Блузма, П. Состав кормв европейской косули в зависимости от лесистости обитаемой територии. *Труды Академии Наук Литовской Сср. Серия В* **1983**, *4*, 84–98.
97. Блузма, П.П.; Балейшис, Р.М. Значение агроценозов для косули в Литве. In *Охотоустройство в Специализированном Лесном Хозяйстве*; LMŪMTI: Kaunas-Girionys, Lithuania, 1983; pp. 64–66.
98. Блузма, П.; Балейшис, Р. Европейская косуля в агроценозах Литвы. In *Млекопитающие в культурном ландшафте Литвы*; Mokslas: Vilnius, Lithuania, 1990; pp. 138–145.
99. Tauginas, J. Žmogaus ūkinės veiklos miške įtaka elninių žvėrių gausumui [Impact of human economic activities in the forest on the abundance of cervids. *Girios* **1977**, *4*, 15–16. (In Lithuanian)

100. Priedytis, A.; Tauginas, J. Feeding effectiveness of granules, made from summer cutting waste of aspen, in the winter ration of roe deer. *Liet. TSR Moksl. Akad. Darbai. Ser. C* **1981**, *3*, 149–158.
101. Bluzma, P.; Baleišis, R.; Balčiauskas, L. Численность и распределение оленьих (Cervidae) на зимних пастбищах различных ландшафтов Литвы [Abundance and distribution of cervids in winter stands of various landscapes of Lithuania]. In *Млекопитающие В Культурном Ландшафте Литвы*; Ekologijos Institutas: Vilnius, Lithuania, 1990; pp. 122–130. (In Russian)
102. Bluzma, P.; Baleisis, R.; Balciauskas, L. Significance of the stand composition to cervid (Cervidae) winter distribution in the forests of various Lithuanian landscapes. *Ekologija* **1996**, *3*, 61–66.
103. Baleisis, R.; Bluzma, P.; Balciauskas, L. Deers (Cervidae) distribution in winter depending on the amount of dwarf shrubs in the forest areas of different landscapes in Lithuania. *Ekologija* **1995**, *2*, 45–48.
104. Baranauskas, K.; Balčiauskas, L.; Mažeikytė, R. Vilnius city theriofauna. *Acta Zool. Litu.* **2005**, *15*, 228–238. [CrossRef]
105. Brazaitis, G.; Pėtelis, K.; Žalkauskas, R.; Belova, O.; Danusevičius, D.; Marozas, V.; Narauskaitė, G. Landscape effect for the Cervidaes Cervidae in human-dominated fragmented forests. *Eur. J. Forest Res.* **2014**, *133*, 857–869. [CrossRef]
106. Прусайте, Я.А.; Блажис, А.С.; Балейшис, Р.М. Влияние эксплуатации различной интенсивности на численность и прирост популяции европейской косули. In *II съезд Всесоюзного Териологического Общества*; USSR: Nauka, Moskva, 1978; pp. 251–252.
107. Flajšman, K.; Pokorny, B.; Chirichella, R.; Bottero, E.; Mattioli, L.; Apollonio, M. I can produce more offspring as you can imagine: First records on exceptionally large litters in roe deer in central/southern Europe. *Eur. J. Wildl. Res.* **2017**, *63*, 42. [CrossRef]
108. Fruzinski, B.; Kaluzinski, J.; Baksalary, J. Weight and body measurements of forest and field roe deer. *Acta Theriol.* **1982**, *27*, 479–488. [CrossRef]
109. Pėtelis, K. The estimation of trophy maturity on roe deer field ecotype. *Miškininkystė* **1997**, *1*, 123–127.
110. Markov, G.; Zvychaynaya, E.; Danilkin, A.; Kholodova, M.; Sugar, L. Genetic diversity and phylogeography of roe deer (*Capreolus capreolus* L.) in different biogeographical regions in Europe. *CR Acad. Bulg. Sci.* **2016**, *69*, 579–584.
111. Olano-Marin, J.; Plis, K.; Sönnichsen, L.; Borowik, T.; Niedziałkowska, M.; Jędrzejewska, B. Weak population structure in European roe deer (*Capreolus capreolus*) and evidence of introgressive hybridization with Siberian roe deer (*C. pygargus*) in northeastern Poland. *PLoS ONE* **2014**, *9*, e109147. [CrossRef] [PubMed]
112. Plis, K.; Niedziałkowska, M.; Borowik, T.; Lang, J.; Heddergott, M.; Tiainen, J.; Bunevich, A.; Šprem, N.; Paule, L.; Danilkin, A.; et al. Pan-European phylogeography of the European roe deer (*Capreolus capreolus*). *Ecol. Evol.* **2022**, *12*, e8931. [CrossRef] [PubMed]
113. Niedziałkowska, M.; Plis, K.; Marczuk, B.; Lang, J.; Heddergott, M.; Tiainen, J.; Danilkin, A.; Kholodova, M.; Zvychaynaya, E.; Kashinina, N.; et al. Genetic diversity and complex structure of the European Roe Deer population at a continental scale. *J. Mamm.* **2024**, *105*, 73–84. [CrossRef]
114. Pūraitė, I.; Paulauskas, A.; Sruoga, A. Analysis of genetic diversity of roe deer (*Capreolus capreolus* L.) in Lithuania using RAPD and allozyme systems. *Biologija* **2013**, *59*, 29–38. [CrossRef]
115. Pūraitė, I.; Paulauskas, A.; Sruoga, A.; Prakas, P. Mitochondrial DNA variation in roe deer population from Lithuania. *Balk. J. Wildl. Res.* **2014**, *1*, 13–19. [CrossRef]
116. Pūraitė, I.; Paulauskas, A.; Sruoga, A. Genetic analysis of roe deer (*Capreolus capreolus* L.) using DNA markers. *Вестник Охотоведения* **2014**, *11*, 226–230.
117. González, S.; del Rio, M.L.; Díez-Baños, N.; Martínez, A.; Hidalgo, M.d.R. Contribution to the Knowledge of Gastrointestinal Nematodes in Roe Deer (*Capreolus capreolus*) from the Province of León, Spain: An Epidemiological and Molecular Study. *Animals* **2023**, *13*, 3117. [CrossRef] [PubMed]
118. Arnastauskienė, T. *Coccidia and Coccidioses of Domestic and Wild Animals in Lithuania*; Mokslas: Vilnius, Lithuania, 1985; pp. 1–176.
119. Arnastauskienė, T. Coccidial fauna of roe deer (*Capreolus capreolus*) in Lithuania. *Acta Parasitol. Litu.* **1974**, *12*, 27–37.
120. Kazlauskas, Y.; Arnastauskene, T. Helminths of wild artiodactyls and hares common to man and domestic animals in Lithuania. *Acta Parasitol. Lith.* **1969**, *9*, 61–67.
121. Падаига, В.; Марма, Б. Зараженость косуль паразитами в звисимости от плотности популяции обитания. *Зоологичеслий журнал* **1970**, *9*, 283–287.
122. Shlikas, A.; Kazlauskas, Y. Survival of Trichuris eggs from roe-deer and hares in Lithuania. *Acta Parasitol. Litu.* **1970**, *10*, 75–80.
123. Kazlauskas, J.; Pužauskas, R. On the factors affecting the distribution of roe-deer helminths in Lithuania. *Acta Parasitol. Litu.* **1974**, *12*, 87–91.
124. Kutkienė, L. The species composition of European roe deer (*Capreolus capreolus*) Sarcocystis in Lithuania. *Acta Zool. Litu.* **2001**, *11*, 97–101. [CrossRef]
125. Malakauskas, M.; Grikienienė, J. Sarcocystis infection in wild ungulates in Lithuania. *Acta Zool. Litu.* **2002**, *12*, 372–380. [CrossRef]
126. Januškevičius, V.; Januškevičienė, G.; Paulauskas, V.; Petkevičius, S. Prevalence and intensity of Sarcosporidia infection in wild cloven-hoofed hunted for food in Lithuania. *Vet. Zootec.* **2016**, *73*, 27–31.
127. Rudaitytė-Lukošienė, E.; Delgado de Las Cuevas, G.E.; Prakas, P.; Calero-Bernal, R.; Martínez-González, M.; Strazdaitė-Žielienė, Ž; Servienė, E.; Habela, M.A.; Butkauskas, D. *Sarcocystis* spp. diversity in the roe deer (*Capreolus capreolus*) from Lithuania and Spain. *Parasitol. Res.* **2020**, *119*, 1363–1370. [CrossRef] [PubMed]
128. Prakas, P.; Rudaitytė, E.; Butkauskas, D.; Kutkienė, L. Sarcocystis entzerothi n. sp. from the European roe deer (*Capreolus capreolus*). *Parasitol. Res.* **2017**, *116*, 271–279. [CrossRef] [PubMed]

129. Ražanskė, I.; Radzijevskaja, J.; Paulauskas, A.; Sruoga, A. Babesia spp. in questing ticks and ticks parasitizing cervids. In *The Vital Nature Sign*; Vytautas Magnus University: Kaunas, Lithuania, 2013; p. 31.
130. Paulauskas, A.; Morkutė, K.; Ražanskė, I.; Radzijevskaja, J. Infection of tick-borne pathogens of games in Lithuania. In *5th International Hunting and Game Management Symposium*; Kusza, S., Jâvor, A., Eds.; Licium-Art Kft.: Debrecen, Hungary, 2016; pp. 10–12.
131. Paulauskas, A.; Ražanskė, I.; Radzijevskaja, J.; Olav, R.; Morkutė, K. Prevalence and diversity of *Babesia* spp. In *wild cervids. In German Society for Parasitology: Program and Abstracts*; University of Göttingen: Göttingen, Germany, 2016; p. 115.
132. Vasaitytė, A.; Klepeckienė, K.; Ražanskė, I.; Radzijevskaja, J.; Paulauskas, A. Investigation of *Bartonella* spp. In *domestic and wild animals from Lithuania. In Smart Bio: ICSB 2nd International Conference, Abstracts Book*; Vytautas Magnus University: Kaunas, Lithuania, 2018; p. 195.
133. Spancerniene, U.; Buitkuviene, J.; Grigas, J.; Pampariene, I.; Salomskas, A.; Cepuliene, R.; Žymantienė, J.; Stankevicius, A. Seroprevalence of hepatitis E virus in Lithuanian domestic pigs and wildlife. *Acta Vet. Brno* **2017**, *85*, 319–327. [CrossRef]
134. Spancerniene, U.; Grigas, J.; Buitkuviene, J.; Zymantiene, J.; Juozaitiene, V.; Stankeviciute, M.; Razukevicius, D.; Zienius, D.; Stankevicius, A. Prevalence and phylogenetic analysis of hepatitis E virus in pigs, wild boars, roe deer, red deer and moose in Lithuania. *Acta Vet. Scand.* **2018**, *60*, 13. [CrossRef] [PubMed]
135. Padaiga, V. *Miškas ir Elniniai [Forest and Cervids]*; (In Lithuanian). Mintis: Vilnius, Lithuania, 1969; pp. 1–122.
136. Падайга, В.И. Зимние пространственное распределение оленей в лесных охотничьих угодиях. Тр. Литовского НИИЛХ **1973**, *14*, 337–351.
137. Sabalinkiene, G.; Simkevicius, K.; Petelis, K.; Stankeviciute, J.; Talijunas, D. High densities of Cervidae effect to forest regeneration in mixed broadleaf forest in south part of Lithuania. *Agrofor* **2016**, *1*, 95–102. [CrossRef]
138. Saveljev, A.P.; Lissovsky, A.A.; Kozlov, Y.A. Comparative analysis of the lists of hunting mammals of the countries of the Baltic region and Belarus. *Russian J. Theriol.* **2020**, *19*, 65–70. [CrossRef]
139. Žemulis, F. Medžioklė Tarybiniais Laikais ir Dabar [Hunting in Soviet Times and Now]. 2012. Available online: https://www.delfi.lt/grynas/aplinka/medziokle-tarybiniais-laikais-ir-dabar.d?id=57145619 (accessed on 17 February 2024).
140. Padajga, V.I. *Methods of Regulating Deer Populations in Intensive Forestry*; LitNIILH: Kaunas, Lithuania, 1970; pp. 1–32.
141. Padaiga, V. Экологические основы управления численностью оленьих в Литовской ССР [Ecological Fundamentals for the Management of Cervid Number in the Lithuanian SSR.] Habil. Ph.D. Thesis, University of Tartu, Tartu, Estonia, 1984.
142. Ribikauskas, V. Medžioklė yra Medžioklė. Available online: https://www.manoukis.lt/mano-ukis-zurnalas/2014/02/medziokle-yra-medziokle/ (accessed on 17 February 2024).
143. Balciauskas, L. Game survey as a mean of sustainable population management. In Proceedings of the International Symposium "Rational Management of Cervids in Forest Habitats", Šiauliai, Lithuania, 28–30 January 2004.
144. Lietuvos Respublikos Medžioklės Įstatymas [Hunting Law of the Republic of Lithuania]. Available online: https://e-seimas.lrs.lt/portal/legalAct/lt/TAD/TAIS.169653/asr (accessed on 15 February 2024).
145. Padaiga, V. *Medžioklės Ūkio Biologiniai Pagrindai*; Žiburys: Vilnius, Lithuania, 1996; pp. 1–210.
146. Melis, C.; Jędrzejewska, B.; Apollonio, M.; Bartoń, K.A.; Jędrzejewski, W.; Linnell, J.D.; Kojola, I.; Kusak, J.; Adamic, M.; Ciuti, S.; et al. Predation has a greater impact in less productive environments: Variation in roe deer, *Capreolus capreolus*, population density across Europe. *Global Ecol. Biogeogr.* **2009**, *18*, 724–734. [CrossRef]
147. Linnell, J.D.; Cretois, B.; Nilsen, E.B.; Rolandsen, C.M.; Solberg, E.J.; Veiberg, V.; Kaczensky, P.; Van Moorter, B.; Panzacchi, M.; Rauset, G.R.; et al. The challenges and opportunities of coexisting with wild ungulates in the human-dominated landscapes of Europe's Anthropocene. *Biol. Conserv.* **2020**, *244*, 108500. [CrossRef]
148. Sönnichsen, L.; Borowik, T.; Podgórski, T.; Plis, K.; Berger, A.; Jędrzejewska, B. Survival rates and causes of mortality of roe deer *Capreolus capreolus* in a rural landscape, eastern Poland. *Mamm. Res.* **2017**, *62*, 141–147. [CrossRef]
149. Cukor, J.; Havránek, F.; Vacek, Z.; Bukovjan, K.; Podrázský, V.; Sharma, R.P. Roe deer (*Capreolus capreolus*) mortality in relation to fodder harvest in agricultural landscape. *Mammalia* **2019**, *83*, 461–469. [CrossRef]
150. Baleišis, R.; Bluzma, P.; Prūsaitė, J.; Vėžys, P. *Medžiojamosios Faunos Apsauga Žemės Ūkio Darbų Metu (Metodiniai Nurodymai)*; Lietuvos TSR Medžiotojų ir Žvejų Draugija: Vilnius, Lithuania, 1980; pp. 1–16.
151. Zažeckis, V.; Baleišis, R. *Kanopinių Žvėrių Atrankinė Medžioklė (Metodiniai Nurodymai)*; Gamtos Apsaugos Komitetas Prie Lietuvos TSR Ministrų Tarybos: Vilnius, Lithuania, 1975; pp. 1–14.
152. Baleišis, R.; Bluzma, P.; Matuzevičius, A.; Zažeckis, V. *Elninių Žvėrių Selekcinė Medžioklė (Metodiniai Nurodymai)*; Lietuvos TSR Medžiotojų ir Žvejų Draugija: Vilnius, Lithuania, 1979; pp. 1–36.
153. Балейшис, Р.М.; Блузма, П.П. Результаты селекционного отстрела оленъих (*Cervidae*) в Литовской ССР. In Териологические Исследования в Литве; Mokslas: Vilnius, Lietuva, 1988; pp. 84–89.
154. Балейшис, Р.М.; Блузма, П.П. Возрастная изменчивость рогов косули в Литве. Бюл. Моск. О-*Ва* Испытателей Природы. Отд. Биол. **1981**, *86*, 3–10.
155. Medžiotojų Selekcininkų Kvalifikacijos Suteikimo Tvarka. 2008. Available online: https://www.e-tar.lt/portal/lt/legalAct/TAR.A053608065EC (accessed on 1 February 2016).

156. Balčiauskas, L.; Varanauskas, R.; Bukelskis, E. Impact of selective hunting on the trop hy size of roe deer: Baltic example. *North-West. J. Zool.* **2017**, *13*, 118–127.
157. Borkowski, J.; Banul, R.; Jurkiewicz-Azab, J.; Hołdyński, C.; Święczkowska, J.; Nasiadko, M.; Załuski, D. There is only one winner: The negative impact of red deer density on roe deer numbers and distribution in the Słowiński National Park and its vicinity. *Ecol. Evol.* **2021**, *11*, 6889–6899. [CrossRef] [PubMed]

MDPI AG
Grosspeteranlage 5
4052 Basel
Switzerland
Tel.: +41 61 683 77 34

Forests Editorial Office
E-mail: forests@mdpi.com
www.mdpi.com/journal/forests

www.ingramcontent.com/pod-product-compliance
Lightning Source LLC
Chambersburg PA
CBHW061608210926
51850CB00009B/2259
* 9 7 8 3 7 2 5 8 4 8 8 7 4 *